AF333768

ICF11 2005

ICF11 2005

Honour and Plenary Lectures Presented at the 11th International Conference on Fracture (ICF11), Held in Turin, Italy, on March 20–25, 2005

Edited by

ALBERTO CARPINTERI

Politecnico di Torino, Torino, Italy

YIU-WING MAI

University of Sydney, Sydney, Australia

and

ROBERT O. RITCHIE

University of California, Berkeley, U.S.A.

Reprinted from *International Journal of Fracture*
Volume 138, Nos. 1–4 (2006)

Springer

A C.I.P. Catalogue record for this book is available from the library of Congress.

ISBN 1-4020-4626-X

Published by Springer
P.O. Box 990, 3300 AZ Dordrecht, The Netherlands

Sold and distributed in North, Central and South America
by Springer
101 Philip Drive, Norwell, MA 02061, U.S.A.

In all other countries, sold and distributed
by Springer
P.O. Box 990, 3300 AZ Dordrecht, The Netherlands

Printed on acid-free paper

All Rights Reserved
© 2006 Springer
No part of the material protected by this copyright notice may be reproduced or utilized in any
form or by any means, electronic or mechanical, including photocopying, recording or by any
information storage and retrieval system, without written permission from the copyright owner.

Printed in the Netherlands

Table of Contents

International Journal of Fracture (2006) 138:1–2
DOI 10.1007/s10704-006-0039-x

© Springer 2006

Editorial*

The present volume of the International Journal of Fracture* presents the contents of the Honour, Opening, Closing and Plenary Lectures given at the 11th International Conference on Fracture (ICF11), Torino, Italy, March 20–25, 2005, the leading international forum on fracture, fatigue, material strength, structural damage and integrity, which every four years is organised in different parts of the world. The related International Society – International Congress on Fracture (ICF) – was founded by Professor Takeo Yokobori in 1965, exactly 40 years ago, and is today – under the actual Presidency of the Editor-in-Chief of this Journal, Professor K. Ravi-Chandar – the premier international body for promoting world-wide cooperation among scientists and engineers on the above mentioned topics. ICF11 has been organised under the High Patronage of the President of the Republic of Italy, under the auspices of the Ministry of Infrastructures and Transportation of the Italian Government, and of the National Science Foundation of Italy (CNR), with the scientific support of the leading continental institutions on the subject of the conference: the European Structural Integrity Society (ESIS), and the American Society for Testing and Materials (ASTM). The Italian Group of Fracture (IGF), the Politecnico di Torino and the Turin Academy of Sciences have taken the role of host organisations.

The Opening Ceremony, introduced and presented by the ICF11 Chairman, Professor Alberto Carpinteri, started with different opening addresses from: the ICF President, Professor Yiu-Wing Mai, the Deputy President of the Turin Academy of Sciences, Professor Sigfrido Leschiutta, and the Deputy Minister for Infrastructures and Transportation of the Italian Government, On. Ugo Martinat. The Opening Ceremony then continued with the presentation of the Doctorate H.C. in Civil Engineering to Professor Benoit B. Mandelbrot and to Professor Grigory I. Barenblatt by the Rector of the Politecnico di Torino, Professor Giovanni Del Tin, and by the Dean of the first Faculty of Engineering of the Politecnico di Torino, Professor Francesco Profumo. The texts of these speeches are reported in the following, as well as that of the closing address by the new ICF President, Professor K. Ravi-Chandar.

The paper by B.B. Mandelbrot proposes an interpretation of roughness based on fractal geometry and describes the implications of such conjecture on fracture and other physical or financial phenomena. G.I. Barenblatt presents the general classification of scaling laws and the basic concepts of physical similarity. The Paris law of fatigue is discussed as an instructive example of incomplete similarity, where the noninteger power is not a material constant. Takeo Yokobori, the ICF Founder President, offers an historical picture of the scientific activities of the International Congress on Fracture and indicates Complexity Science as

---

*This special issue is also available as hardbound volume. ISBN: 1-4020-4626-X, www.springer.com

the cutting edge of advanced research in fracture. G. Maier et al. write a survey on engineering-oriented results obtained from inverse analysis applied to fracture mechanics. W.W. Gerberich and co-workers emphasize how measured elastic and plastic properties in volumes having at least one dimension on the order of 10–1000 nm are length scale dependent up to a factor of three. H. Gao applies fracture mechanics concepts to hierarchical biomechanics of bone-like materials. He answers some questions related to the optimization of strength, toughness and stiffness.

A. Pineau illustrates the local approaches to fracture for the prediction of the fracture toughness of structural steel. It is shown that the ductile-to-brittle transition curve can be well predicted by these approaches. Y. Murakami investigates on the effect of hydrogen on fatigue properties of metals used for fuel cell systems. J.G. Williams et al. review previous studies on impact loaded bi-material strip in shear. A global energy balance solution is given to include a cohesive zone. A.J. Rosakis and co-workers report on the experimental observation of supershear rupture in frictionally-held interfaces. The results suggest that under certain conditions supershear rupture propagation can be facilitated during large earthquakes. H. Abé and co-workers make a survey of some recent achievements for realizing a reliable circuit design against electromigration failure. The growth of voids in the metal lines ultimately results in electrical discontinuity. R. de Borst presents an overview of novel discretization techniques for capturing propagating discontinuities: meshless methods, partition-of-unity property based shape functions, and discontinuous Galerkin methods.

The guest editors of the present volume would like to thank all the authors of the papers for their outstanding scientific contributions. Special thanks are also due to Dr. Giuseppe Lacidogna, Dr. Marco Paggi, and Dr. Simone Puzzi, for their effective cooperation in the assemblage of the final ICF11 scientific programme, in the organization of the Opening Ceremony, and in the production of the present volume.

Alberto Carpinteri, Yiu-Wing Mai, Robert O. Ritchie
Volume Guest Editors

International Journal of Fracture (2006) 138:3–11
DOI 10.1007/s10704-006-7148-4

© Springer 2006

ICF11 Official speeches

Opening Address of the Chairman of the 11th International Conference on Fracture, Professor Alberto Carpinteri

Authorities, Dear Colleagues, Ladies and Gentlemen,

On behalf of the Organising Committee of ICF11, I am very pleased to introduce the 11th International Conference on Fracture, to be held in Turin, Italy, on March 20–25, 2005. ICF11 has been organised under the High Patronage of the President of the Republic of Italy, under the auspices of the Ministry of Infrastructures and Transportation of the Italian Government, and of the National Science Foundation of Italy (CNR), with the scientific support and sponsorship of worldwide leading Institutions in the fields of Fracture, Fatigue, Material Strength and Structural Integrity, like the International Congress on Fracture itself (ICF), the European Structural Integrity Society (ESIS), the American Society for Testing and Materials (ASTM), while the Italian Group of Fracture (IGF), the Politecnico di Torino and the Turin Academy of Sciences have taken the role of host organisations.

The conference is endorsed by a strong scientific programme and by the attendance of senior Scientists and younger Delegates coming from 53 different Countries. Besides the more traditional topics, the scientific programme will cover exciting and new developments such as scaling laws, nanomechanics, smart materials, biomechanics, geophysics and tectonics, infrastructure durability, damage and restoration of historical and monumental buildings.

Twelve Plenary Lectures will be delivered by well-known Speakers during the Opening, Plenary and Closing Sessions. 38 Keynote Lectures will be delivered by leading Scientists in the field of Fracture to characterise the topics of the Mini-Symposia, whereas nearly 1100 presentations are scheduled to take place during 228 Special and Contributed Sessions. This total is obtained by multiplying the 16 parallel sessions by the 14 working time periods of the conference, and adding four further events in the Auditorium. I would like to express my most sincere appreciation to the Organisers of the Special Sessions and Mini-Symposia as well as to the Referees of the papers.

The Lingotto Conference Centre, selected to host ICF11, is one of the largest in Europe, offering first-rate services. In addition, Turin – a very historic and artistic city but also a modern and dynamic one – is ready to receive you in the best way, as it will be for the Olympic Winter Games one year from now. I hope that you will also have an opportunity to visit other beautiful places and cities of Italy during your post-conference tours. I wish you an enjoyable stay in Italy.

Photo 1. Opening Address of Prof. Alberto Carpinteri, Chairman of the 11th International Conference on Fracture.

Photo 2. Opening Address of Prof. Yiu-Wing Mai, President of the International Congress on Fracture.

Opening Address of the President of the International Congress on Fracture, Professor Yiu-Wing Mai

Thank you very much, Professor Alberto Carpinteri. Ladies and Gentlemen, good morning again, welcome to this Quadrennial Conference of the International Congress on Fracture. It is nice to see so many old friends in the audience, but it is even more pleasing to see many new faces, young and dynamic, to join this Congress. It is perhaps timely on this occasion to revisit and reaffirm the aims and the philosophies of the International Congress on Fracture, when it was first founded by Professor Takeo Yokobori and his team of wise men in the 1960's. The three major aims as put up in the next slide have remained even truer today. On the integration of disciplines, the second aim has captured very well those new areas on fracture mechanics of nanomaterials and biomaterials, as well as smart materials used in infrastructure durability, to name but a few. The first aim is particularly valuable if we extend the definition of "public community" to include developing countries with fast economic growth, such as China, the East European block, and some Asian countries, on the necessity and awareness of structural safety, integrity, on codes and standards as enabled by fracture research to avoid disasters and catastrophes.

Over the years, many Presidents of ICF together with their Executive Committees have worked towards achieving those aims and objectives. Professor Yokobori is our Founder President, and he is here amongst us, sitting in the first row. Also present here are the past Presidents, Rob Ritchie and David Taplin. The first ICF was held in Sendai in 1965, and then every four years, in other cities, as shown in this slide. So, started from Sendai, Brighton, Munich, Waterloo, Cannes, New Delhi, Houston, Kiev, Sydney and Honolulu, we are now in Turin. We had eleven ICFs held in different countries spread over the continents. ICF 11 is our 40th anniversary after Sendai in 1965, and we have a very exciting and innovative programme in the next five days, as Professor Alberto Carpinteri already mentioned. I would like to take this opportunity to thank Professor Carpinteri and his team in putting the program together. I hope you will have, over the next few days, a most fruitful and rewarding time to exchange scientific ideas and technical discussions, to meet new and old friends, and to visit this beautiful city of Turin. So welcome again to ICF11 and thank you very much.

Ceremony of Presentation of the Doctorate Honoris Causa in Civil Engineering to Professor B.B. Mandelbrot and to Professor G.I. Barenblatt by Professor Giovanni Del Tin, Rector of the Politecnico di Torino

It is my great pleasure to open the 11th Conference on Fracture, representing the Politecnico di Torino, which was the first Technical University to be founded in Italy, in the year 1859. Our Institution has always been very much involved in studies regarding material strength, structural integrity and, in recent years, fracture mechanics. Today, fracture mechanics is part of several different Master and Doctorate programs in Civil, Mechanical, Aeronautical, Chemical, Materials, Nuclear Engineering. The research work conducted on fracture at the Politecnico by young Post-Doctoral Research Fellows, as well as by senior Scientists, is remarkable and fruitful, as this significant event will also bear out.

This is the first ICF Conference to be held in Italy. Our manufacturing and infrastructural industries are really interested in the topics addressed at the conference, in that they affect the know-how and development of strategic fields: from aerospace to nuclear engineering, from nanotechnology to civil engineering, from geophysics to the maintenance of the architectural and archaeological heritage. The Italian academic environments are particularly proud of having organised such an important conference in Italy, and even prouder are our Colleagues from the Politecnico, for having brought ICF to the capital of Italian Industry.

Fracture mechanics can be considered – today, and even more so into the 21st Century – not solely as an important section of a discipline – Material Strength Theory – but rather as the discipline in itself. All the significant aspects of material and structural behaviour are described and analysed in terms of fracture mechanics concepts: creep, plastic deformation, ductile failure, brittle rupture, dimensional transitions, etc. At the same time, the range of structural types and materials dealt with by fracture mechanics is great and extremely diversified: from nanotubes to bridges, from metallic alloys to biological tissues, from fibre-reinforced materials to foams. In such a broad and global context, the boundaries between traditional disciplines – mathematics, mechanics, physics, material science, engineering – tend to fade and vanish. Today, in its maturity, Fracture Mechanics appears interdisciplinary and complex, and, for these very reasons, one of the most challenging research topics in Science.

Two eminent and very well-known Scientists, who have made outstanding and seminal contributions to the Science of Complexity, will give today the Opening Lectures of the conference, and, at the same time, their *Lectiones* for a Doctorate Honoris Causa in Civil Engineering from the Politecnico di Torino: Benoit B. Mandelbrot and Grigory I. Barenblatt. Their names are very famous in different areas of human knowledge. They both utilised and, in some cases, created new concepts and tools – such as nonlinear dynamics, deterministic chaos, fractal geometry, intermediate asymptotics – to describe and capture different natural phenomena – such as turbulence and fracture. A common aspect in Fluid and Solid Mechanics, in fact, is the transition from stability to instability with increasing specimen size-scale: in the former case, we have a transition from laminar to turbulent flow; in the latter, a transition from ductility to brittleness prevails. Such fundamental trends, that no single phenomenological model will fully explain, can only be accounted for by the more comprehensive and synthetic view of dimensional analysis, fractal geometry, intermediate asymptotics, renormalization group theory.

Photo 3. Presentation of the Doctorate Honoris Causa in Civil Engineering to Prof. Benoit Mandelbrot. From left to right: Prof. Mandelbrot, Prof. Carpinteri (Chairman of ICF11), Prof. Francesco Profumo (Dean of the 1st Faculty of Engineering of the Politecnico di Torino), Prof. Giovanni Del Tin (Rector of the Politecnico di Torino).

Photo 4. Presentation of the Doctorate Honoris Causa in Civil Engineering to Prof. Grigory Barenblatt. From left to right: Prof. Carpinteri, Prof. Barenblatt, Prof. Profumo, Prof. Del Tin.

Ceremony of Presentation of the Doctorate Honoris Causa in Civil Engineering to Professor B.B. Mandelbrot and to Professor G.I. Barenblatt by Professor Francesco Profumo, Dean of the 1st Faculty of Engineering

First of all, I would like to say that the 1st Faculty of Engineering of the Politecnico di Torino is really very proud to confer a honorary degree upon Professor Benoit B. Mandelbrot and Professor Grigory I. Barenblatt. They both have been active in several different areas of Engineering, and even in others that are usually considered external to the Engineering context. For this reason, the Doctorate Honoris Causa in Civil Engineering that today will be conferred upon them may even be regarded as exceedingly specific or limited. On the other hand, they both have made outstanding and admirable contributions to Civil Engineering.

Quoting from the minutes of the meeting of the Council of the 1st Faculty of Engineering of June 18, 2004:

"Benoit B. Mandelbrot, founder of Fractal Geometry and pioneer of Complex Systems Mathematics, is a figure of outstanding achievement in all the areas of Applied Science. In the field of Civil Engineering, he has made substantial contributions to the sectors of Geomorphology, Orography, Topography and Solid Mechanics. In particular, through the identification of the fractal nature of fracture surfaces, he paved the way for the attainment of fundamental results such as the definition of

the role of microstructural disorder and the scale effects on the fracture energy of materials".

Professor Mandelbrot can be viewed as the successor to great mathematicians of the past (Cantor, Sierpinski, Menger, von Koch, Hausdorff, Besicovitch, Weierstrass, Peano). They considered as monstrous, pathological or strange the irregular and self-similar sets that represent the rough and jagged character of natural objects. Usually, these sets are characterised by noninteger physical dimensions, intermediate between isolated points and continuous lines, or between lines and surfaces, or between surfaces and volumes.

Quoting again from the same official resolution of the 1st Faculty of Engineering:

"Grigory I. Barenblatt, a scientist characterised by an outstanding range of interests, is an international figure of great stature in different areas of Physics, Mathematics and Engineering. Always grounded in brilliant mathematical insights, his theories have become an absolute term of reference in the literature. In the field of Civil Engineering he has made fundamental contributions to the sectors of Hydraulics, Solid Mechanics and Fracture Mechanics, including, in particular, the cohesive crack model, then successfully applied to most of the materials of technological significance".

Professor Barenblatt extended and generalised the concepts of dimensional analysis, giving a solution to different problems of mathematical physics and an explanation to empirical formulas with noninteger and irrational exponents. He gave a systematic organisation to these problems and showed how his theory of intermediate asymptotics – as well as fractal geometry and renormalization group theory – always lead to power-laws or scaling laws in space and time. Among the problems he addressed we should mention: turbulence, diffusion in porous media, explosions, nonlinear wave propagation, fracture, fatigue. He can be considered as the successor to legendary figures such as Theodor von Karman and G.I. Taylor.

Laudatio to Benoit B. Mandelbrot by Professor Alberto Carpinteri

This Opening Ceremony of ICF11 and the conferring of Honorary Degrees on Benoit B. Mandelbrot and Grigory I. Barenblatt is a consequence of the direction indicated by our Founder President, Professor Takeo Yokobori, to devote future research efforts on the strength and fracture of materials as Complexity Science and Engineering.

"Why is geometry often described as cold and dry? One reason lies in its inability to describe the shape of a cloud, a mountain, a coastline, or a tree. Clouds are not spheres, mountains are not cones, coastlines are not circles, and bark is not smooth, nor does lightning travel in a straight line". With these words Benoit B. Mandelbrot opened his well-known essay "The Fractal Geometry of Nature" (1982). In the natural world, he observed and surveyed irregularities, tortuosities and discontinuities, that appeared on any scale and could not be described and analysed through Euclidean geometry or classical infinitesimal calculus. In this connection, Mandelbrot disagrees with Leibniz's idea: "Natura non facit saltus".

His paradigm of Fractal Geometry, with self-similarity and noninteger physical dimensions of the related sets, strives to reproduce natural phenomena, with their intrinsic roughness and disorder, in space and time.

After studying in Lyon (France), Mandelbrot entered the Ecole Normale in Paris, and then continued his studies at the Ecole Polytechnique (1944) under the supervision of Paul Lévy, an eminent Scientist in Statistical Mechanics. After completing his studies in Paris, Mandelbrot went to the United States, where he visited the California Institute of Technology (1947), and then spent one year at Princeton University, Institute of Advanced Study, under the guidance of John von Neumann (1953). In 1958 he moved to the U.S. on a permanent basis and began his long standing and most fruitful collaboration with IBM at their world renowned laboratories in Yorktown Heights (NY State). This environment allowed him to explore a wide variety of fields and ideas.

Mandelbrot's decision to make contributions to several scientific domains was a deliberate choice, which he made at a young age. During his long scientific life, he has investigated problems in the fields of linguistics, communication errors, computer graphics, fluid dynamics, geophysics, cosmology, finance and economics, and so on. After ritiring from IBM, he found similar opportunities at Yale University (1999), where at present he is Sterling Professor of Mathematical Sciences (first tenured University position). He has also been Visiting Professor at Harvard University and many other prestigious Institutions in the world. Professor Mandelbrot has received numerous honours, including fifteen (15) Honorary University Degrees. The principal motivation is that he seeks a measure of order in physical, biological and social phenomena that are characterized by abundant data but extreme sample variability. In 1994, he delivered the "Vito Volterra Lecture" at the Accademia Nazionale dei Lincei in Rome.

In 1984, with D.E. Passoja and A.J. Paullay, Mandelbrot published his famous paper entitled "Fractal character of fracture surfaces of metals" on the journal *Nature* (Vol. 308, pp. 721–722).

The fractal character of fracture surfaces (invasive fractals of dimension > 2) as well as of cross-sectional ligaments (lacunar fractals of dimension < 2) enabled our research group to recognise a fundamental reason for scale effects on fracture energy and tensile strength.

When the scale of observation approaches zero (as at the micro- or nano-scale), nominal fracture energy tends to zero, whereas nominal tensile strength tends to infinity.

It is therefore with a sense of scientific gratitude that today I can read again a sentence from Professor Mandelbrot's fundamental essay, which encouraged me to work along these lines:

"Almost every case study we perform involves a *divergence syndrome*. That is, some quantity that is commonly expected to be positive and finite turns out either to be infinite or to vanish. At first blush, such misbehavior looks most bizarre and even terrifying, but a careful reexamination shows it to be quite acceptable...as long as one is willing to use new methods of thought" (*The Fractal Geometry of Nature*, W.H. Freeman and Company, New York, 1982, page 19).

Laudatio to Grigory I. Barenblatt by Professor Alberto Carpinteri

"A very common view is that these scaling or power-law relations [in mathematical physics and in engineering] are nothing more than the simplest approximations to the available experimental data, having no special advantages over other approximations. It is not so. Scaling laws give evidence of a very deep property of the phenomena under consideration – their *self-similarity*: such phenomena reproduce themselves, so to speak, in time and space... These powers appeared generally speaking to be certain transcendental numbers rather than simple fractions as for classical self-similarities".

This sentence is taken from the Preface to the well-known book by Professor Barenblatt entitled "Scaling, Self-similarity and Intermediate Asymptotics" (1996), where the Author distinguishes between complete and incomplete self-similarity.

In the former and ideal case, the power-law solutions can be obtained through classical dimensional analysis, whereas in the latter, and more realistic, case, the solutions can be obtained in a more sophisticated way and present anomalous and irrational exponents. This should be considered as an original and fundamental justification for the so-called "empirical" formulas of Applied Science and Engineering.

Grigory I. Barenblatt took his Ph.D. from Moscow University in 1953, under the guidance of Andrej N. Kolmogorov, a famous Scientist in Statistical Mechanics. He became Professor at the same University in 1962. From 1961 to 1975 he was Head of the Division of Plasticity at the Institute of Mechanics. From 1975 to 1992 he was Head of the Theoretical Department at the Institute of Oceanology of the USSR Academy of Sciences.

Since 1990 he has been Visiting Professor at many Western Universities:

Université de Paris VI, Pierre et Marie Curie (1990);
University of Minnesota, Minneapolis (1991);
Rensselaer Polytechnic Institute, Troy-NY State (1991);
University of Rome "Tor Vergata" (1992);
University of Cambridge, G.I. Taylor Professor of Fluid Mechanics (1992–1994) – this was the first time this title was assigned (Emeritus since 1994);
Universidad Autonoma de Madrid (1993);
University of Illinois at Urbana Champaign (1995);
Stanford University, Timoshenko Visiting Professor (1996–1997);
University of California at Berkeley, Professor in Residence (1996–);
Lawrence Berkeley National Laboratory (1997–).

Professor Barenblatt has received numerous honours. He is a Foreign Member of different National Academies, in the US, Denmark, Poland, the UK. He received a Doctorate Honoris Causa from the Royal Institute of Technology in Sweden, and, among many others, two prestigious awards from Italy (both in 1995): the Lagrange Medal from Accademia Nazionale dei Lincei in Rome, and the Panetti Prize from Accademia delle Scienze of Turin.

In 1959 Grigory I. Barenblatt published a series of papers titled "On equilibrium cracks formed in brittle fracture", with different sub-titles. A fundamental mathematical model of elastic body was introduced, simulating cracks and explicit cohesive closing forces in the crack tip region. Such forces cancel the stress-singularity and make the model stable.

In the following years, G.I. Barenblatt considered self-similarity and scaling laws in fracture and fatigue. In particular, as he will clarify in his *Lectio*, he focused on scaling in fatigue crack growth, with reference to the Paris & Erdogan Law, which is a power-law with an anomalous exponent, m.

As regards Fracture Mechanics, the critical vision of G.I. Barenblatt is very similar to that of B.B. Mandelbrot: fracture is a complex and multiscale phenomenon, that we can fully understand and dominate only by making use of appropriate and refined conceptual tools.

Photo 5. Closing Address of Prof. Robert Ritchie, Member of the Organizing Committee.

Photo 6. Closing Address of Prof. Krishnaswamy Ravi-Chandar, the new President of the International Congress on Fracture.

Closing comments

I am honored and pleased to be elected as President of the International Congress on Fracture for the next 4 years. I share this honor with members of the Executive Committee, Professors Alberto Carpinteri, Bhushan Karihaloo and Robert Goldstein as Vice Presidents, Professors Leslie Banks Sills, Elisabeth Bouchaud, Teruo Kishi, Shou-wen Yu, and Robert McMeeking as Directors, Professor Toshimitsu Yokobori, Jr. as the Secretary General and Professor David Taplin as the Treasurer. Professor Yiu-Wing Mai and his team are to be thanked for their excellent service to ICF during the past 4 years. Also, I would like to thank Professor Alberto Carpinteri and his colleagues for their excellent organization of the 11th International Conference on

Fracture, here in Turin; this conference has been the largest in the series of ICFs and a very stimulating conference.

It would not be an exaggeration to claim that advances in fracture control represent one of the crowning achievements of the 2nd half of the 20th Century. We have developed a deeper understanding of material strength and durability, developed better materials and models, improved methods of analysis, experimentation and simulation, developed strategies for structural integrity assessment and lifetime prediction, etc. So much so, that our scaling laws have enabled us to do better than nature! We have altered the landscape (mostly for the better) and built structures that are large and strong to accommodate our needs; while our methods of locomotion are perhaps not as efficient as nature's own, they certainly are more capable, flexible and reliable. Witness the fact that a thousand delegates from around the world were able to assemble in this venue for the sake of scientific knowledge and information. The safety and reliability of these endeavors are ensured by the fracture theories advanced by this community.

All of these accomplishments have had such an impact that ours may appear to be an aging discipline – sadly, there appears to be an overall decline in funding for fracture research within the United States and other countries are likely to follow soon. Indeed there is a growing sense that there is nothing more to be done in fracture research! However, this 11th quadrennial International Conference on Fracture has pointed out the fallacy of such pessimism. Fracture research is vibrant and it touches vast areas of human endeavor; fracture occurs over a wide range of scales in diverse applications as highlighted by the many plenary lectures: Professor Bill Gerberich on nanoscale problems, Professor Huajian Gao on biophysics applications, Professor Ares Rosakis on geophysical applications, Professor Hiroyuki Abe on microelectronics, Professor Murakami on the effects of fracture on a hydrogen economy, just to mention a few topics. The plenary lectures and the focused symposia at this meeting highlighted all of these areas that have just begun to be explored and remain as fertile ground for us to plough further. So, I am excited about being President of ICF at this juncture and hope to be able to promote the organization and the discipline in these interesting times.

The next conference – the 12th International Conference on Fracture – will be held in July 2009 in Ottawa, Canada, under the able leadership of Drs. Bill Tyson and Mimoun Elboujdaini, and I hope to see all of you there!

Krishnaswamy Ravi-Chandar
President, International Congress on Fracture
The University of Texas at Austin
Austin, TX 78712-0235, USA
E-mail: kravi@mail.utexas.edu

International Journal of Fracture (2006) 138:13–17
DOI 10.1007/s10704-006-0037-z

© Springer 2006

Fractal analysis and synthesis of fracture surface roughness and related forms of complexity and disorder

BENOIT B. MANDELBROT
Sterling Professor of Mathematical Sciences, Yale University, New Haven, CT 06520-82383 (E-mail: benoit.mandelbrot@yale.edu)

Abstract. Roughness is, among human sensations, just as fundamental as color or pitch, or as heaviness or hotness. But its study had remained in a more primitive state, by far. The reason was that both geometry and science were first drawn to smooth shapes. Thus, color and pitch came to be measured in cycles per seconds, that is, were reduced to sinusoids, in other words to uniform motions around a circle – the epitome of a smooth shape. A study of roughness had necessarily to wait until specific mathematical tools had been discovered and, much later, suitably interpreted. Fractal geometry began when I reinterpreted the flight from nature that had led mathematicians to conceive of notions like the Holder exponent, the Cantor set, or the Hausdorff dimension. They boasted of these notions being 'monstrous' but in fact I turned them over into everyday tools of science. I also added further tools that – taken together – made roughness quantitatively measurable for the first time. Acquiring a quantitative measure is the step that moves a field into maturity. And this move instantly led to a striking conjecture. In 1984, 'Nature' published an article I wrote with D. A. Passoja and A. J. Paullay on metal fractures. We found that the traditional measures of their roughness range all over. To the contrary, their fractal roughness varies very little not only between samples but also between materials. Last time I checked the "universality" had been extended but not explained. The new intrinsic measure created a major intellectual mystery. The first major new tool that I added to those contributed by the likes of Holder, Cantor, and Hausdorff was multifractality, for both measures and functions. I was motivated by the urge to model the intermittence of turbulence but my first full paper (in 1972) also noted that the same techniques ought to apply to the intermittence in the variation of financial prices. An ancient adage claimed that the City of London is as unpredictable as the weather. I found unexpectedly quantitative truth to this adage by showing that both phenomena can be tackled with essentially the same tools. Roughness is everywhere therefore fractal geometry has little fear of running out of problems. This address will sketch the fractal geometry of roughness and explore some new developments relevant to this Congress.

1. Introduction

Throughout my long working life, I have constantly been seeking the thrill of transforming old toys into new tools. Think of Kepler who changed ellipses from geometric toys into tools that describe the planets' motions. What I have always sought is the thrill of identifying analogous problems; problems that take a 'useless mathematical device' and change it into the defining tool solving a very practical problem.

High among my list of 'Keplerian' thrills is the use of the 'Holder-Lipschitz exponent' and 'fractal dimension' to provide – at long last – a numerical measure of the ubiquitous but elusive notion of roughness. When I began to prepare my contribution to these Proceedings, I was hoping to provide a survey of this match. But lack of time has defeated this effort and the best that can be done consists in a very general text and two very specific figures. Figure 1 is the original one and Figure 2 is more recent.

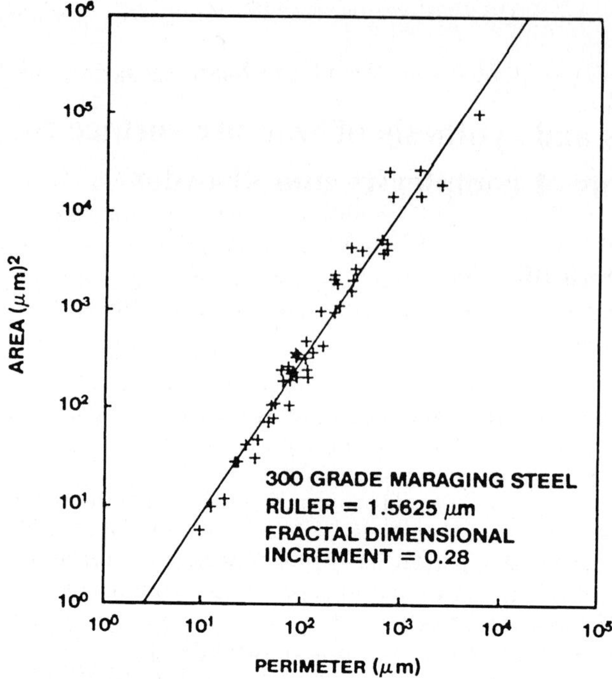

Figure 1. Original evidence of the fractality of metal fractures. (From Benoit B. Mandelbrot, Dann E. Passoja & Alvin J. Paullay, Fractal character of fracture surfaces of metals. *Nature* **308** (1984) p 721).

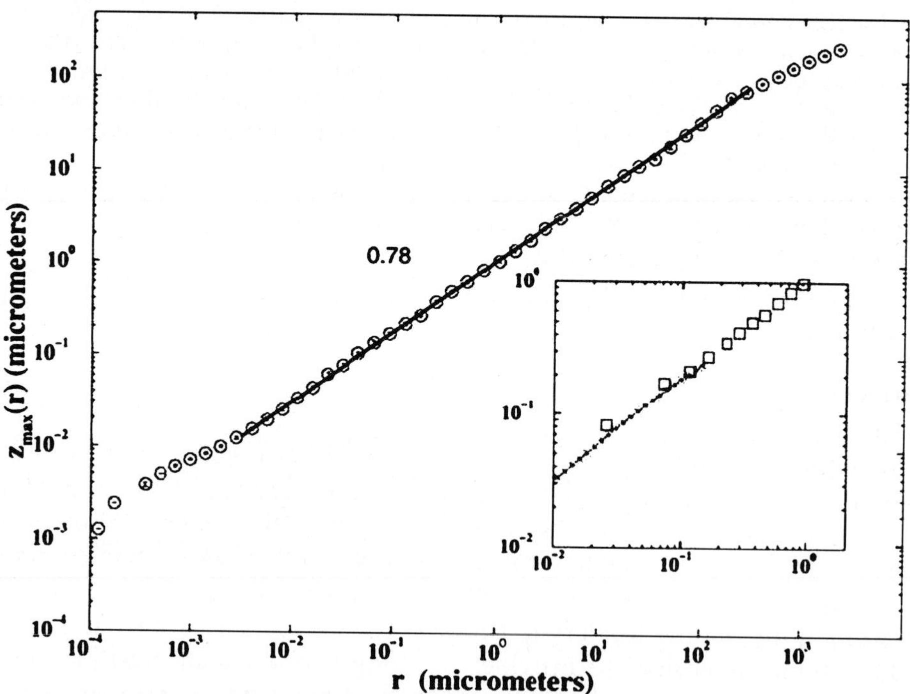

Figure 2. Recent confirming evidence of the fractality of metal fractures. (Figure 10 of Elizabeth Bouchaud, Scaling properties of cracks. *Journal of Physics: Condensed Matter* **9** (1997) p 4336). The experimental points obtained with two techniques gently collapse onto the same curve with overlap extending approximately from 10 nm to 1 m.

2. Toward a geometry of roughness, one of the basic sensations of Man, ubiquitous in nature and culture

Why is Euclid's standard geometry so widely and so often described as grey, dull, cold, and dry? One reason lies in its inability to describe the shape of a cloud, a mountain, a coastline, or a tree. Clouds are not spheres, coastlines are not circles, and bark is not smooth, nor does lightning travel in a straight line. Mt. Fuji excepted, of course, mountains are not cones. More generally, everyday patterns of Nature are irregular and fragmented while standard geometry favors the smooth and simple. The level of complexity of much in Nature is not merely higher but altogether different. For practical purposes, natural patterns show an infinite number of distinct scales of length. Identical remarks apply to many aspects of culture.

More specifically, Poincaré remarked that there are questions that one chooses to ask and other questions that ask themselves. The latter can be mundane. • How to measure and compare the roughness of ordinary objects such as broken stone, metal, glass, or rusted iron? • How long is the coast of Britain? • What shape is the Earth, more precisely a mountain, a coastline, a river, or a dividing line between two rivers' watersheds? That is, can the term 'geometry' deliver what it seems to promise? • How to define the speed of the wind during a storm? • What shape is a cloud, a flame, or a welding? • What is the density of galaxies in the universe?

In this list, the first few demonstrate that a question that has long remained without response tends to be abandoned to children. Let's move on to related questions. • How to characterize the boundary of a plane random walk? Analogous questions are constantly being added. • How to characterize the boundary between two basins of attraction in a chaotic dynamical system? • How to measure the variation of the flow of messages on the Internet? • How to measure the volatility of the prices quoted on financial markets?

All these questions challenge us to change our view of nature, to tame forms that standard geometry considers as formless. My broad and specific response to this challenge was to conceive and develop a new geometry of nature and culture.

Fractals' key feature is that they satisfy a form of 'dilation-reduction' invariance called 'scaling' which expresses that the degree of their irregularity and/or fragmentation is identical at all scales. Fractals can be curves, surfaces, or disconnected 'dusts,' and some are so oddly shaped that we formerly lacked terms for them. For example, Earth's relief, clouds, and real physical fractures are natural fractal surfaces. I showed how to imitate them by computer-generated fractals. Price records are multifractal functions somehow created by culture. I showed how to imitate them by computer-generated multifractals. Imitation, it should not be forgotten, is the first step to understanding.

3. Where does the fractal theory of roughness stand among the other branches of physics?

Let me argue that my work has belonged all along to a suitably broadened view of the scope of physics. The classic branches of physics arose directly from the desire to describe, understand, and control the basic messages that the brain receives from the senses. The transformation of a mess of sensations into a quantitative science followed about the same process for many sciences. In most instances, three stages stand out reasonably clearly: elaboration of a rich descriptive vocabulary, elaboration of a 'narrative' with a reassuring message of how everything holds together, and a proper quantitative science. The latter only begins with the identification of a numerical measure for some common ordinary feature of everyday experience.

Quantitative measures of mass and size came early and mark the dawn of history. Visual signals led very early on to the vocabulary of bulk and shape, and of brightness and color; auditory signals, to the vocabulary of loudness and pitch, and to scales known to musicians since time immemorial. However, optics and acoustics did not fully arise as sciences until two hundred years ago, when concrete notions represented by ancient words were made quantitative, for example, when color and pitch could be measured by frequencies. Similarly, the first step in the theory of the perennial sensation of "hot" was well-recorded by history. It was the 'invention' by Galileo of the notion of bodies of uniform hotness, together with a proper intrinsic quantitative measure of hotness: the temperature.

Judged against this background, the sense of smooth vs rough, which is a priori equally essential, remained in a primitive state. There was no rich descriptive vocabulary and nothing comparable to temperature had been invented, until I identified suitable exponents in remote mathematical esoterica due to Hölder and Lipschitz around 1870, and Hausdorff in 1919. To accomplish this has been an awesome and humbling privilege.

Previously, the notion called either roughness or volatility was measured by a mean square deviation from a 'norm.' For fluctuations in a perfect gas in equilibrium, this measurement happens to be formally proportional to temperature! But in the context of roughness or volatility, it proved an inappropriate borrowing.

Instead, starting almost from scratch, I had to create a new tool-box specifically geared towards the study of forms of roughness that possess certain geometric scaling invariances. Each invariance intrinsically introduces one or more numerical invariants. I reinterpreted one as the first of many quantitative measurements of roughness.

Later, many additional intrinsic measurements were also brought up by fractal and multifractal geometry; it even made a set's 'degree of emptiness' into a concrete and useful notion.

Inevitably, a growing toolbox raises a question: To what extent are the quantitative concepts exact or only approximate? For the sake of comparison, the science of sound showed healthy opportunism by centering on the idealized pure sounds of strings and pipes. Those pure sounds proved to be useful 'icons' because they are quantitatively measurable and mathematically manageable. In addition, their study suffices for song and music, and clarifies even the sounds that it fails to characterize or fully explain, such as drums and concert halls.

More generally, in the long road from raw sensation to science, a key moment was marked by the successful identification of a proper 'compromise' between simplicity and breadth of applicability. Fractal dimension is the first agreed-upon quantity able to measure pure roughness the way temperature measures uniform hotness. This was analogous to the question of applicability of the notion of a system of uniform temperature. A growing and long list of widely diverse examples shows fractal or multifractal forms of roughness/fragmentation to be ubiquitous to an extraordinary degree in nature and culture.

Specific applications cannot be discussed here further but let me note that my early illustrations of a multifractal measure has provided excellent models of the variations on financial prices and on the internet.

A coexistence emerges in physics, between the usual smooth or smoothly varying phenomena, and the fractal ones. This raised a general issue. In responding, I observed that numerical solutions of very familiar partial differential equations easily yield close approximations to fractals. For example, this happens under new but natural conditions that include movable singularities or boundaries. They occur for the Laplace equation in the case of interacting galaxies under Newton attraction, and for diffusion-limited aggregates. This and other motivations led me to put forward a bold and general conjecture described in chapter 11 of *The Fractal Geometry of Nature*. Solving the usual partial differential equations of physics can yield either the familiar and expected smoothness, or fractality.

Reference

My 1982 book, *The Fractal Geometry of Nature*, remains the basic reference but hundreds of more specialized books are now available. On my web homesite www.math.yale.edu/mandelbrot, they are listed in a continually updated "Chronicle of Books".

International Journal of Fracture (2006) 138:19–35
DOI 10.1007/s10704-006-0036-0

© Springer 2006

Scaling phenomena in fatigue and fracture

G. I. BARENBLATT
*Department of Mathematics, University of California and Lawrence Berkeley National Laboratory
Berkeley, CA 94720-3840, USA (E-mail: gibar@math.berkeley.edu)*

Received 1 March 2005; accepted 1 December 2005

Abstract. The general classification of scaling laws will be presented and the basic concepts of modern similarity analysis – intermediate asymptotics, complete and incomplete similarity – will be introduced and discussed. The examples of scaling laws corresponding to complete similarity will be given. The Paris scaling law in fatigue will be discussed as an instructive example of incomplete similarity. It will be emphasized that in the Paris law the powers are not the material constants. Therefore, the evaluation of the life-time of structures using the data obtained from standard fatigue tests requires some precautions.

Key words: fracture – advanced similarity analysis, fatigue – advanced similarity analysis, scaling laws in fatigue, scaling laws in fracture, incomplete similarity, pan's law, fracture – cracks, fatigue – cracks, dynamic cracks.

1. Introduction

Signore Presidente, Ladies and Gentlemen, it is my pleasant duty to thank the Organizing Committee and personally the Chairman, Professor Alberto Carpinteri, for the honor rendered to me by the invitation to deliver this lecture.

Scaling laws

$$y = C x_1^{\alpha_1} \dots x_n^{\alpha_n} \tag{1}$$

often appear in modeling phenomena in nature, engineering, and society. The reason of their importance is that the scaling laws reveal a deep feature of processes: *self-similarity*. For processes developing in time self-similarity means that the phenomenon is reproducing itself in scales, which vary in time:

$$\mathbf{U}(\mathbf{r}, t) = \mathbf{U}_0(t) f\left(\frac{\mathbf{r}}{\ell(t)}\right). \tag{2}$$

Establishing scaling laws and self-similarity was always considered as an important, sometimes crucially important step in construction of engineering and physical theories. In the pre-computer era they were considered as special, "exact" solutions illuminating complicated models, elegant, sometimes useful, but nevertheless restricted in their value, elements of theories. Later, when computers entered into play, the role of such solutions did not diminish, just the contrary. However, the general attitude to them changed: they started to attract attention mainly as *"intermediate asymptotics"* – an important element of physical or engineering theories describing the behavior of systems when the influence of accidental details of the initial and/or boundary conditions already disappears, but the system is still far from the ultimate state of equilibrium.

Establishing the scaling laws and self-similarities was of special importance in studies of turbulence and structural strength. Remarkably both these branches of natural and engineering science were started by two great Italians: Leonardo da Vinci, born in a suburb of Florence, and Galileo Galilei, born in Pisa: both these places are not far from Torino where we are assembled these days.

Turbulence is the fluid flow with randomly accumulating and dissipating vortices.

Structural strength is inseparable from *fatigue*. Fatigue in a broader sense is the deformation and flow of solids with accumulating defects, terminating by *fracture*.

Generally in scientific circles turbulence is considered as a major challenge of applied mathematics and classical physics. Among those who made seminal contributions to studies of turbulence were giants – A.N. Kolmogorov and W. Heisenberg, as well as great applied mathematicians of the 20th century – L. Prandtl, Th. von Kármán, and G.I. Taylor.

I have studied both structural strength and turbulence for 50 years, and I can say that in the circles of physicists and applied mathematicians the problem of structural strength, and in particular the problem of fatigue, remains overshadowed by turbulence. Meanwhile, fatigue and fracture present not less thrilling fundamental problems than turbulence, not to speak about practical importance. Moreover, the phenomenon of fatigue is in principle even more complicated! Indeed, stop the turbulent flow of air and/or water, and the fluid becomes indistinguishable from that at the beginning of the motion. This is not the case for fatigue. And there is something worse. In turbulence, we have experimental reasons to believe that the fluid, like air and water remains Newtonian even in the most complicated flows up to the scales of smallest vortices. At the same time even for quasi-brittle solids the very possibility of using any constitutive equations for the materials near the defects is doubtful!

Parallels between turbulence and fatigue are instructive – also in another aspect. Yes, turbulence is generally considered as a great challenge, and during more than a century an army of engineers and scientists led by geniuses attacked it. But let's ask ourselves with full sincerity: what was achieved during this time as far as the creation of a pure, self-contained theory of turbulence based on first principles like, e.g., the theory of elasticity is concerned? The answer is disappointing – nothing! And it is clear now that such a theory will not be constructed in real time – such was in particular the opinion of A.N. Kolmogorov. He claimed that the practical way is to construct models based on special hypotheses relying on the results of experimental studies. Clearly such models could be valid for special classes of flows only. It can be expected that the same path is to be followed in structural strength and fatigue studies.

I want to emphasize that practically all significant results in turbulence were obtained using similarity considerations and scaling laws. The value of these tools and their technique should be properly understood, and they should be used in everyday practice by engineers and researchers working in fatigue and fracture.

It is well known that the subject of scaling in structural strength attracted the attention of engineers and researchers starting from Galileo. There exist multiple treatises, monographs, and papers, where the scaling problems in structural strength are treated. I want to mention especially recent monographs and papers by Bažant and his school (Bažant, 1984, 2002, 2004a,b; Bažant and Planas, 1998; Bažant and Yavari, 2005) – and our President Carpinteri and his school (Carpinteri, 1989, 1994, 1996, 1997; Carpinteri and Massabo, 1996; Carpinteri et al., 2002, 2003). We all learned a lot from their remarkable works. We have a special section to discuss the problems of scaling in structural engineering, where their contributions will find a descent place.

I will speak about the results related to a different approach, developed independently starting from the late 1950s – early 1960s in my close collaboration with my unforgettable colleague and friend, Ya. B. Zeldovich (Barenblatt and Zeldovich, 1971, 1972). My students and close colleagues now eminent Professors G.I. Sivashinsky and L.R. Botvina participated in various parts of these works.

2. Scaling laws obtained by dimensional analysis

Naturally I will pay basic attention to fatigue and fracture. However, I want to make a short *intermezzo*: to demonstrate one of the milestones not only of fluid mechanics, but of engineering science as a whole achieved by similarity methods. I mean the famous scaling law obtained first by Taylor (1941) for the radius r_f of the shock wave formed after an atomic explosion (Figure 1):

$$r_f = \left(\frac{Et^2}{\rho_0} \right)^{1/5}. \tag{3}$$

Here E is the energy of the explosion, t the time after the explosion, and ρ_0 is the air density before the explosion. Scaling law (3) was obtained by Taylor using *dimensional analysis* (I will demonstrate this technique on examples from fracture

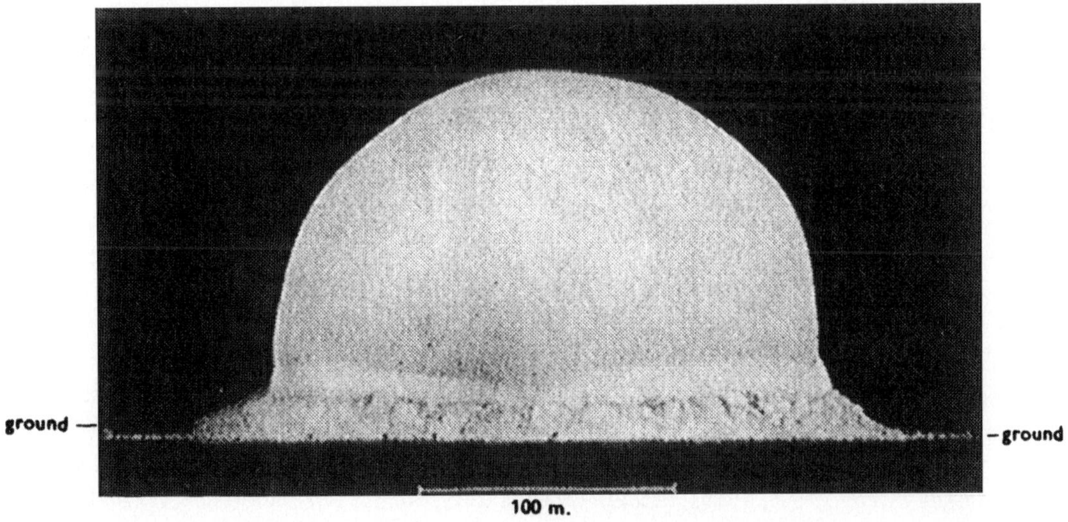

Figure 1. Photograph of the fireball of the atomic explosion.

and fatigue later, the explanation of it with details and examples can be found in Barenblatt, 1996, 2003) I emphasize specially: of crucial importance was not the formal application of rather simple rules of dimensional analysis, but the preliminary idealization of the problem, which was invented by Taylor: he made two basic assumptions – the explosion is instantaneous and concentrated at a point. Furthermore, the initial atmospheric pressure p_0 is negligible in comparison with the pressure behind the shock wave. This basic idealization allowed Taylor to assume that the radius of the shock wave r_f can depend on E, t, and p_0 only. After this assumption scaling law (3) followed from dimensional analysis immediately. Scaling law (3) obtained excellent experimental confirmation (Taylor, 1950, Figure 2). I bring your attention again to the fact that this remarkable final result was obtained without solving complicated equations.

In fracture mechanics also there are well-known scaling laws, which can be obtained after a plausible idealization in Taylor style using dimensional analysis. I want to mention here some of them. The first example is the symmetric wedging of a brittle or quasi-brittle thick plate (plain strain) by a thin wedge (Barenblatt, 1959) (see Figure 3a).

We introduce a natural idealization. First we assume that the spearheads of the wedge are concentrated on opposite lines, so that the action of the wedge is represented by two equal and oppositely directed concentrated tractions P, per unit length of the plate thickness (Figure 3b). Further, we assume that the plate is infinite. Thus, the length of the crack ℓ can depend only on P, the Poisson ratio v, and a characteristic of fracture toughness, the *cohesion modulus K*. I emphasize here the known difference between the cohesion modulus K, introduced by the author in (Barenblatt, 1959), and determined under conditions of stable crack propagation, and another fracture toughness characteristic K_{Ic}, introduced practically simultaneously by Irwin (1960) and determined by the beginning of the unstable crack propagation. In the case under consideration the crack is stable, therefore, a relation for the crack length ℓ is valid

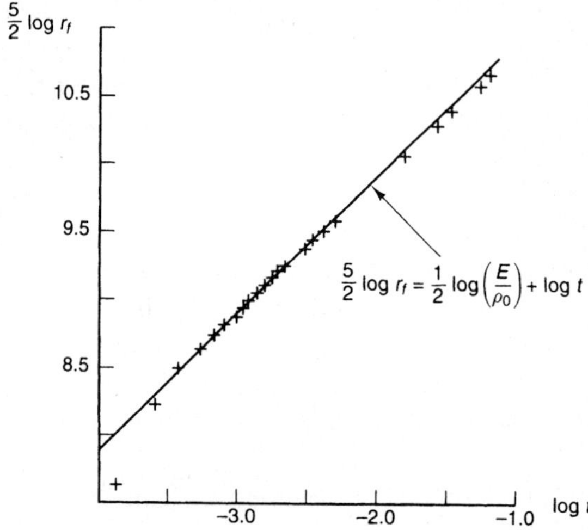

Figure 2. The experimental results confirm scaling law (3) (Taylor, 1950).

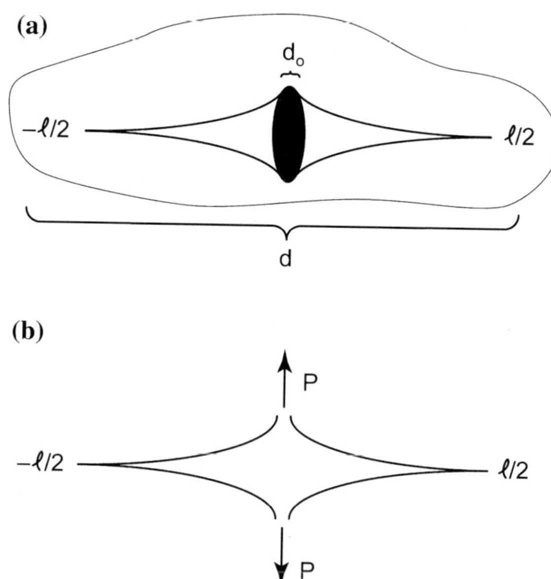

Figure 3. The wedging of a thick plate. (a) Original problem. (b) Idealized problem.

$$\ell = f(P, K, \nu). \tag{4}$$

The dimensions of the quantities entering relation (4) are obviously as follows:

$$[\ell] = L, \qquad [P] = FL^{-1}, \qquad [K] = FL^{-3/2}, \qquad [\nu] = 1,$$

where L and F are the dimensions of the length and force. From the arguments only P and K have independent dimensions, and only one quantity of the dimension of length can be formed of them: P^2/K^2. Therefore, the dimensionless quantity $\ell/(P^2/K^2)$ can depend on the Poisson ratio ν only, and we obtain

$$\ell = \mathrm{Const}(\nu)\frac{P^2}{K^2}. \tag{5}$$

An analytic calculation (Barenblatt, 1959) shows that Const is equal to one. Scaling law (5) also obtained a reliable experimental confirmation, and relation (5) served for experimental evaluation of the cohesion modulus (see the comprehensive book by Panasyuk, 1968).

The cohesion crack model was introduced by Barenblatt (1959). Before that the analogous problem was considered by Barenblatt (1956) for the case when the propagation of the crack is resisted by uniform compression q, like in rock massives (also an instructive illustrative example is achieved by a ruler inserted between the pages of a horizontally lying book). I want to remember here my unforgettable colleagues S.A. Christianovich and Yu. P. Zheltov who started (Zheltov and Christianovich, 1955) to consider such problems in connection with hydraulic fracturing of oil strata. In this case the scaling law is different:

$$\ell = \frac{2}{\pi}\frac{P}{q} \tag{6}$$

so that the length of the crack is proportional to the first, not the second, degree of the load.

A remarkable scaling law was obtained using the dimensional analysis by Roesler (1956), and Benbow (1960) for the cone crack formed under small diameter punch in a block of fused silica (Figure 4). Roesler and Benbow also replaced the problem by an idealized one: point indentor punched into an infinite block. The scaling law, also obtained by dimensional analysis, for the diameter of the base of the cone crack D under the load P has the form:

$$D = \mathrm{Const}(\nu) \left(\frac{P}{K} \right)^{2/3} .$$

(7)

This scaling law also obtained remarkable experimental confirmation (Figure 5).

When considering the examples presented above, a natural question can arise: in fact, for instance, for the problem of wedging of a thick plate, there appear two additional length scales: the width of the spearhead of the wedge d_0, which we equal to zero, and the size of the block d, which was assumed to be infinite. However, in

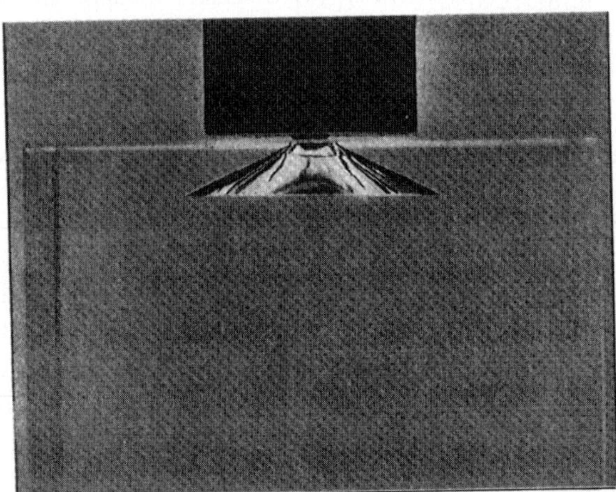

Figure 4. Conical crack in a fused silica block (Benbow, 1960).

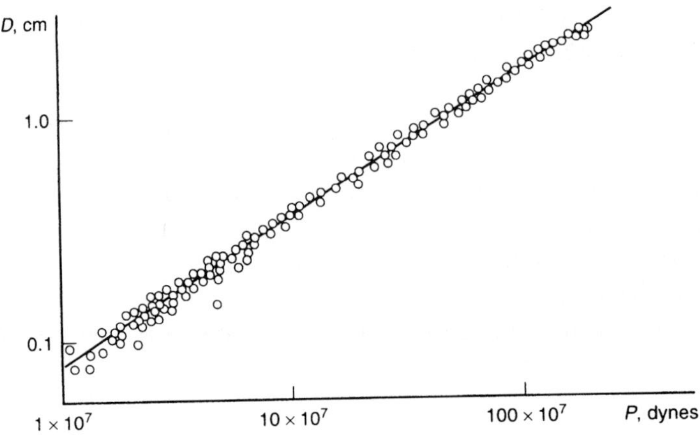

Figure 5. The experimental data confirm scaling law (7) (Benbow, 1960).

fact, both these length scales are finite, and if we take them into account in relation (4), the dimensional analysis will lead to a different, and completely non-constructive result:

$$\ell = \left(\frac{P}{K}\right)^2 \Phi(\Pi_1, \Pi_2, \nu),$$

$$\Pi_1 = \frac{d_0}{\left(\frac{P^2}{K^2}\right)}, \qquad \Pi_2 = \frac{d}{\left(\frac{P^2}{K^2}\right)}. \tag{8}$$

In fact, scaling law (5) is an *"intermediate-asymptotic law"* valid when

$$d_0 \ll \ell \ll d. \tag{9}$$

The advantage of the problem of wedging of a plate is that it can be proved rigorously that there exists a finite, non-zero limit of the function $\Phi(\Pi_1, \Pi_2, \nu)$, at $\Pi_1 \to 0$ and $\Pi_2 \to \infty$. Therefore, at sufficiently small Π_1, and sufficiently large Π_2 the intermediate asymptotic scaling law (5) works. Apparently, excellent experimental confirmation of scaling laws (3) and (7) shows that the same situation happens in Taylor and Roesler–Benbow problems, although up to the present time nobody has proved it rigorously.

But what is the practical meaning of this fact? It means that if we take a different larger or smaller wedge, and a larger or smaller block from the same material, so that $d_0' = \lambda d_0$, $d' = \mu d$, where λ and μ are certain positive numbers of the order of unity, nothing will change at this intermediate stage! Similarly, if we take a smaller or larger punch and a smaller or larger block in the Roesler–Benbow problem the propagation of the crack at the intermediate stage will not change. Also, in Taylor problem of very intense explosion the moderate variation of the charge size and of the atmospheric pressure will change noting in the propagation of the shock wave.

These are simple examples of the invariance of the intermediate asymptotics with respect to the *renormalization group*, a concept of extreme importance in the whole of this subject. The details concerning the renormalization group and its connection with intermediate asymptotics can be found in the books of Goldenfeld (1992) and Barenblatt (1996, 2003).

3. Classification of scaling laws: complete and incomplete similarity

A natural question arises. There exist many scaling laws in engineering science and physics. Whether the idyllic situation demonstrated above on three examples is always the case, and all scaling laws can be obtained using dimensional analysis after a simple natural idealization-like point explosion or pressing along a line in an infinite space. The answer is no, and there should be no illusions: as a rule the scaling laws and self-similarities cannot be obtained by dimensional analysis alone, and such "natural" idealization does not exist. In particular, this is true for a basic scaling law in fatigue: the Paris law; we will see it later.

To clarify this point, let's consider the scaling laws from a more general viewpoint. Namely, consider a certain physical relation, i.e., a relation, valid for all observers having different sizes of units of measurement of the same physical nature:

26 *G.I. Barenblatt*

$$a = f \overbrace{(a_1, \dots, a_k, b_1, b_2 \dots)}^{n \text{ arguments}}. \tag{10}$$

Here a is the quantity, possibly a vector, to be determined, $a_1, \dots, a_k$ – the arguments, having independent dimensions, like say length, time, density whereas the dimensions of a, b_1, b_2 – are dependent, and can be expressed via the dimensions of $a_1, \dots, a_k$:

$$[a] = [a_1]^p \dots [a_k]^r; \qquad [b_1] = [a_1]^{p_1} \dots [a_k]^{r_1} \dots . \tag{11}$$

Dimensional analysis allows one to reduce relation (10) to a dimensionless form:

$$\Pi = \Phi \overbrace{(\Pi_1, \Pi_2 \dots)}^{n-k \text{ arguments}},$$
$$\Pi = \frac{a}{a_1^p \dots a_k^r}, \qquad \Pi_1 = \frac{b_1}{a_1^{p_1} \dots a_k^{r_1}}, \dots \tag{12}$$

The advantage of relation (12) in comparison with the original relation (10) is that the number of arguments of function Φ in (12) is less than the number of arguments of function f in the original equation (10). Moreover, according to (12) the basic function f entering a physical relation possesses an important property of *generalized homogeneity*, i.e., it can be represented via a function Φ of a lesser number of arguments:

$$f = a_1^p \dots a_k^r \Phi \left(\overbrace{\frac{b_1}{a_1^{p_1} \dots a_k^{r_1}}, \dots}^{n-k \text{ arguments}} \right). \tag{13}$$

When we idealized the problems considered in the previous section, we did not intend to consider the limiting cases $\Pi_1 = 0$, $\Pi_2 = \infty$. What we really did – we obtained the asymptotics (*intermediate asymptotics*), i.e., the asymptotic relations which appear when $\Pi_1, \Pi_2, \dots$ are small or large, but not equal to their limiting values.

If there exists a finite non-zero limit C of the function Φ at $\Pi_1, \Pi_2 \dots \to 0, \infty$, then at sufficiently small (large) $\Pi_1, \Pi_2 \dots$ the function Φ can be replaced by its limiting value C, and the scaling law is valid:

$$a = f = C a_1^p \dots a_k^r \tag{14}$$

so that the arguments $b_1, b_2 \dots$ are not represented in the resulting relation, and all powers $p, \dots, r$ can be obtained by a simple procedure of dimensional analysis, as it was demonstrated in the problems considered in the previous section.

Colleagues, we say often, teaching our students, and reducing the relations to a dimensionless form, that if a dimensionless argument is small or large, it can be neglected. Generally speaking this statement is erroneous; it is correct only when the finite non-zero limit of the function Φ in (12) does exist, which is *a priori* unknown. And what will happen if such a finite non-zero limit of the function Φ does not exist, which is in fact the general case? In this case, we cannot say anything in general terms, if, of course, we do not know the solution, in which case the similarity approach is superfluous.

However, the following remarkable possibility should not be missed. Let the finite non-zero limit of the function Φ not exist, but at small (large) $\Pi_1, \Pi_2, \ldots$ let the function Φ possess in its turn the property of generalized homogeneity in its dimensionless arguments. Here, we consider the simplest case (for the general discussion the reader can be addressed to the books of Barenblatt, 1996, 2003), which we will need further:

$$\Phi = \Pi_1^\alpha \Phi_1 + \cdots , \tag{15}$$

where this time the function Φ_1 has a finite non-zero limit at $\Pi_1, \Pi_2, \ldots \to 0, \infty$. Substitute (15) to (12), and we obtain that in this case again at large (small) values of $\Pi_1, \ldots$, a scaling law of type (14) is valid:

$$a = f = C a_1^{p-\alpha p_1} \ldots a_k^{r-\alpha r_1} b_1^\alpha . \tag{16}$$

There is however, a substantial difference between scaling laws (14) and (16):

(a) The exponents $p - \alpha p_1 \ldots r - \alpha r_1$ cannot be obtained by using dimensional analysis alone, because the power α in (15) is unknown. It should be obtained using some additional information, including sometimes experimental or computational data.

(b) The argument b_1 does not disappear from scaling law (16), it continues to influence the phenomenon, although it enters only in multiplicative combination with other parameters.

(c) The generalized homogeneity of function Φ, contrary to the case of function f in the original relation (10), does not follow from a fundamental physical principle of equality of all observers having different magnitude of the units of measurement of the same physical nature. Just the contrary: this is a property only of the special phenomenon under consideration.

In the case when the scaling law of type (14) is valid we speak about the *complete similarity* of the phenomenon in the parameters $\Pi_1, \Pi_2, \ldots$ (cf. Reynolds number similarity in turbulence, the generally accepted term for the situation when the influence of the Reynolds number disappears).

In the second case, when the scaling law of type (16) is valid we speak about *incomplete similarity*. Clearly, incomplete similarity is a much more general case than complete similarity, although it is still a very special case of asymptotic laws at small (large) values of parameters $\Pi_1, \Pi_2 \ldots$. Discovery of incomplete similarity in gas dynamics, turbulence and fatigue marked important steps in the development of these disciplines. Mandelbrot fractals (1975, 1977, 1982) also present a remarkable example of incomplete similarity. Incomplete similarity is also related to the asymptotic invariance of the mathematical model to a renormalization group, but in this case the renormalization group is more complicated. The details can be found in the books (Barenblatt, 1996, 2003).

4. Paris law – an example of incomplete similarity

An instructive and important example of incomplete similarity is represented by the Paris scaling law in fatigue.

The standard fatigue experiment is performed as follows. A specimen (notched or slotted bar or plate) is loaded by a combination of static tensile and a pulsating tensile load of constant frequency and amplitude. At the tip of the notch a fatigue crack is formed, and its propagation, i.e., its length as a function of the number of cycles n is recorded. The Paris law is specified for multi-cycle fatigue tests, when the number of cycles before failure is of the order of many millions.

Processing the experimental data of such tests (Paris and Erdogan, 1963) (see also the preceding paper Paris et al., 1961) revealed the following scaling law

$$\frac{\mathrm{d}\ell}{\mathrm{d}n} = A(\Delta N)^m,$$ (17)

which is being established after a certain initial stage. Here $\mathrm{d}\ell/\mathrm{d}n$ is the crack speed averaged over the cycle; $\Delta N = N_{\max} - N_{\min}$ is the stress-intensity-factor amplitude, A and m are empirically obtained constants. Under the conditions of the present experiment $N = C\sigma\sqrt{\ell}$, where σ is the pulsating bulk stress, and the constant C is a form-factor, which can be evaluated using the technique of the theory of elasticity. The Paris law (17) has found subsequently multiple confirmations for different materials (see, e.g., Figure 6) and now is considered as one of the fundamental laws of structural strength engineering science.

And here an important and very practical question arises: whether A and m are material constants, or may they be different for different specimens? The constant m was found to vary in a wide range from slightly more than two to ten and even more. In engineering practice, scaling law (17) is used for an important prediction of the life-time of the structure, i.e., the number of cycles before failure n_F. Relation (17) can be rewritten in the form:

$$\frac{\mathrm{d}\ell}{\mathrm{d}n} = AC^m(\sigma_{\max} - \sigma_{\min})^m \ell^{m/2}.$$ (18)

By integration we obtain:

$$\frac{1}{\ell_0^{\frac{m}{2}-1}} - \frac{1}{\ell^{\frac{m}{2}-1}} = G(m-2)n,$$ (19)

where $G = AC^m(\sigma_{\max} - \sigma_{\min})^m/2$, and ℓ_0 – the initial crack length. Remember that in multicycle fatigue the number of cycles before the failure is very high, so that it is possible to neglect in the evaluation of the life-time the number of cycles corresponding to the preliminary stage when the Paris law (17) still does not hold.

So, because the intermediate self-similar stage where the Paris law is valid holds during the basic part of the fatigue fracture process, the estimate for the life-time, i.e., for the number of cycles before the failure n_F can be obtained from (19) assuming $\ell \gg \ell_0$ and neglecting the second term on the left-hand side of (19). We obtain

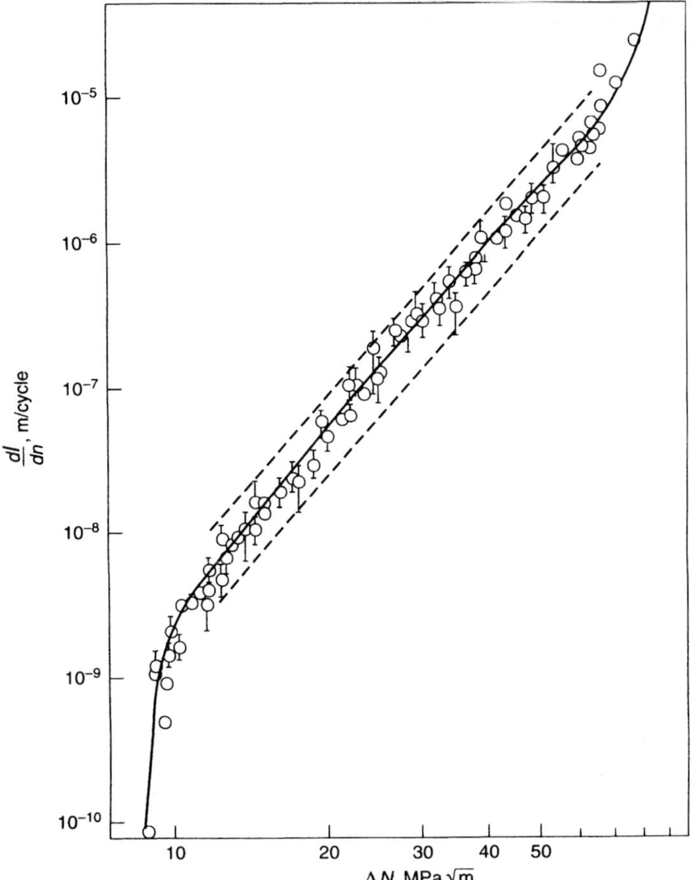

Figure 6. Experimental data for fatigue crack growth in the aluminum alloy confirm Paris law (17) in the major part of the crack velocity range (Botvina, 1989).

$$n_F = \frac{1}{(m-2)G\ell_0^{\frac{(m-2)}{2}}}.$$ (20)

Clearly the life-time n_F sharply depends upon the parameter m, the exponent entering the Paris law, and also upon the pre-power coefficient A, entering expression (17) of this law. Therefore, it is worthwhile to analyze the Paris scaling law (17) on the basis of the procedure outlined above to understand whether these coefficients are indeed the material constants.

We assume that the shape of the loading cycle is fixed. Then the dependent quantity, the mean crack velocity averaged over the cycle $d\ell/dn$ can depend in principle upon the following arguments: the stress-intensity-factor amplitude $\Delta N = N_{\max} - N_{\min}$, stress-intensity-factor asymmetry $R = N_{\max}/N_{\min}$ (as a reminder, $N_{\max}$ and $N_{\min}$ are the maximum and minimum values of the stress intensity factor over the cycle), and, what is specially important, the characteristic length scale of the specimen h, e.g., its diameter, or thickness. The number of cycles n does not enter the list of the governing parameters because it is the intermediate self-similar stage under consideration. Also, important material properties should be included in the list of arguments: the yield stress σ_Y (analysis of the fracture surface shows that the local yield takes place at least at a certain part of the cycle), and a fracture toughness parameter. As the fracture toughness property it is reasonable to take Irwin's parameter K_{Ic} because the crack extension goes by jumps, i.e., unstably. Thus, we assume that there exists a relation of the form:

$$\frac{d\ell}{dn} = f(\Delta N, R, \sigma_Y, K_{Ic}, h). \tag{21}$$

The corresponding dimensions of the parameters entering (21) are as follows:

$$[N] = [K_{Ic}] = FL^{-3/2}, \qquad [\sigma_Y] = FL^{-2}, \qquad [h] = L, [R] = [n] = 1 \tag{22}$$

Here again F is the dimension of force, L is the dimension of length. The dimension of determined quantity $d\ell/dn$ is therefore equal to L.

We assume as the governing parameters with independent dimensions the parameters Δ_N and σ_Y. The dimensional analysis gives by a standard procedure

$$\frac{d\ell}{dn} = \left(\frac{\Delta N}{\sigma_Y}\right)^2 \Phi\left(\frac{\Delta N}{K_{Ic}}, R, Z\right), \tag{23}$$

where the dimensionless parameter

$$Z = \frac{\sigma_Y \sqrt{h}}{K_{Ic}} \tag{24}$$

is the square-root of the ratio of characteristic specimen length scale h to the fracture yield length scale σ_Y^2/K_{Ic}^2.

The evaluations show that the dimensionless parameter $\Pi_1 = \Delta N/K_{Ic}$ is small. And here, we have an appropriate case to apply the technique of the analysis of asymptotic scaling laws, *"advanced similarity analysis"*, discussed in Section 2. Thus, if a finite non-zero limit of the function Φ in (23) at $\Delta N/K_{Ic} \to 0$ does exist, i.e., if there exists a *complete similarity* in the parameter $\Delta N/K_{Ic}$, we obtain the scaling law:

$$\frac{d\ell}{dn} = \left(\frac{\Delta N}{\sigma_Y}\right)^2 \Phi_1(R, Z) \tag{25}$$

so that the parameter m in the Paris scaling law (17) is equal to 2, however, the constant A appears to be non-universal, depending on the specimen size. The analysis of the experimental data shows that $m = 2$ is practically never the case: for some aluminum alloys m is close to 2, but nevertheless always larger than 2. For the vast majority of cases m is substantially larger than 2.

Assuming incomplete similarity we obtain (Barenblatt and Botvina, 1981) (see also Carpinteri, 1996):

$$\Phi = \left(\frac{\Delta N}{K_{Ic}}\right)^{\alpha(R,Z)} \Phi_1(R, Z) \tag{26}$$

exactly the form of the experimentally observed Paris law (17) with the following expressions for the parameters of this law:

$$A = \frac{\Phi_1(R, Z)}{\sigma_Y^2 K_{Ic}^{\alpha}}, \quad m = 2 + \alpha(R, Z). \tag{27}$$

The most important conclusion of the analysis just performed is that the parameters A and m of the Paris law *are not the material characteristics*. Besides the asymmetry of the cycle R they should depend on the specimen length scale.

This conclusion is of high importance, and it had to be checked experimentally. Indeed, it obtained a persuasive experimental confirmation. Botvina, processing the data by Heiser and Mortimer (1972) (see Barenbalatt and Botvina, 1981), and Ritchie (2005), processing the data by Ritchie and Knott (1973) (Figures 7 and 8), showed that the dependence of m upon Z, i.e., upon the specimen size can be substantial. Therefore, using in practical structural design of the results of the standard fatigue experiments performed on small specimens can be dangerous: the real life-time of the structure can be overestimated.

I think that this example has a wider meaning. Power laws are often used in engineering practice as material properties. Characteristic examples – the calculation of J-integrals for plastic materials assuming the power law constitutive equation, or the evaluation of the life-time of polymeric structures assuming the power-type dependence of the fracture toughness on the crack-tip velocity. *In fact, the universality of constitutive relations should be carefully checked in the specimens of various sizes, otherwise the predictions of strength could be unreliable.* This is one of many results obtained in the last decades and based on the concept of incomplete similarity. There are many other results, in particular, in turbulence, where this approach has led to a reconsideration of seemingly well established views.

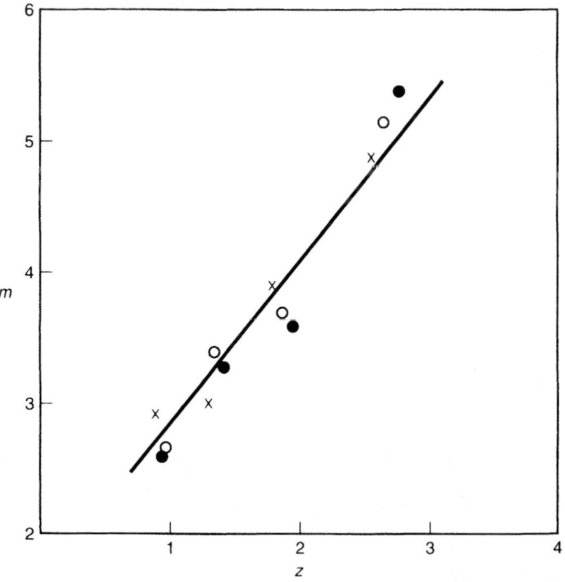

Figure 7. The dependence of the exponent in Paris law (17) on the similarity parameter Z for 4340 steel (Barenblatt and Botvina, 1981).

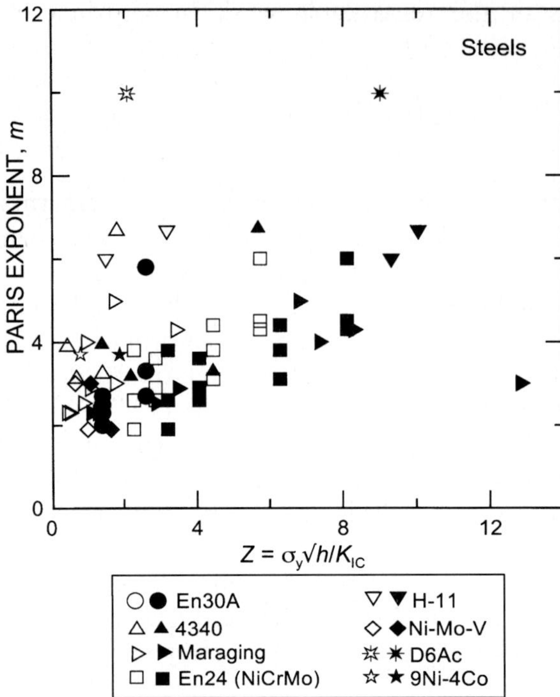

Figure 8. The dependence of the exponent in Paris law (17) on the similarity parameter Z for various steels (Ritchie, 2005).

5. Conclusion

I want to mention in the end of my lecture a very recent result of Broberg (2004), related to dynamic crack propagation and at the same time – indirectly – to incomplete similarity. This is a fresh result, and I do not think that its meaning is understood completely.

It is known that in various fields of applied mathematics over the last century, there appeared the Korteweg–de Vries equation

$$\partial_t u + u\partial_x u + \beta\partial^3_{xxx}u = 0, \tag{28}$$

where t is the time, x the space coordinate, and $\beta > 0$ is a constant. Equation (28) has special solitary wave solutions of the traveling wave type: $u = u(x - \lambda t + c)$:

$$u = \frac{u_0}{\cos h^2\left(\sqrt{\frac{u_o}{12\beta}}\zeta\right)}, \quad \zeta = x - \lambda t + c, \quad u_0 = 3\lambda \tag{29}$$

so that the wave is uniformly propagating with velocity λ and amplitude 3λ.

Transforming the variables $x = \ln\xi$, $t = \ln\tau$, and $c = -\ln A$ is reducing the traveling wave solution (28) to a simple self-similar form:

$$u = \frac{12\lambda}{2 + (\eta)^{\sqrt{\lambda/\beta}} + (\eta)^{-\sqrt{\lambda/\beta}}}, \quad \eta = \frac{\xi}{A\tau^\lambda}. \tag{30}$$

One of the results of the remarkable work by Gardner et al. (1967) was that the traveling wave solutions of type (29) are intermediate asymptotics of the solutions corresponding to hump-like shapes of the initial conditions. The velocity λ and the amplitude $u_0 = 3\lambda$ depend on the initial conditions of the original problem for which the traveling wave is an intermediate asymptotic. The self-similar interpretation (30) of the traveling wave shows that this is exactly the case of the incomplete similarity where the power cannot be determined by dimensional analysis.

Broberg (2004) found recently, using the numerical cell method which he invented, that the situation similar to traveling-wave solutions of the Korteweg–de Vries equation happens also for dynamic cracks, so that for steady dynamic cracks the velocity of propagation and the amplitude are determined by initial conditions. Contrary to the case of Korteweg–de Vries solitary waves where the velocity can be an arbitrary positive number, the velocities of dynamic cracks (experimental fact!) does not exceed approximately one third of the Rayleigh velocity. What is the reason of that is yet unknown – discovery of that reason would be of fundamental interest for dynamic fracture.

I want to say in conclusion that in our time *multiscale phenomena*, where the parameters of the same dimensions but of largely different magnitude enter the model of the phenomenon simultaneously, attract more and more attention. (The examples presented above are characteristic ones.) Nanoscience, which is very popular now is one of the characteristic fields, too. Scaling and incomplete similarity will play a decisive role in studying structural strength multi-scale models, turbulence, and other branches of applied mathematics and engineering science.

Thank you for your attention.

Acknowledgements

This work was supported by the Director, Office of Science, Advanced Scientific Computing Research, U.S. Department of Energy under Contract No. DE-AC03-76SF00098.

References

Barenblatt, G.I. and Botvina, L.R. (1981). Incomplete self-similarity of fatigue in the linear range of crack growth. *Fatigue of Engineering Materials and Structures* **3**, 193–212.

Barenblatt, G.I. and Zeldovich, Ya.B. (1971). Intermediate asymptotics in mathematical physics. *Russian Mathematical Surveys* **26**(2), 45–61.

Barenblatt, G.I. and Zeldovich, Ya.B. (1972). Self-similar solutions as intermediate asymptotics. Ann. Rev. Fluid Mech. **4**, 285–312.

Barenblatt, G.I. (1956). On certain problems of the theory of elasticity, which arise in the theory of the hydraulic fracture of the oil stratum. *Journal of Applied Mathametics in Mechanical (PMM)* **20**(4), 475–486.

Barenblatt, G.I. (1959). On the equilibrium cracks formed in brittle fracture. *Journal of Applied Mathametic Mechanics (PMM)* **(3)** 434–444, **(4)** 706–721, **(5)** 893–900.

Barenblatt, G.I. (1996). *Scaling, Self-Similarity, and Intermediate Asymptotics.* Cambridge University Press, Cambridge.

Barenblatt, G.I. (2003). Scaling, Cambridge University Press, Cambridge.

Bažant, Z.P. and Planas, J. (1998). *Fracture and Size Effect in Concrete and Other Quasibrittle Materials.* CRC Press, Boston, London, New York, Washington, D.C.

Bažant, Z.P. and Yavari, A. (2005). Is the cause of size effect on structural strength fractal or energetic-statistical? *Engineering Fracture Mechanics* **72**, 1–31.

Bažant, Z.P. (1984). Size effect in blunt fracture: concrete, rock, metal. *Journal of Engineering Mechanics ASCE* **110**(4), 518–535.

Bažant, Z.P. (2002). *Scaling of Structural Strength.* Hermes Penton Science, London.

Bažant, Z.P. (2004a). Probability distribution of energetic-statistical size effect in quasibrittle fracture. *Probabilistic Engineering Mechanics* **19**, 307–319.

Bažant, Z.P. (2004b). Scaling theory for quasibrittle structural failure. *Proceedings of the US National Academy of Sciences* **101**(37), 13400–13407.

Benbow, J.J. (1960). Cone cracks in fused silica. *Proceeding of the Physics Society* **B75**, 697–699.

Botvina, L.R. (1989). *The Kinetics of Fracture of Structural Materials.* Nauka, Moscow.

Broberg, K.B. (2004). Significance of morphological changes at a propagating crack edge. *International Journal of Fracture*, **130**, 723–742.

Carpinteri, A., Chiaia, B. and Cornetti, P. (2002). A scale-invariant cohesive crack model for quasi-brittle materials. *Engineering Fracture Mechanics* **69**, 207–217.

Carpinteri, A., Cornetti, P., Barpi, F. and Valente, S. (2003). Cohesive crack model description of ductile to brittle size-scale transition: dimensional analysis vs. renormalization group theory. *Engineering Fracture Mechanics* **70**, 1809–1839.

Carpinteri, A. and Massabo, R. (1996). Bridged versus cohesive crack in the flexural behaviour of brittle-matrix composites. *International Journal of Fracture* **81**, 125–145.

Carpinteri, A. (1989). Cusp catastrophe interpretation of fracture instability. *Journal of the Mechanics and Physics of Solids* **37**(5), 567–582.

Carpinteri, A. (1994). Scaling laws and renormalization groups for strength and toughness of disordered materials. *International Journal of Solids and Structures* **31**(3), 291–302.

Carpinteri, A. (1996). Strength and toughness in disordered materials: complete and incomplete similarity. In *Size-Scale Effects in the Failure Mechanisms of Materials and Structures* Edited by Carpinteri A. E. & F.N. Spon, London, et. al., 3–26.

Carpinteri, A. (1997). *Structural Mechanics – A United Approach.* E. & F.N. Spon, London, et. al.

Gardner, C.S.J., Greene, J.M., Kruskal, M.D. and Miura, R.M. (1967). A method for solving the Korteweg-de Vries equation. *Physics Review Letters* **19**, 1095–1097.

Goldenfeld, N.D. (1992). *Lectures on Phase Transition and the Renormalization Group.* Perseus Publishing, Reading.

Heiser, F.A. and Mortimer, W. (1972). Effects of thickness and orientation on fatigue crack growth rate in 4340 steel. *Met. Trans.* **3**, 2119–2123.

Irwin, G.R. (1960). Fracture made transition for a crack traversing a plate. *Transactions of the ASME Series.* **D82**, 417–425.

Mandelbrot, B. (1975). *Les Objects Fractals: Forme, Hasard et Dimension.* Flammarion, Paris.

Mandelbrot, B. (1977). *Fractals, Form, Chance and Dimension.* W.H. Freeman and Co., San Francisco.

Mandelbrot, B. (1982). *The Fractal Geometry of Nature.* W.H. Freeman and Co., San Francisco.

Panasyuk, V.V. (1968). *Limiting Equilibrium in Brittle Bodies with Cracks.* Naukova Dumka, Kiev.

Paris, P.C. and Erdogan, F. (1963). A critical analysis of crack propagation laws. *Journal of Basic Engineering Transactions ASME, Series* **D85**, 528–534.

Paris, P.C. and Gomez, M.P. and Anderson, W.P. (1961). A traditional analytic theory of fatigue. *The Trend in Engineering* **13**, 9–14.

Ritchie, R.O. and Knott, J.F. (1973). Mechanisms of fatigue crack growth in low alloy steel. *Acta Metallurgica* **21**, 639–648.

Ritchie, R.O. (2005), Incomplete self-similarity and fatigue crack growth. *International Journal of Fracture* **132**, 197–203.

Roesler, F. (1956). Brittle fracture near equilibrium. *Proceeding of the Physics Society* **B69**, 981–992.

Taylor, G.I. (1941) The formation of a blast wave by a very intense explosion. Report RC-210, Civil Defense Research Committee.

Taylor, G.I. (1950). The formation of a blast wave by a very intense explosion. II. The atomic explosion of 1945. *Proceeding of the Royal Society* **A201**, 175–186.

Zheltov, Yu.P. and Christianovich, S.A. (1955). On the hydraulic fracture of oil stratum. *Izvestiya, USSR Academic Science Technical Science* **5**, 3–41.

International Journal of Fracture (2006) 138:37–45
DOI 10.1007/s10704-006-0058-7

© Springer 2006

ICF contribution to fracture research in the second half of the 20th century[*]

Takeo Yokobori
Academician The Japan Academy, 7-32, Uenokoen, Taitoku, Tokyo, Japan, #110-0007

Received: 1 March 2005; accepted 1 December 2005

Abstract. Historical explanation and some remarks for future have been described on The International Congress on Fracture (abbreviated as ICF), including the International Journal and International Cooperative Research as relevance. The aim of them in all concerns the systematized atom–nano–meso (–in new words)—macroscopic researches on strength and fracture, nonlinearly (especially, say, as complexity system science and engineering).

1. Introduction

For instance, First Announcement of Ninth International Conference on Fracture describes: Founded by Professor T. Yokobori in 1965, the International Congress on Fracture (ICF) is today the premier international body for promotion of worldwide cooperation amongst scientists and engineers concerned with the mechanics and mechanisms of fracture, fatigue and strength of solids. Over the years, ICF has made considerable progress in providing an international forum for highlighting individual and national accomplishments in the overall field of fracture.......

It is a great honor to the author.

In the present article, the author will concern ICF contribution to fracture research in the second half of the 20th century as requested. Nevertheless I believe it is important to reflect historical relevance, which is also included in this article.

2. The ICF (The International Congress on Fracture) and relevance in retrospect

2.1. THE ORIGIN OF THE CONCEPTS

It was the author's question on the days of young student at high school and university: why the machines and the structures cannot be designed entirely in terms of atomic or physical term without any ambiguous parameters?

"Remarkable progress had been made in each different discipline of the single subject of the strength and fracture of solids, for example, the fields of solid-state physics, metallurgy, continuum mechanics, environmental studies and statistical theory." There is, however, a major gap between the microscopic and mac-

---

[*]Full paper lectured at ICFII by the author will be published in Strength, Fracture and Complexity, an International Journal, Vol. 4 No. 4 (2006) IOS Press

roscopic understanding of strength and fracture of materials, and, thus, a long way before such complexity system problems are fully solved. In order to fill up this gap, it is essential to integrate the many disciplines mentioned above into the field of strength and fracture of materials as a consolidated subject. Also cross-fertilization provided from a common forum to investigators on different materials and different approaches would be important in establishing unifying principles, thus contributing advances in this complexity science. Finally it is intended to direct the attention to the problem area for which further progress is needed. ..." This is written in PREFACE of Proceedings of The First International Conference on Fracture (Yokobori 1966). Practical applications, which become extremely important, will be naturally concerned and be solved by this interdisciplinary way.

The author had long this concept since the 1940's, felt keenly the necessity of realizing the idea, and published the book on interdisciplinary approach to Strength, Fracture and Fatigue, 1955 (Yokobori 1955). In writing the book the following three concepts were in mind:

(1) For each single materials

    (a) With respect to failure and fracture, such as, yielding, brittle fracture, ductile fracture, fatigue, creep and creep fracture, corrosion fatigue, stress corrosion, hydrogen embrittlement, and their multiplication, etc.

    (b) By interdisciplinary, nonlinearly coupled micro and macro, and as time sequence phenomena

(2) Furthermore, comparative studies on various materials, such as metallic, inorganic materials, polymers, composite and biomaterials, etc.

These concepts were succeeded in his later books (Yokobori 1964, 1974). Simply speaking, the concept may be called synthetical view as a whole.

In 1956 the author initiated Japan National Symposium on Strength, Fracture and Fatigue and in 1960 the Fracture Committee, that is, 129 Committee, JSPS (Japan Society for the Promotion of Science), respectively.

In 1961 over 3 months continued, the author visited at many scientists and engineers concerned from door to door throughout Europe and USA. The author canvassed the opinions on his proposal for the following three subjects for interdisciplinary research on strength and fracture:

(1) The creation of an international journal on fracture. By that time, there had been none in the world. Furthermore, the author had long been thinking it is needed to give a position to fracture in the classification of science and engineering.

(2) Holding an international conference on fracture

(3) The coordination of an international cooperative research system.

In 1964 the Research Institute for Strength and Fracture of Materials, Tohoku University and in 1965 The Japanese Society for Strength and Fracture of Materials was established, respectively by the author.

2.2. On international journal of fracture

With respect to an international journal of fracture, through the process as mentioned in Section 2.1, the main Regional Editors in International Board of Editors had been established by the author's negotiation. The author succeeded in publication by P. Noordhoff LTD. Groningen, The Netherlands, who published the English version (Yokobori 1955) of the author's Japanese book 1955 edition, immediately before. Taking into account the situation Prof. Yokobori asked Prof. M. L. Williams to be in charge of Editor-in-Chief. Thus the first issue started 1965 under the name of international journal of fracture mechanics, according to Prof. M. L. Williams preferring to *fracture mechanics* rather than *fracture*. At any rate, a simple background and the point is briefly written in Forward of the first issue (Williams 1965).

2.3. ON THE INTERNATIONAL CONGRESS ON FRACTURE

In parallel with the international journal, the Organising Committee for an International Conference on Fracture of this nature had been established as follows: Chairman: Takeo Yokobori; Secretaries: Tadashi Kanazawa, Shigeyasu Koda, Takeshi Kunio, Hideji Suzuki and Shuji Taira; Members: Tadashi Ishibashi, Yuichi Kawada, Kozo Kawata, Kaizo Monmma, Yo Okada, Kiyoshi Okano, Y. Suezawa and Teruyoshi Udoguchi; from outside Japan, B. L. Averbach, J. Friedel, P. Haasen, A. K. Head, N. J. Petch, M. L. Williams and S. N. Zhurkov.

The 19 countries that participated in the Conference are: Australia, Canada, Czechoslovakia, France, Germany, Greece, Hungary, India, Ireland, Israel, Japan, Korea, The Netherlands, Poland, Spain, Sweden, United Kingdom, USA and USSR.

In order to foster research in the mechanics and phenomena of fracture, understanding of fatigue and the problems on strength of materials related to fracture and to promote means whereby results or the work may be publicly communicated, the establishment of an international organization was discussed and decided upon during the Sessions. Among attendants from outside Japan were N. P. Allen, A. S. Argon, C. Atkinson, B. L. Averbach, L. F. Coffin, Jr., D. L. Davidson, J. E. Dorn, F. Erdogan, W. N. Findley, N. E. Frost, W. W. Gerberich, J. J. Gilman, J. Glucklich, P. Haasen, G. T. Hahn, A. K. Head, C. C. Hsiao, D. Hull, A. S. Kobayashi, E. Krempl, E. Kroner, R. F. Landel, H. Liebowitz, H. Liu, N. P. Louat, S. S. Manson, F. A. McClintock, A. J. McEvily, C. J. McMahon, Jr., J. D. Morrow, T. Mura, R. W. Nichols, Y. A. Ossipian, P. C. Paris, E. R. Parker, P. L. Pratt, R. M. N. Pelloux, J. R. Rice, A. R. Rosenfield, G. C. Sih, E. Smith, K. U. Snowden, A. M. Sullivan, J. L. Swedlow, A. S. Tetelman, J. Weertman, V. Weiss, A. A. Wells, E. T. Wessel, A. R. C. Westwood, M. L. Williams, V. F. Zackay, S. N. Zhurkov. The organization was named the International Congress on Fracture (Permanent Body). Prof. Takeo Yokobori was designated to become the Chairman of the Provisional Committee of the Organization. Member: B. L. Averbach, J. Fiedel, P. Haasen, A. K. Head, A. Kochendörfer, R. W. Nichols, N. J. Petch, M. L. Williams, S. N. Zhurkov; and Secretaries, ex-officio: T. Kawasaki, and J. L. Swedlow. Hereafter this conference is called The First International Conference on Fracture.

It was decided by acclamation in The Committee that the name of the international standing organization be The International Congress on Fracture and the symbol ■ designed by Mrs. Yokobori and, used for Sendai Conference be that of this standing organization for the nice memory of the origin. This is the reason why the word "Congress", but not "Institute" nor "Association" was used. Also it was decided that the initials ICF is used as the abbreviated name of The Congress. It should notice International Conference on Fracture held every 4 years also is called as ICFX abbreviated. Yokobori would like to propose to denote The ICF for the Organization (Congress), and, on the other hand, ICF for each Conference every 4 years.

With respect to some regulation, Yokobori prepared the manuscript of Proposed Statutes of the International Congress on Fracture consulting with many peoples concerned, taking into account Constitution of IIW (The International Institute of Welding) and The Statutes of IAEE (The International Association for Earthquake Engineering). In making the manuscript of Proposed Statutes of ICF the following three points were taken into account:

- The first point is to keep the initial aim of ICF as possible.
- The second point is that the Congress will keep a truly international character.
- The third point is that Congress should be given active co-operation by many organizations, societies and individuals in the field concerned.

Many copies had been distributed early summer in 1967 to the societies and the groups concerned over 26 countries, asking opinions and comments. The answers came from many of them, and these all approved the PROPOSED STATUTE essentially.

As is described in Minutes (ICF FOUND. EC. M-1, 1969; ICF Council M-1, 1969), the modification for opening the membership to individuals was accepted.

The following paragraph was described in "SOUVENIR, Sixth International Conference on Fracture, December 4–10, 1984, New Delhi, India (Yokobori 1984).

"...

....

The Provisional Committee Members who had contributed in founding ICF was nominated as Founder Members. Founder Members are B. L. Averbach, J. Friedel, P. Haasen, A. K. Head, T. Kawasaki, A. Kochendörfer, R. W. Nichols, N. J. Petch, J. L. Swedlow, T. Yokobori, M. L. Willams, S. N. Zhurkov.

Approving the Statutes, The International Congress on Fracture was established at the Council Meeting, Brighton April 16, 1969. Prof. T. Yokobori who took the initiative in founding ICF was by acclamation elected as the First President of ICF. The Council then proposed candidates for the three Vice-Presidents and were unanimously chosen: Y. N. Rabotnov, B. L. Averbach and R. W. Nichols. During the Third International Conference on Fracture held at Munich in April 1973, the Council elected Prof. Yokobori as the Founder–President of the ICF, in his life time, appreciating his great achievements for initiating, and with this respect, the STATITES was amended at The COUNCIL Meeting, Munich, April 12, 1973 (ICF COUNCIL M-2, 1973).

....

...."

On the way to the initiation of The ICF and the relevance, the spilt history will be published elsewhere (Yokobori 2006).

2.4. On co-ordination of an international co-operative research system

Some spilt history will be published elsewhere (Yokobori 2006).

3. The contribution of ICF and relevance to fracture research in historical aspects

There are many contributions and it is difficult to describe them in limited pages. A few will be mentioned.

It was not until a few years after the author published the book (Yokobori 1955) that Griffith Theory being introduced by this book was cited in Japan in 1958 in Transaction of The Japan Society of Mechanical Engineers. In addition, as a turning point with ICF1, 1965, Sendai, the concept and the school initiated by G. Irwin 1956 has been introduced into Japan. These situation have been a remarkable contribution of ICF and the relevance to a rapid development of fracture mechanics in Japan.

A short historical view in relation to the matter is shown in Table 1.

It is most impressive to the author that the paper on the introduction of the J-integral concept, a path-independent integral by J. R. Rice was published in 1968 a few years after ICF1, 1965 in which he personally participated.

Over the world, by Quadrennial Conference including interim conferences there has been remarkable contribution along the aim of The ICF for 40 years as shown in Table 2. The author wishes to express hearty thanks to all these persons who served for their much efforts.

Among many achievements of ICFX, the following program may be very important: At ICF4, 1977, Waterloo, Pannel Discussions on Fracture, Education and Society (Taplin 1977a), and on Fracture, Politics and Society (Taplin 1977b), respectively were held (Prof. D. M. R. Taplin, Chairman). The author would like to respect that such program was held about 28 years ago, because the author believes the matter, especially the latter matter is now being, and be much more in future very important problem.

Next to mind comes ICF10, 2001 Honolulu (Chairman, Prof. T. Kishi, Co-chairs, Prof. R. O. Ritchie, Prof. R. Ravi-Chander, and Prof. Toshimitsu Yokobori, Jr.: President, Prof. R. O. Ritchie) (Table 2). This is one of the most large conference among ICFXs in the scale of total participants approximately 700 from 50 countries, and presented papers approximately 700. The author deems a great honour to make an Honour Lecture, and, in that, proposed the fracture research as complexity system science, emphasizing its importance.

It is quite timely and of much significance that fracture research as complexity system science is highlighted in ICF11, 2005, Turin under Prof. A. Carpinteri, The Chairman (Table 2), say, with Honour Lecture by Prof. B. B. Mandelbrot on the complexity system science. I sincerely hope that ICF11 will make outstanding contribution to the fracture science and engineering.

4. Prospect in ICF

4.1. STRENGTH AND FRACTURE AS COMPLEXITY SYSTEM SCIENCE — A NEW PARADIGM

Recently it has been revealed by many scientists (Tazaka 1998; Tsuda 1998; Kishida 2000) in various field of science and engineering, say, pure science, electronics, informatics, economics, medicine that complexity system science as a new paradigm for complex phenomena or complex function of a system consisting of many

Table 1. A short historical view of the proposal of the line of the concept for the criterion of fracture as compared in Japan with the world[a].

Country	World		Japan	
Aspect Year	Mechanical	Physical	Mechanical	Physical
1920	Griffith theory		Strength of materials approach	
1931				Morphological study: Terada and Hirata
1937	Neuber hypothesis			
1939	Weibull theory with relation to Griffith crack			
1940–1941		Furth theory (melting)		
1941				Reaction rate process theory of fracture for glass: Omori
1943–1947	Mechano-physico-metallurgical studies with relation to Griffith crack: Zener, Hollomon et al	Rate process theory of fracture Fiber: Tobolsky & Eyring, 1943 Metals: Machlin & Nowick, 1947		
1948	Griffith–Irwin–Orowan formula	Nucleation theory of rupture of liquid, Fisher		
1948–1949				Stochastic theory of fracture, Glass: Hirata, 1948 Metals: Yokobori, 1949
1952–1954		Crystal dislocation model for fracture: Keohler, Mott, Stroh, Cottrell	Rate process theory and nucleation theory of fracture of solid, especially combining micro and macro stress concentration: Yokobori, Micro and macro mechanical approach to fracture combining crack and crystal dislocation: Yokobori	
1955			Interdisciplinary book: Strength, Fracture & Fatigue of Materials, Gihodo (In Japanese): Yokobori, [Translation English Edition, 1965]	

Table 1. Continued.

Country	World		Japan	
Aspect Year	Mechanical	Physical	Mechanical	Physical
1956			Creep research: TAIRA	
1956–1958	Fracture mechanics: Irwin			
1957			Double tension test: Yoshiki & Kanazawa	
1958			Brittle fracture studies: Kihara	
1965		The First International Conference on Fracture, Sendai (Interdisciplinary) initiated by T. Yokobori		
1968	J. integral: Rice			

[a]The table is corrected one to the Takeo Yokobori's table in Fracture Mechanics (1978), Univ. Press of Virginia, P. 9.

various elements may go to replace the present day paradigm based on so-called law of causality.

Some of the characteristics of complexity system are as follows (Tazaka 1998; Tsuda 1998).

(1) Nonlinearity
- Nonlinear system in which the constituent elements themselves are constantly changing and, that, the joint between them are not neglected.
- The relations between each constituent element also are dynamically changing.
- The phenomena are beyond the separation between human learning and science & engineering.

(2) Emergence-chaos
- Phase change, structure change.
- Step by step a quantitative change translates into qualitative change.
- Microscopic characteristics translate into macroscopic characteristics.
- In this case, the same shape will reveal between the part and the bulk.— Fractal.

(3) Self-organization

On the situation in mind, the author and colleagues held an international conference, 2001, Sendai in order to get a clue to solve Strength and Fracture Problems as Complexity System Science and Engineering (Yokobori 2003). At that stage, the conference did not apparently quite fit with this category. The author, however, believed that the stimulating motive is very important and hoped the approach will grow

Table 2. Quadrennial Conference ICF X Chairman and The International Congress on Fracture President.

Quadrennial Conference	Location	Date	Organizing Chairman	President
ICF-1	Sendai (Japan)	12–17 Sept. 1965	T. Yokobori	T. Yokobori (Sept. 1965–Apr. 1969) (Provisional)
ICF-2	Brighton (UK)	13–18 Apr. 1969	R.W. Nichols	T. Yokobori (Apr. 1969–Apr. 1973)
ICF-3	Munich (Germany)	8–13 Apr. 1973	A. Kochendörfer	B.L. Averbach (Apr. 1973–Jun 1977)
ICF-4	Waterloo (Canada)	19–24 Jun 1977	D.M.R. Taplin	R.W. Nichols (Jun 1977–Apr. 1981)
ICF-5	Cannes (France)	29 Mar.–3 Apr. 1981	D. Francois	D.M.R. Taplin (Apr. 1981–Dec. 1984)
ICF-6	New Delhi (India)	4–10 Dec. 1984	P. Rama Rao	D. Francois (Dec. 1984–Mar. 1989)
ICF-7	Houston (USA)	20–24 Mar. 1989	K.S. Salama	P. Rama Rao (Mar. 1989–Jun 1993)
ICF-8	Kiev (Ukraine)	8–14 Jun 1993	V.V. Panasyuk	J.F. Knott (Jun 1993–Apr. 1997)
ICF-9	Sydney (Australia)	1–5 Apr. 1977	Co-chairs B.L. Karihaloo K.W. Mai	R.O. Ritchie (Apr. 1997–Dec. 2001)
ICF-10	Honolulu (USA)	2–6 Dec. 2001	T. Kishi Co-chairs R.O. Ritchie K. Ravi-Chander A.T. Yokobori, Jr.	Y-W. Mai (Dec. 01)
ICF-11	Turin (Italy)	20–25 Mar. 2005	Alberto Carpinteri	

more and more significantly in near future. Message from Sir Alan Cottrell says that the conference is scientific challenge for the future, and the author was very much encouraged. The concept of complexity system science is closely related to or essentially or virtually similar to Bio-informatics.

In this way, the author hope the research towards such direction will be prevailing in ICF at the 21 century.

4.2. SYNERGETIC RESEARCH ON SAFETY FOR ACCIDENTS ASSOCIATED WITH FRACTURE OF ARTIFICIAL STRUCTURES

Recently the world appears to be changing in every aspect. At this point we should look for another new lines of fracture research. On the day of rapidly developing science and technology, it is most important to ensure the safety related to failure of artificial structures. In the lecture "The Problems on Safety Science and Engineering" at 41st Japan Academy Public Lecture Meeting, Ueno, Tokyo (Yokobori 2004), the author emphasized and proposed as follows: "The remarkable development of recent science and technology has reflected, on the other hand, on the serious aspects, such as, the matter related to uncertainty or unsafely among the people. For instance, there often occur many accidents associated with various artificial structures. With respect to so-called human error, most of all analysis concern mainly man–machine interactions. On the other hand, the author had long being made a classification in more wide range including almost human themselves (Yokobori 1987). Thus the matter involves various field of science and technology, such as, physics, chemistry, mechanics, metallurgy, materials science and engineering, psychology, economics, politics, sociology, medicine, etc. Therefore the synergetic research is inevitable, and such research project will be desirable as research network at least in Japan. It also belongs to a category of complexity system science." It will be nice among the countries concerned in The ICF.

Finally, most thanks should be expressed to many persons for the tremendous co-operation and help on The ICF and relevance. Without these, there were no today's prosperous ICF.

The data in the historical description in Retrospect were quoted from and based on the Minutes of The ICF, Proceedings and Journals, etc. The author will make corrections if any.

References

Kishida, T. (2000). Gakushikai Kaiho 827:105 (in Japanese).

Taplin, D.M.R. (1977a). The Fourth International Conference on Fracture, Panel Discussions, Fracture, Education and Society

Taplin DMR (1977b) The Fourth International Conference on Fracture, Panel Discussions, Fracture, Politics and Society

Tazaka H (1998) Fukuzatsukei no Chi, Kodan Co. (in Japanese)

Tsuda I (1998) Gakushikai Kaiho 821:106 (in Japanese)

Williams ML (1965) Forward. Int J Fract Mech 1(1):1

Yokobori T (1955) Zairyo-Kyodo Gaku, pp. 300, Gihodo: (1965) Translated Edition. In: Crisp JDC (ed) (Translated by Inoue M. and Matsuo S.) The strength, fracture and fatigue of materials. P. Noordhoff, Groningen, The Netherlands

Yokobori T (1964) Zairyo-Kyodo Gaku, First edition p 292 Iwanami Shoten: (1968) Translated Edition. In: Crisp JDC (ed) (Translated by Matsuo S) An interdisciplinary approach to fracture and strength of solids. Wolters-Noordhoff Sci. Pub., Groningen, The Netherlands

Yokobori T (1966) Preface. In: Yokobori T, Kawasaki T, Swedlow JL (eds) Proc. The First International Conference on Fracture, Vol 1. Sendai, Japan

Yokobori T (1974) Zairyo-Kyodo Gaku, 2nd edn, Iwanami shoten:(1978) Translated Edition in Russian, Pisarenko GS (ed) Dumka, Kiev

Yokobori T (1984) Souvenir, ICF6. New Delhi, India

Yokobori T (1987) J Jpn Soc Mech Engrs 90(827):1255–1258 (In Japanese)

Yokobori T (ed) (2003) Strength, fracture and complexity, an international journal, vol 1 and vol 2. IOS Press

Yokobori T (2004) 41st Japan academy public lecture meeting, Oct. 2004. Preprints, The Japan Academy, Ueno-cōen, Taito-ku, Tokyo

Yokobori T (2006) Strength, fracture and complexity, an international journal. Vol. 4 No. 4 IOS Press (To be published)

International Journal of Fracture (2006) 138:47–73
DOI 10.1007/s10704-006-7153-7

© Springer 2006

Inverse analyses in fracture mechanics

G. MAIER*, M. BOCCIARELLI, G. BOLZON and R. FEDELE
Department of Structural Engineering, Technical University (Politecnico) of Milan, Milan, Italy
Author for correspondence. (E-mail: giulio.maier@polimi.it)

Received 1 March 2005; accepted 1 December 2005

Abstract. The present purpose is a survey of some engineering-oriented research results which may be representative of the main issues in the title subject. Some recent or current developments are pointed out in the growing area of fracture mechanics centered on the calibration of cohesive fracture models for quasi-brittle materials, by approaches which combine experimentation, experiment simulation and minimisation of the discrepancy between measured and computed quantities. Specifically, reference is made herein to the following topics in calibration of fracture constitutive models: (a) deterministic characterisation of concrete-like materials by traditional three-point-bending tests (TPBTs), supplemented by optical measurements; (b) wedge-splitting tests (WST) and extended Kalman filter (EKF) for the stochastic estimation of fracture parameters; (c) in situ parameter identification for the local diagnosis of possibly deteriorated concrete dams on the basis of flat-jack tests; (d) fracture properties of ceramic materials and coating-substrate interfaces identified through indentation tests, imprint mapping and inverse analysis in micro-technologies.

Key words: Inverse analysis, in situ experiments, interface models, non-destructive testing, parameter identification.

1. Introduction

The quantitative assessment of constitutive parameters by inverse analysis exhibits at present growing scientific interest and practical usefulness, as material models become more realistic and complex, and computational tools more and more powerful. Inverse analysis for the calibration of materials or structural models consists of: data collection from experimental tests in laboratory or in situ; computer simulation of the tests by employing the model to calibrate; minimisation, with respect to the sought parameters, of a suitable norm which quantifies the discrepancy between experimental data and the corresponding values provided by the simulation. Several aspects of the inverse problem theory and its applications are treated systematically in the recent literature, see e.g., Bui (1994); Stavroulakis (2000); Stavroulakis et al. (2003); Mróz and Stavroulakis (2005). A noteworthy by-product of the experimental test simulation is the sensitivity analysis of measurable quantities with respect to the parameters to identify: in fact, sensitivity matrices may corroborate the identifiability of the sought parameters and orient the optimisation of the experiment design, see Kleiber et al. (1997).

In fracture mechanics, like in other fields, inverse analysis exhibits peculiar features and represents an intersection of diverse disciplines which provide the following contributions: mathematical minimisation of generally non-convex, sometimes non-smooth discrepancy functions, with possible ill-posedness requiring special provisions, such as Tikhonov regularisation; sophisticated experimental mechanics for accurate measurements; statistical methods for processing measurement and model errors in order to assess consequent uncertainties on the estimates; large-scale simulations and computations; recourse to soft-computing, such as neural networks, in view of routine industrial use.

This paper is intended to provide a critical survey of the title subject through brief discussions of particular but representative practical problems in structural engineering and micro-technologies, recently tackled by the authors' team with reference to fracture properties of quasi-brittle materials described by traditional constitutive models, like those dealt with, e.g., by Karihaloo (1995) and Bažant and Planas (1998). More specifically, the present paper will address timely issues of model calibration concerning primarily concrete dams and coated materials at the micro-scale. The case histories referred to herein are intended to evidence that also in engineering applications of fracture mechanics, progress may be fostered by synergistic combinations of experiments and their simulations for inverse analyses.

Many existing structures, particularly concrete dams built decades ago, are deteriorated by physico-chemical ageing processes (primarily by alkali–silica reaction), and/or by past extreme loading, such as earthquakes. In order to assess present mechanical properties crucial for structural integrity, two kinds of methodologies are currently adopted in engineering practice: (a) global (static or dynamic) diagnostic analyses, intended to identify possibly deteriorated elastic moduli as damage indices (see, e.g., Ardito et al., 2004; Maier et al., 2004); (b) local diagnosis based either on in situ 'non-destructive' tests or on specimen extraction and laboratory tests.

Techniques of the latter local kind are considered herein (Sections 2 and 3) for the characterisation of concrete from a fracture mechanics standpoint. In Section 2, two conventional laboratory procedures for the identification of quasi-brittle fracture parameters are referred to, namely: three-point bending test (TPBT), supplemented by electronic-speckle pattern interferometry (ESPI), combined with batch, deterministic inverse analysis (fairly detailed descriptions available in: Bolzon et al., 1997b; Bolzon and Maier, 1998); wedge-splitting test (WST) associated to sequential, stochastic extended Kalman filters (EKF) (Bolzon et al., 2002). In both cases, a mode I cohesive crack model with a bilinear softening branch is adopted with four fracture parameters to identify, in accordance with CEB-FIP Model Code (1993); see also: Alvaredo and Torrent (1987); Brühwiler and Wittmann (1990); Guinea et al. (1994).

The cohesive models and, hence, the experiments are analytically described as linear complementarity problems (LCPs) and, therefore, the discrepancy minimisation exhibits the format of 'mathematical programming with equilibrium constraints' (MPEC), at present a recurrent mathematical construct in econometry and management and numerically solvable by new *ad hoc* algorithms, e.g., see Nash (1951), Cottle et al. (1992), Luo et al. (1996). However, for the fracture parameter identification based on TPBT, traditional numerical procedures are employed, such as first-order (gradient) algorithms and direct-search genetic algorithms.

As for WST data, extended Kalman-Bucy filter is adopted herein for stochastic, sequential parameter identification, leading both to parameter estimates and to a quantification (through a covariance matrix) of the estimate uncertainties due to random noises in the experimental measurements (see: Bittanti et al., 1984; Bui, 1994).

The peculiar LCP description is exploited in order to make cost-effective the repeated computations of sensitivity matrices required by the filtering process (Bolzon et al., 2002).

In the context of local structural diagnosis by in situ tests, Section 3, two novel procedures for fracture parameter identification are currently being investigated with reference to concrete dams. Both techniques are emerging from the combination of traditional in situ tests with inverse analyses, and are centred on: pressurisation and dilatometric measurements of in-depth drilled holes; flat-jacks employed as usual on the surface of the construction, but in a new fashion with respect to the present practice. The latter technique is outlined in this paper, whereas the former is dealt with in Fedele et al. (2005) as for elastic moduli, and in Maier et al. (2004) as for fracture properties. The numerical study of these innovative methodologies entails comparative assessments of gradient-based algorithms and artificial neural networks (ANNs). From the engineering practice standpoint, it is advantageous to perform once-for-all large-scale computations by algorithms of the former kind, in order to 'train' the implemented ANNs to be employed later, repeatedly and routinely, in situ (see e.g., Waszczyszyn, 1999).

Surface engineering is an emerging discipline that is becoming more and more important in technologies leading to products such as electronic packages, magnetic recording media, optical devices and tribological protection of mechanical components. Typical representative applications of coatings on bulk materials are as follows: thin layers deposited on silicon substrates in micro-electro-mechanical systems (MEMS) to improve their electric performance; hard surface coatings (ceramics, diamond-like-carbon materials) deposited on cutting tools in order to improve wear and corrosion resistance; metal substrates coated by ceramic layers or metal/ceramics functionally-graded materials (FGM), to enhance thermal barrier properties at high temperature in turbine blades, diesel and jet engines, nuclear fusion equipment, rockets and space shuttles; see, e.g.,: Spearing (2000); Cetinel et al. (2003); Zhi-He et al. (2003). The fruitful use of thin coatings in advanced technologies, and the increase of their performance and life-time, require an accurate characterisation of their mechanical properties after the deposition on the substrate, since this process can alter their characteristics with respect to the original bulk state. Properties like elastic modulus, yield strength, fracture energy and interfacial adhesion, and also residual stresses, play a crucial role on the correct functioning and long-term performance of coatings, avoiding cracking or delamination from the substrate. However, most existing techniques intended to assess, e.g., adhesion, are qualitative only (Ohring, 1991; Volinsky et al., 2002).

A spreading methodology for material characterisation at the micro and nano-scales is based on indentation tests. Fracture induced in brittle materials by indenters with sharp corners, and delamination between film and substrate caused by indentation tests, have been investigated by several researchers, see e.g.,: Xiaodong et al. (1997); Abdul-Baqi and Van der Giessen (2001, 2002); Carpinteri et al. (2001); Li and Siegmund (2004).

The model calibration methodology proposed in Bolzon et al. (2004) and Bocciarelli et al. (2005), is based on indentation test, imprint mapping and inverse analysis. This method is applied herein in Section 4 to the identification of material parameters both for the fracture characterisation of ceramic materials at the micro-scale and for delamination in film-substrate systems. A pyramidal indenter is considered in the former case, so that the development of cracks starting from the pyramid edges turns out to be one of the dominant irreversible phenomena during indentation. The novelty is represented by the subsequent mapping of the residual imprint through an atomic force microscope (AFM), and by the use of these deformation measures, besides the traditional indentation curves, for the parameter estimation through discrepancy minimisation. The inverse analysis technique in fracture mechanics via micro-indentation and imprint mapping combined, is shown to provide fracture characterisation of both bulk materials and thin film coatings.

In Section 5, some limitations, open problems and future prospects are briefly presented of the research issues surveyed in this paper as typical and representative of the broad and growing title subject.

2. Calibration of cohesive crack models by laboratory tests on cementitious materials

2.1. PIECEWISE-LINEAR COHESIVE CRACK MODELS

Fracture propagation in quasi-brittle materials like ceramics, concrete and rocks, is usually simulated to structural engineering purposes by cohesive crack models: specifically, it is assumed that the process is concentrated in a displacement-discontinuity locus ('process zone') while the surrounding material is still undamaged and elastic, see e.g.: Wittmann and Hu (1991); Karihaloo (1995); Bažant and Planas (1998). Such an assumption (Barenblatt, 1962; Hillerborg, 1991) is conceptually and mathematically similar to modelling of inelastic behaviour of beams and frames by the softening plastic hinge notion (Maier, 1968; Bolzon and Corigliano, 1997a; Jirasek and Bažant, 2001; Cocchetti and Maier, 2003).

For overall analyses focusing on fracture of concrete dams, mode I often turns out to be reasonable even in the absence of clear motivations such as symmetry. In fact, in homogeneous media, crack advancement direction tends to re-establish mode I; on the other hand, mixed mode modelling is still rather controversial and, in the absence of a reliable quantification, modes II and III can often be interpreted by the rough idealisation of no sliding displacements due to aggregate asperity; see, e.g., Bolzon et al. (1994a). However, the parameter identification methods considered in what follows with reference to opening mode crack alone are quite general. In particular, if the complementarity construct is preserved in passing from mode I to mixed mode (e.g., by piece-wise linearisation, see Figures 1a and 1b), then only changes in the number of variables and in the computational effort intervene, not in conceptual and mathematical terms; similar unified framework is provided by piece-wise linear material models also to direct elasto-plastic analyses in rates, in finite steps and under the assumption of holonomic path-independent behaviour; see: Maier and Comi (2000); Tin-Loi and Xia (2001b); Cocchetti et al. (2002); Bolzon and Cocchetti (2003).

The relationship of tensile traction p vs opening displacement w depicted in Figure 1a exhibits a bilinear softening (sloping down) branch, governed by four parameters (like p_c, p_b, k, h, indicated there). This bilinear model turns out to be flexible enough to match experimental results satisfactorily to many practical purposes (see e.g.: Alvaredo and Torrent, 1987; CEB-FIP Model Code, 1990; Guinea et al., 1994).

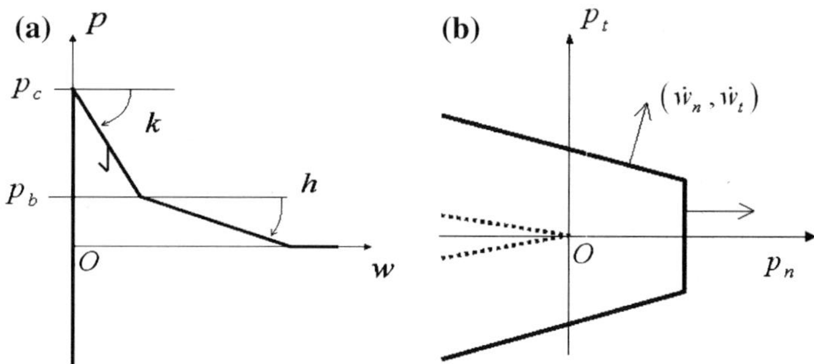

Figure 1. Piecewise-linear cohesive crack models: (a) mode I with four parameters to identify; (b) mixed mode (original yield locus in solid lines; dotted lines for residual strength locus after fracture).

The above cohesive crack model in its 'holonomic' interpretation (reversible, path-independent) can be analytically described by a now popular mathematical construct, namely as a LCP. This description reduces the computational burden of direct and inverse analyses, especially when a 'sifting' of the potentially active yielding modes is carried out in order to reduce the number of variables (Cocchetti and Maier, 2003).

The analytical description of the local behaviour depicted in Figure 1a can be cast into the following LCP format relating normal traction p to opening displacement w:

$$\boldsymbol{\varphi} = -p_c \mathbf{v}_1 + p_b \mathbf{v}_2 + [k\mathbf{m}_1 + h\mathbf{m}_2]\boldsymbol{\lambda} + p\mathbf{n} \leq \mathbf{0}, \quad \boldsymbol{\lambda} \geq \mathbf{0}, \quad \boldsymbol{\varphi}^T \boldsymbol{\lambda} = 0 \tag{1}$$

where: p_c and p_b denote tensile strength ('critical traction') and 'break-point traction', respectively; k and h govern the slopes of the two linear softening branches; $\boldsymbol{\varphi} = \{\varphi_1, \varphi_2, \varphi_3\}^T$ and $\boldsymbol{\lambda} = \{\lambda_1, \lambda_2, w\}^T$ are vectors which collect the yield functions φ_i and, respectively, opening displacement w and auxiliary variables λ_1 and λ_2; $\mathbf{n}$, $\mathbf{v}_1$, $\mathbf{v}_2$, $\mathbf{m}_1$, $\mathbf{m}_2$ are vectors and matrices of non-dimensional constant entries (precisely: 0, 1 or -1); for details, see: Maier (1970); Bolzon et al. (1994b, 2002); Maier and Comi (2002); Tin-Loi and Que (2001a).

For any given traction p, the LCP in Equation (1), endowed with a non-symmetric indefinite matrix, provides its solution in terms of displacements w in accordance with Figure 1a, namely: $w = 0$ for $p < 0$; an unbounded continuous set for $p = 0$; two solutions for $0 < p < p_c$; one for $p = p_c$; no solution for $p > p_c$.

The four parameters p_c, p_b, k and h (henceforth gathered in vector $\mathbf{x}$), visualised in Figure 1a, are to be identified. Clearly, any alternative set of 4 parameters in one-to-one correspondence with the above mentioned ones is eligible for identification (e.g., the fracture energy, i.e., the area enclosed by the plot and the displacement axis, might replace the traction at the break point), but the selected ones have the advantage of entering linearly in the formulation. The elastic moduli concerning the bulk material (Young's E and Poisson's ν) are supposed to be *a priori* known from conventional uniaxial tests.

2.2. FRACTURE TESTS

Three-point bending tests under displacement control are routinely and economically performed in industrial environments, primarily to evaluate the fracture energy, see e.g., RILEM (1985), though with some controversial result (Guinea et al., 1992;

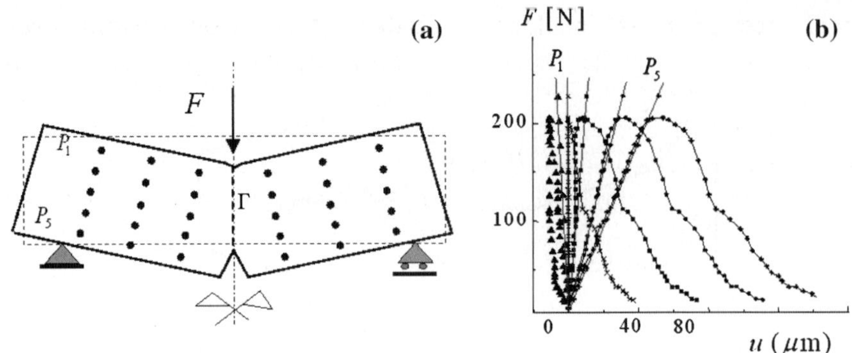

Figure 2. Three-point bending test: (a) specimen and monitored points; (b) horizontal relative displacements u measured by laser interferometry.

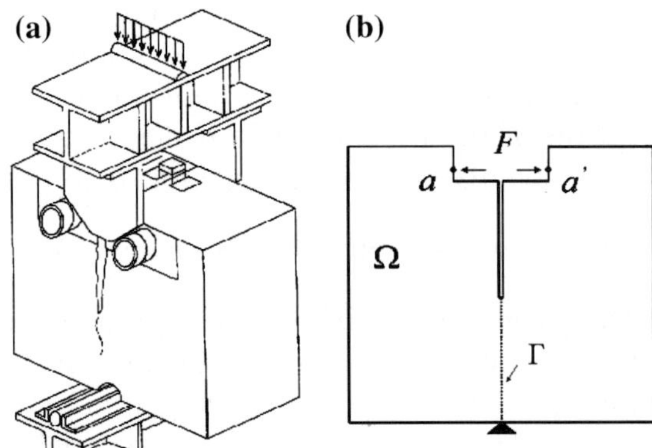

Figure 3. Wedge-Splitting Test: (a) equipment (from Denarié et al., 2001); (b) specimen and measured quantities (schematically).

Planas et al., 1992; Elices et al., 1992). The identification through TPBTs of the four parameters in the above cohesive crack model for concrete-like materials turned out to be little robust if based only on the measurement of the reaction force F vs the imposed vertical displacement (Figure 2a), see Bolzon et al. (1997b). Additional input data for the inverse problem in point consist of a set of displacements measured on the specimen surface and acquired by ESPI (Figure 2b). This technique uses laser to produce fringe patterns which are recorded by a video-camera and transformed into digitalised images. The relative displacements of symmetric point couples are obtained from the spacing between fringes. Sensitivity analysis provides orientation to the choice of the spacing between the points to be monitored on the specimen surface, see Bolzon and Maier (1998).

Wedge-splitting test, schematically represented in Figure 3 with a prismatic specimen, is alternative to more traditional tests like the TPBT, and turns out to be especially suitable to determine fracture properties of concrete with large-size aggregates, like dam concrete.

The main potentially advantageous features of WST are as follows: usual mechanical testing machines can be employed; the experiment can be performed on prismatic or cylindrical specimens cast in place, or on cores taken from existing structures; the influence of self-weight may be regarded as negligible for usual sizes; additional provisions for multiaxial tests, dynamic loading and hydraulic fracture can be integrated in relatively easy ways into the basic WST set-up. This kind of tests turns out to be extensively used since about 15 years, with specimen size ranging from 5 cm up to 3.2 m, on concrete, mortars, advanced cementitious materials and rocks (see e.g.: Brühwiler and Wittmann, 1990; Denariè et al., 2001).

A typical WST set-up between the actuator and the specimen consists of a beam with wedges, plates equipped with roll bearings and a rounded support. A notch is cast or cut into the specimen, in order to force a straight path of crack propagation along the symmetry plane Γ (Figure 3b). The descending wedge causes a controlled, stable crack growth in the specimen. The available experimental data consist of the imposed relative displacement (IRD) between points a and a' and of the consequent horizontal splitting force F, see Figure 3b. Embedded optical fibres endowed with Bragg grating might provide additional experimental data in terms of strains (see Denariè et al., 2001), but are not considered herein.

2.3. TEST SIMULATIONS

The computer simulations adopted here for both the above mentioned TPBT and WST are based on a two-dimensional, plane-strain interpretation of the specimen and on a discrete cohesive crack idealisation with the bilinear tensile softening shown in Figure 1a. All the dissipative phenomena are supposed to be concentrated on the locus Γ of possible displacement discontinuities, along the symmetry axis of the specimen.

In both experimental techniques, the design of specimen and equipment, and the monotonic growth in time of the imposed displacement (proportional to a monotonically increasing factor t) guarantee regularly progressive fracture processes. Regular progression means here: without manifestations of irreversibility in terms of crack closure (see Bolzon et al., 1995); stability, in the sense of no snap-back occurrence (see, e.g.: Carpinteri, 1989; Cen and Maier, 1992); uniqueness of incremental solutions (Bolzon et al., 1997c). Therefore, the cohesive crack model can be regarded, to the present calibration purposes, as path-independent (or holonomic), and can be described as a LCP, Equation (1).

The space discretisation is conveniently performed by means of mixed finite elements (FE) and generalised variables (in Prager's sense). This kind of mixed FE modelling preserves the energy meaning of the dot product of work-conjugate variables, and the essential features of the involved operators in passing from the continuum to the discrete formulation, see, e.g.: Comi et al. (1992); Bolzon (1996); Bolzon and Corigliano (1997a). An alternative, cost-effective approach to the computer simulation of systems with localised nonlinearities like the two present ones, is provided by boundary element methods (BEMs), in particular by the symmetric Galerkin BEM, like in Maier et al. (1993), Frangi and Maier (2002).

Also the overall FE or BE holonomic analysis of the TPB and WS tests can be formulated as a LCP, as follows (details in Bolzon et al., 1997b, 2002):

$$\boldsymbol{\Phi} = -p_c \mathbf{V}_1 + p_b \mathbf{V}_2 + [k\mathbf{M}_1 + h\mathbf{M}_2 + \mathbf{Z}'] \boldsymbol{\Lambda} + t\,\mathbf{P}^E \leq 0, \quad \boldsymbol{\Lambda} \geq 0, \quad \boldsymbol{\Phi}^T \boldsymbol{\Lambda} = 0 \qquad (2)$$

where: $\mathbf{V}_1, \mathbf{V}_2, \mathbf{M}_1, \mathbf{M}_2$ and $\mathbf{\Lambda}^T = \left\{ \mathbf{\Lambda}_1^T \ \mathbf{\Lambda}_2^T \ \mathbf{W}^T \right\}$ are the overall counterparts, in the discrete model of the whole system, of vectors and matrices denoted by the relevant lower-case symbols in Equation (1) and concerning local quantities; matrix $\mathbf{Z}' = diag\left[\mathbf{0} \ \mathbf{0} \ \mathbf{Z}\right]$ contains as diagonal block the (symmetric, negative semi-definite) influence matrix $\mathbf{Z}$ which relates, in linear elasticity, vector $\mathbf{W}$ governing the relative displacements along Γ to the vector $\mathbf{P}$ governing the consequent tractions across the fracture itinerary surface Γ; vector $\mathbf{P}^E$ describes the tractions on Γ due to the reference, for $t=1$, external action) on the undamaged specimen (namely: imposed vertical displacement in TPBT; imposed horizontal relative displacement in WST). The dimension of the LCP Equation (2) is equal to $3\mathrm{n}_d$, n_d being the number of the space-discretisation nodes on Γ_d.

In view of the linear background of the modelled system outside Γ_d, measurable quantities in the cracked specimen can be computed as follows:

$$\mathbf{Y}^{\mathrm{comp}} = t\mathbf{Y}^E + \mathbf{G}\mathbf{W}(t; \mathbf{x}) \tag{3}$$

where: vector $\mathbf{Y}^E$ gathers the measurable quantities in the computed elastic response of the sound specimen to unitary external action; vector $\mathbf{W}$ contains nodal displacements obtainable as a solution of LCP Equation (2) and, hence, depends on the four sought parameters collected in vector $\mathbf{x}$; matrix $\mathbf{G}$ consists of elastic influence coefficients.

Ad hoc mathematical programming techniques (e.g.: Dirkse and Ferris, 1995; Facchinei and Pang, 2003) can be employed to solve LCP as forward operator, namely the direct fracture problem Equation (2), in order to compute the specimen response in terms of the measurable response quantities (horizontal splitting force F in WST; vertical force F and relative horizontal displacements u measured by ESPI in TPBT, Figure 2).

The LCP mathematical construct which describes both experiments according to Equations (2) and (3) can be exploited to obtain in a computationally convenient way the derivatives of the measurable quantities $\mathbf{Y}$ with respect to the parameters to identify, gathered in the so-called 'sensitivity matrix' $\mathbf{L}$, see Bolzon et al. (2002).

2.4. DETERMINISTIC, BATCH PARAMETER IDENTIFICATION

In the deterministic framework of the traditional least-square methodology (see e.g.: Bui, 1994; Stavroulakis et al., 2003), all the data are compared simultaneously to the computed ones, in a 'batch' fashion. The inverse problem of identifying the model parameters $\mathbf{x}$ consists of minimising the discrepancy between the quantities (gathered in vector $\mathbf{Y}_i^{\mathrm{exp}}$) measured experimentally at different instants $t_i, i=1\ldots n_t$, during the test, and the corresponding computed quantities, say $\mathbf{Y}_i^{\mathrm{comp}}$. This discrepancy, say ω, is quantified as follows:

$$\omega(\mathbf{x}) = \sum_{i=1}^{n_t} \left(\mathbf{Y}_i^{\mathrm{exp}} - \mathbf{Y}_i^{\mathrm{comp}}(\mathbf{x})\right)^T \mathbf{D}_i^{-1} \left(\mathbf{Y}_i^{\mathrm{exp}} - \mathbf{Y}_i^{\mathrm{comp}}(\mathbf{x})\right) + (\mathbf{x} - \mathbf{x}_0)^T \mathbf{D}_0^{-1} (\mathbf{x} - \mathbf{x}_0) \tag{4}$$

Matrices $\mathbf{D}_i$ and $\mathbf{D}_0$ are suitably chosen weights, also with the role of making non-dimensional and hence comparable the relevant quadratic forms, but primarily intended to confer more weight to less uncertain data, and, hence, usually represented by the covariance matrix $\mathbf{C}_i$ of the i-th supply of data. In Equation (4) vector $\mathbf{x}_0$ denotes an initialisation vector of *a priori* parameter estimates (often arising from the judgement of a hypothetical 'expert'). The last term in Equation (4) plays a regularisation role for the possible ill-posedness of the optimisation problem, in the spirit of Tikhonov 'convexification', see e.g., Bui (1994).

In view of Equations (2)–(4), the parameter identification problem can be formulated as follows:

$$\min_{\mathbf{x},\boldsymbol{\Lambda}_i} \{\omega(\mathbf{x}, \boldsymbol{\Lambda}_i)\}, \text{ subject to } \boldsymbol{\Phi}_i(\mathbf{x}, \boldsymbol{\Lambda}_i) \leq \mathbf{0}, \quad \boldsymbol{\Lambda}_i \geq \mathbf{0}, \quad \boldsymbol{\Phi}_i^T \boldsymbol{\Lambda}_i = 0, (i = 1 \ldots n) \tag{5}$$

Noteworthy is the peculiar circumstance that the constraints are represented by a complementarity relationship, in particular here a LCP (but elsewhere a nonlinear complementarity problem, NLCP, when the adopted discrete crack model is nonlinear). As a consequence, the optimisation problem Equation (5) exhibits special mathematical and computational features. Clearly, complementarity constraints imply non-convexity and non-smoothness. Such kind of problems belongs to the class of MPEC (e.g., Facchinei and Pang, 2003), and turns out to be remotely rooted in the theory of non-cooperative games in econometrics (Nash, 1951).

Traditional penalty approaches, such as sequential unconstrained minimisation techniques (SUMT) (see e.g., Fiacco and McCormick, 1968), might turn out to be cost-effective for the numerical solution of MPEC, Equations (4) and (5). Recently, smoothing techniques have been proposed for other mechanical problems formulated as MPECs, apparently with remarkable computational benefits, by Pang and Tin-Loi (2001) and by Tin-Loi and Que (2002b).

In the field of inverse analysis, a MPEC was solved first in plasticity by an *ad hoc* two-phase method by Maier et al. (1982). The first phase consists of the minimisation of a non-convex quadratic function under linear inequality constraints only, by a standard algorithm; the second phase is a decomposition procedure which implies a sequence of quadratic programs leading to the global minimum in a finite number of operations.

For the quasi-brittle fracture parameter identification by TPBT mentioned in Subsection 2.2, problem (5) was numerically solved in Bolzon et al. (1997b), Bolzon and Maier (1998), by sequential quadratic programming (SQP algorithm) and by a conventional genetic algorithm. As well known, genetic algorithms consist of a large number of direct solutions (based on parameters generated at each step, starting from a random population of points in the parameter space) and, thus, they circumvent any mathematical difficulties, including the occurrence of local minimum points (see e.g., Tin-Loi and Que, 2002a). However, they turned out to increase dramatically the computational effort.

2.5. Stochastic, sequential parameter identification

The EKF procedure basically consists of a time-stepping sequence of estimations, which starts from an priori estimates (Bayesian approach) and exploits a flow of experimental measurements, accompanied by statistical data on their uncertainties. Restarting from the final results, the whole sequence can be repeated until convergence is achieved ('global iterations'). At each step, the estimates obtained at the previous instant are updated (and generally improved) by using the new supply of experimental data, and the relevant uncertainties are processed in order to update those of the resulting estimates. Kalman filter is, therefore, a sequential stochastic algorithm (see e.g.: Bittanti et al., 1984; Bui, 1994), at difference from the batch deterministic parameter identification techniques mentioned in Subsection 2.4.

Reference is made here again to the PWL cohesive crack model specified in Subsection 2.1, Equation (1), and to the identification of its four parameters, Figure 1a, through experimental data by WST (Subsection 2.2, and Figure 3). The whole procedure proposed for quasi-brittle fracture behaviour characterisation and expounded in Bolzon et al. (2002) can concisely be outlined as follows. Some fundamentals and pertinent operative details on Kalman filter applications to the present kind of inverse mechanical problems are available there and in Bittanti et al. (1984).

The initialisation consists of a vector $\hat{\mathbf{x}}_0$ gathering *a priori* estimates of the sought parameters and a covariance matrix $\hat{\mathbf{C}}_0$ which quantifies their degree of uncertainty ($\hat{\mathbf{C}}_0$ is diagonal in the frequent absence of correlations). At the measurement instant $t_i, i = 1 \ldots n_t$, i.e., under the IRD at the i-th stage of the WST, the updating equations of the EKF can be formulated as follows:

$$\begin{cases} \hat{\mathbf{x}}_i = \hat{\mathbf{x}}_{i-1} + \mathbf{K}_i (\mathbf{Y}_i^{\text{exp}} - \mathbf{Y}_i^{\text{comp}}(\hat{\mathbf{x}}_{i-1})) \\ \hat{\mathbf{C}}_i = \hat{\mathbf{C}}_{i-1} - \mathbf{K}_i \mathbf{L}_{i-1} \hat{\mathbf{C}}_{i-1} \end{cases} \tag{6}$$

$$\mathbf{K}_i = \hat{\mathbf{C}}_{i-1} \mathbf{L}_{i-1}^T (\mathbf{L}_{i-1} \hat{\mathbf{C}}_{i-1} \mathbf{L}_{i-1}^T + \hat{\mathbf{C}}_i^{\text{exp}})^{-1}, \quad \mathbf{L}_{i-1} = \left. \frac{\partial \mathbf{Y}_i^{\text{comp}}}{\partial \mathbf{x}^T} \right|_{\mathbf{x} = \hat{\mathbf{x}}_{i-1}} \tag{7}$$

In these formulae, capped symbols $\hat{\mathbf{x}}_{i-1}$ and $\hat{\mathbf{C}}_{i-1}$ indicate the mean value vector and the covariance matrix of the estimates at the previous instant; $\mathbf{L}_{i-1}$ denotes the 'sensitivity matrix', i.e., the matrix of the derivatives of the forward operator with respect to the parameters to identify; $\mathbf{K}_i$ is the so-called 'gain matrix'; $\mathbf{Y}_i^{\text{exp}}$ and $\hat{\mathbf{C}}_i^{\text{exp}}$ represent the experimental data supplied by the equipment at instant t_i and their covariance matrix, which quantifies the measurement 'noise', respectively. The final estimation now includes mean values and covariance matrix of the sought parameters: the capacity of processing experimental uncertainties and quantifying the resulting uncertainties of the identified parameters represents an advantageous peculiar feature of the EKF methodology.

The repeated computation of the sensitivity matrix $\mathbf{L}_{i-1}$, for $i = 1 \ldots n_t$, generally represents a significant burden for large-size problems usually arising in practical engineering situations. Such burden can be drastically reduced in the present fracture mechanics problem, due to LCP formulation conferred to the forward operator, Equations (2) and (3), as demonstrated in Bolzon et al. (2002): in fact, closed-form analytical formulations of the derivatives can be found instead of costly finite-difference approximations. In general, every entry of any sensitivity matrix quantifies the influence of a sought parameter, here $x_j (j = 1 \ldots 4)$, on a measurable quantity (here $Y_i, i = 1 \ldots n$): therefore it can usefully orient the design of the experiment, since, obviously, information of a quantity practically uninfluenced by a parameter cannot contribute to its estimation. Matrices $\mathbf{L}_{i-1}$, as key ingredient of the filtering method, reflect the linearisation over the step $t_{i-1} \to t_i$ of the forward operator. At present, the numerical consequences of this linearisation can be mitigated by the so-called 'unscented' version of the Kalman filter, see e.g., Norgaard et al. (2000).

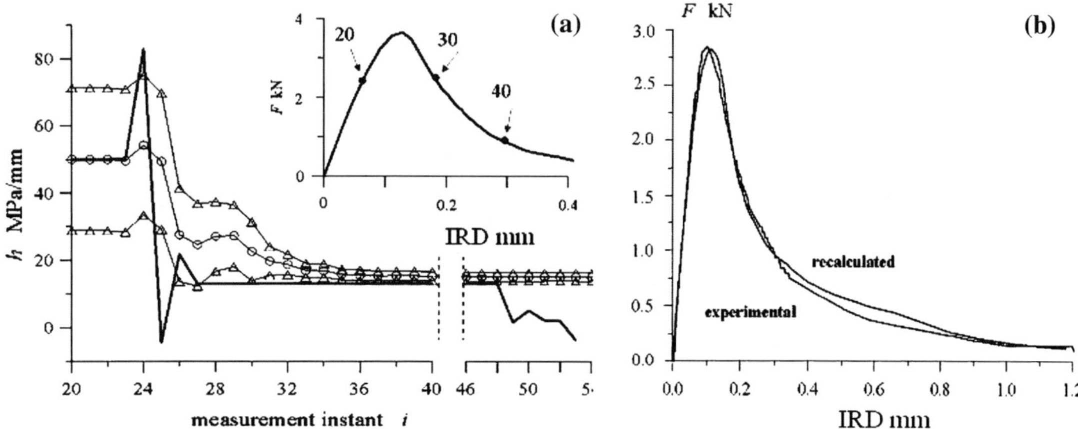

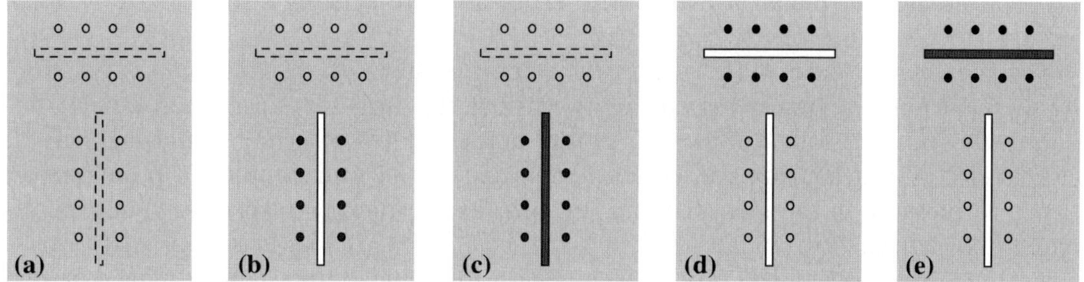

Figure 4. Kalman filter estimation of parameter h, on the basis of pseudo-experimental data (correct value $h = 13$ MPa/mm): (a) mean values (marked by o) and 99% confidence limits (marked by $\triangle$) vs measurement instant index i, with noisy data; thick line visualises mean values in the absence of noise; (b) experimental reaction force F vs imposed relative displacement IRD plot, compared to the simulated response computed by the identified model.

Figure 5. Steps in Phase A of the innovative flat-jack inverse-analysis procedure.

Every identification procedure is preliminarily assessed by means of pseudo-experimental (i.e., computer generated) data. As an example of such validation exercises, presented in Bolzon et al. (2002), Figure 4a visualises the convergence of the estimated mean value of parameter h to the correct value assumed for generating the pseudo-experimental data. The confidence range (99% with Gaussian distribution of probability density) shrinks progressively as new data are processed, until convergence is achieved. The experimental plot of force F vs conjugate displacement input and its counterpart computed by using the calibrated cohesive crack model, are comparatively shown in Figure 4b.

3. In situ assessment of stresses and concrete properties in large dams

Flat-jack tests are frequently employed in civil engineering to estimate stress states and elastic Young moduli, especially in monumental masonry buildings and large concrete structures like dams. The technique consists of the following stages (see e.g., Goodman, 1980, with reference to rock mechanics): a slot is cut on the structure surface, and consequent relative displacements of couples of points across the slot are measured; a flat-jack is put inside the slot and is pressurized until the deformation due to the cut generation is recovered; the pre-existing stress normal to the cut is assumed equal to the product of the fluid pressure inside the jack and of a corrective constant established in laboratory for each jack; a second slot, parallel to the first, is cut; both slots are pressurized by flat-jacks and this gives rise approximately to a uniaxial compression test; relative displacements of points between the two slots are measured at different pressure levels, and elastic stiffness is assessed from the stress-strain curve thus obtained.

This traditional technique exhibits severe limitations, namely: it provides a single component of the elasticity tensor (and often also one component of the stress tensor); the region between the two flat-jacks is dealt with like a laboratory specimen under uniaxial compression, neglecting the links with the structure; fracture properties are not considered.

In order to achieve a more extensive set of information about stress states and material properties in concrete dams, a novel methodology based on conventional flat-jack equipment and inverse analysis is proposed and outlined in what follows. The new technique encompasses two different phases (see Figures 5 and 6, where slots and measurement marks are black when active): phase A (steps a–e) concerns the identification of material elastic properties and pre-existing stress; phase B (steps f and g) focuses on fracture properties.

Specifically, in Phase A (Figure 5) the operative steps are as follows: (a) positioning of measurement bases; (b) cutting a vertical slot and measuring consequent displacements of, say, four couples of points across it; (c) inserting a flat-jack in the vertical slot and measuring consequent displacements of four couples of points across it, due to pressure in the jack; (d) and (e) same as (b) and (c), respectively, but involving the horizontal slot.

A three-dimensional FE model of a region near the outer surface of a dam has been generated by a commercial computer code (Abaqus, 1998), in order to simulate the above outlined experiment and to compute experimentally measurable quantities as functions of the parameters to identify (now elastic moduli and stresses). The inverse analysis procedure rests here on the usual least-square formulation and on the trust-region algorithm (TRA), based on gradient recurrent computation, see e.g., Coleman and Li (1996). For the present least-square identification problems, in the

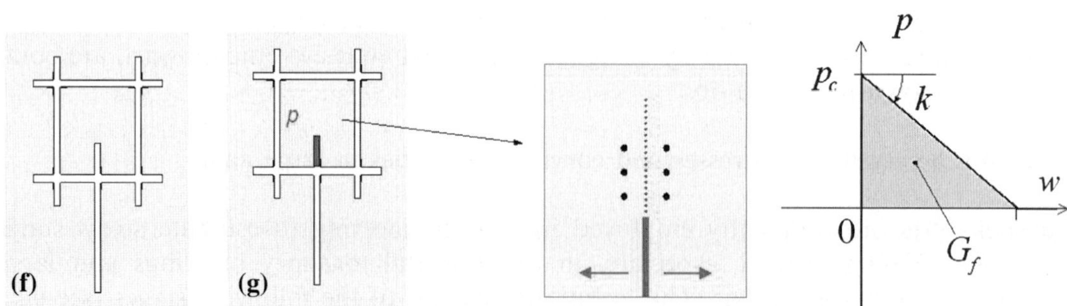

Figure 6. Steps of Phase B (fracture test) of the flat-jack, inverse analysis procedure; on the right the adopted cohesive crack model.

Authors' experience, TRA turns out to be computationally preferable with respect to other first order and direct search (zero order) algorithms.

The diagnostic procedure in point leads first to the identification of Young moduli E_V and E_H in vertical (subscript V) and horizontal (subscript H) directions. In view of the casting and compacting processes, concrete in a dam should be considered as an orthotropic and transversally (horizontally) isotropic material, hence endowed with five independent elastic moduli. However, in the present preliminary illustrative exercise, only the two Young moduli are regarded as parameters to identify, by using measurements in steps (c) and (e) as input for the inverse analysis. Clearly, these measures are independent from the pre-existing stress state as long as the pressure is limited and the system behaves linearly; this circumstance allows to identify E_V and E_H without knowing the stress. The so assessed elastic properties of the material are exploited together with the experimental data achieved from steps (b) and (d), in order to estimate the normal stresses σ_H and σ_V, pre-existing in the dam (possibly due to unknown expansive pathological ageing processes, such as alcali-aggregate reaction, AAR). Tangential stress τ_{HV} can be identified as well by measuring also relative displacements almost tangentially to the slot, i.e., between basis marks in diagonal directions. Clearly, plane stress hypothesis is realistic near the free surface of the dam.

An alternative set of quantities to be used as input for the inverse analysis consists of local deformations measured by strain-gages placed near the slots. This alternative turns out to be practically advantageous because electric signals from strain-gages can be automatically (rather than manually) gathered, interpreted and saved on site. In both cases, i.e., measuring either relative displacements or local strains or both, numerical tests on the inverse analysis based on the TRA led so far to encouraging results, even when random noise affects (pseudo-) experimental input data.

Phase B is based on the pressurization leading to fracture of a sort of specimen generated by additional cuts, and on the measurement of relative displacements near the mouth of the crack propagation path, see Figure 6. This phase can be subdivided in two steps (Figure 6): (f) additional cuts are made so that a region of material becomes almost isolated from the rest of the dam and free of stresses except for the rear zone; (g) a special flat-jack is inserted in the notch which is part of the previous last vertical slot, and opening displacements are measured by a set of instruments along the crack path, at different increasing pressure levels. The jack employed in the last step (g) should be small and possibly *ad hoc* designed, but all the other instruments may be those used traditionally.

A three-dimensional FE model has been built up (see Figure 7), in order to simulate the fracture process, and to compute the measurable relative displacements as functions of the two parameters in the adopted popular cohesive crack model shown in Figure 6. The quasi-brittle fracture model is simplified with respect to the one dealt with in Section 2, in view of the peculiar features of dam engineering situations now considered. The suitability of the geometry designed for the fracture test (and visualized, together with the FE mesh, in Figure 7) has been validated by nonlinear analyses resting on a continuum elastic-plastic-softening constitutive model for concrete (Fenves model, in the computer code Abaqus, 1998): such preliminary analyses pointed out that only negligible plastic strains arise outside the envisaged crack propagation plane.

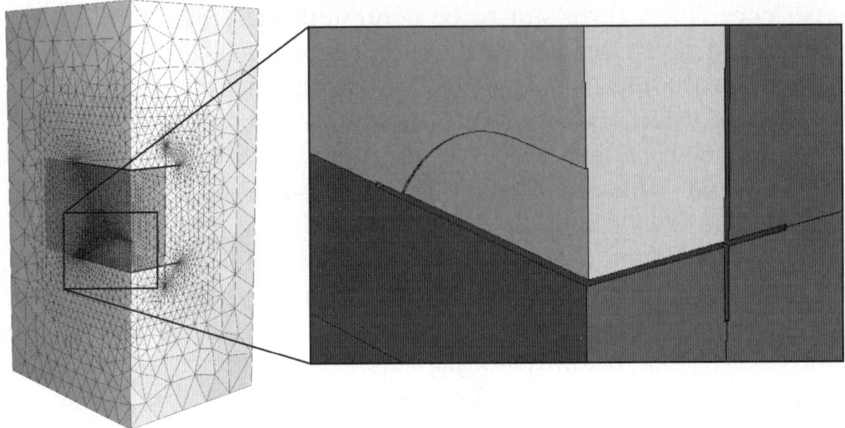

Figure 7. Finite element model of the proposed flat-jack test for fracture parameters identification. The zoomed picture visualizes the semicircular notch.

The identification procedure by inverse analysis exploits the measurable data obtained in step (g), i.e., a set of opening IRD at different pressure levels. The chosen cohesive crack model (Figure 6) is defined by two independent parameters out of three (tensile strength p_c, fracture energy G_f and slope k of the post-peak segment). Comparative inverse analyses have been performed for the identification first of p_c and G_f, second of p_c and k, starting from the same set of input data, generated by adding to the computed pseudo-experimental data a random noise uniformly distributed on a suitable range.

As a typical numerical test on the new technique, with noise in the range ±5%, identification of p_c and G_f led to −2.30% error on p_c and +3.80% error on G_f, the identification of p_c and k to almost the same error (−2.30%) on p_c and −8.05% error on k estimate. Such kind of results corroborates the identifiability of fracture param eters by the proposed method (contractivity of uncertainties is observed in passing from input data to output estimates) and evidences the importance of a proper choice of the parameters to identify. Figure 8a visualizes the convergence of the TRA along the itinerary of steps shown in Figure 8b (parameters are normalized with respect to their 'exact' values). The contour plot in Figure 8b of the objective function ω on the (p_c, G_f) plane evidences that this discrepancy function turns out to be nonconvex.

In dam engineering practice the characterization of possibly deteriorated concrete sometimes must be performed repeatedly on the same large dams. For future routine applications it would be desirable to mitigate the computational burden of inverse analysis based on a large FE model, like the present one shown in Figure 7 (with 138,000 degrees of freedom). Soft-computing by ANNs (see e.g.: Haykin, 1999; Waszczyszyn, 1999) may satisfy this requirement and might eventually lead to local diagnoses carried out in situ routinely and economically.

A preliminary orientative employment of ANNs has been investigated for Phase B (fracture test) of the flat-jack technique proposed herein. The task is to estimate the tensile strength p_c and the fracture energy G_f on the basis of a set of 16 experimental data, namely from the relative displacements of three couples of points measured by conventional instruments at five pressure levels.

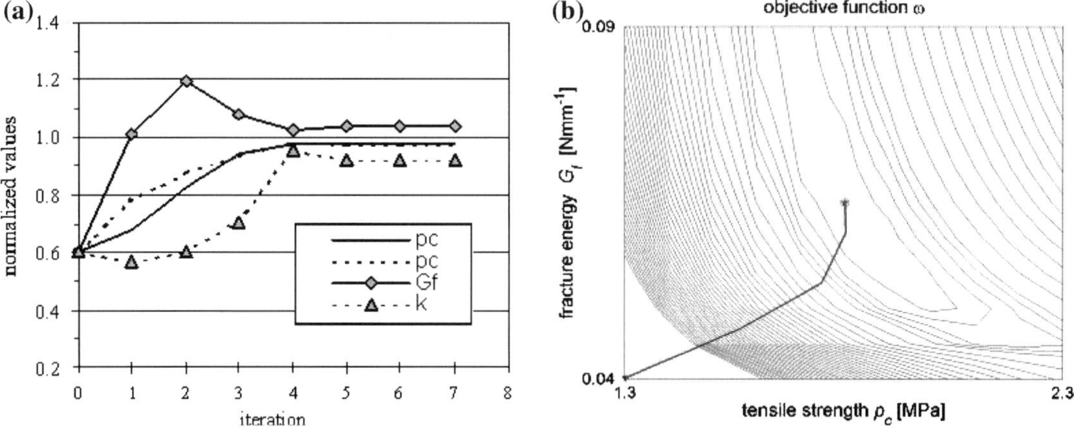

Figure 8. (a) Convergence of the Trust Region algorithm for the identification of the fracture parameters p_c and G_f (solid line) and of p_c and k (in dashed lines); (b) map of the discrepancy function and iteration path to its absolute minimum.

To the above specified purpose, an 'architecture' 16-4-2 (16 input neurons, one hidden layer with 4 neurons and 2 output neurons) was selected by a trial-and-error approach for the feed-forward, back-propagation network. The neurons in the hidden layer are assigned an hyperbolic tangent sigmoid as 'transfer function', the neurons in the output layer a linear 'transfer function'. In order to calibrate the 72 weights and 6 biases of the network, the implemented ANN has been trained by 363 patterns (i.e., pairs of input and output vectors), generated once-for-all by direct analyses through the FE model in Figure 7 on the basis of equally spaced points in a selected region of the space (p_c, G_f, E). In all the training inputs perturbations have been added randomly extracted from a range of $\pm10\%$ of the current value for the opening displacements and $\pm5\%$ for the elastic modulus, with uniform probability density distribution. The training process (i.e., the minimization of the non-dimensional mean-square error by means of a back-propagation algorithm) was stopped after 600 iterations (or 'epochs', in the ANN jargon).

The trained network has been tested by 72 patterns generated by values of randomly chosen p_c, G_f and E, and by the relevant measurable displacements again affected by $\pm10\%$ ($\pm5\%$ for E) uniformly distributed random noise. The relative errors of the two fracture parameters in the ANN learning and testing processes are shown in Figure 9: the neural network turns out to approximate fairly satisfactorily the present complex relationship, originally materialized by the large FE model (Figure 7), between the two variable vectors of the fracture parameters (and Young's modulus E) and the relevant measurable displacements. In this illustrative example of computerized in situ flat-jack fracture experiment, pre-existing stresses are assumed to be *a priori* known.

A rather extensive investigation of the potentialities of ANNs in local diagnostic analyses of concrete dams can be found in Fedele et al. (2005) with reference to dilatometric tests in parallel deep holes drilled in the dam. This novel technique combines inverse analysis with experiments which are traditional in rock mechanics (see e.g., Goodman, 1980). The dilatometric test is complementary, clearly not alternative,

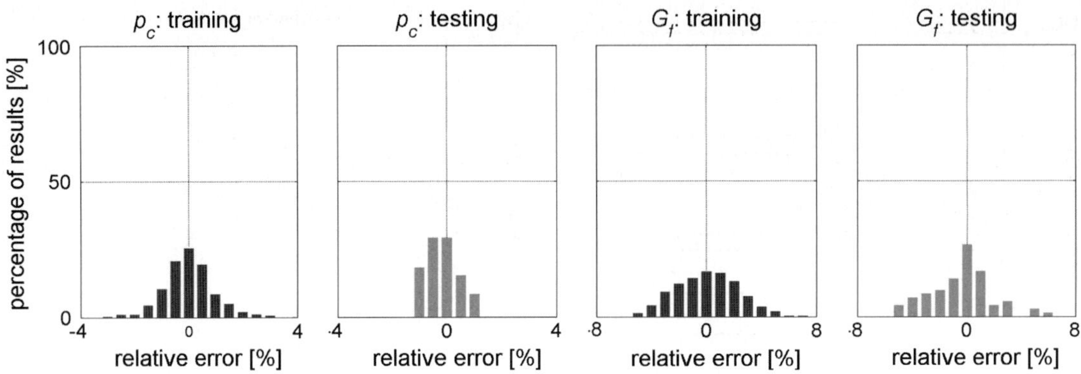

Figure 9. Relative errors of two fracture parameters identified by a neural network, with 363 learning and 72 testing patterns.

to the flat-jack procedure discussed in what precedes. Such dilatometric diagnostic method was partly anticipated by the 'pressure tunnel' studied in Gioda and Maier (1980) and has recently been extended to fracture mechanics, Maier et al. (2004).

4. Indentation, imprint mapping and inverse analysis combined in micro-technologies

A spreading methodology for mechanical characterisation of materials at the micro- and nano-scales is based on indentation tests. Fracture processes in quasi-brittle materials and delamination in film-substrate systems are easily induced in a non-destructive manner by, e.g., pyramidal (Vickers) or conical (Rockwell) indenters, and can effectively be simulated using interface and cohesive crack models to parameter identification purposes.

4.1. Fracture models and test

The cohesive model adopted in what follows was originally proposed for mode I fracture by Rose et al. (1981) for metals and for bimetallic interfaces, then it was extended to two- and three-dimensional mixed-mode fracture by Xu and Needleman (1994), Camacho and Ortiz (1996), Ortiz and Pandolfi (1999). Its formulation is centred on the following 'potential function':

$$\phi(\delta) = ep_c\delta_c\left(1 - \left(1 + \frac{\delta}{\delta_c}\right)\exp\left(-\frac{\delta}{\delta_c}\right)\right) \tag{8a}$$

$$\text{where: } \delta = \sqrt{\beta^2\left(w_{t1}^2 + w_{t2}^2\right) + w_n^2} \tag{8b}$$

In the above formulae, e is Neper number (basis of natural logarithms), p_c indicates the maximum cohesive normal traction and δ_c represents a characteristic opening displacement. It can be easily shown that the quantity which, in this model, plays the role of mode I fracture energy is $G_F = ep_c\delta_c$. Equation (8b) defines δ as an effective displacement discontinuity across the interface, β being a parameter which quantifies different weights for sliding (w_{t1} and w_{t2}) with respect to opening (w_n) displacements. The non-holonomic behaviour in this interface model is governed by the maximum attained effective displacement jump $\delta_{\max}$, which represents the only

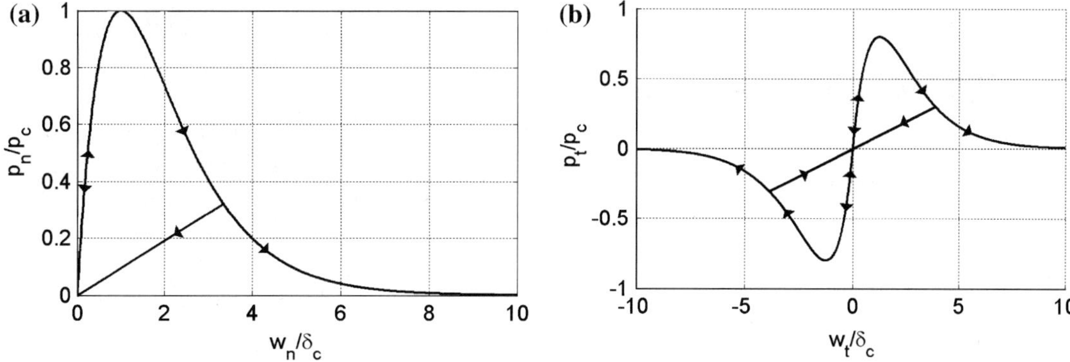

Figure 10. Interface model for mode I (a) and mode II (b) fracture, the latter for $\beta = 0.8$.

internal variable in the model. The relationship which describes the evolution of this variable reads:

$$\dot{\delta}_{\max} = \begin{cases} \dot{\delta} & \text{if } \delta = \delta_{\max} \quad \text{and } \dot{\delta} \geq 0 \\ 0 & \text{otherwise} \end{cases} \tag{9}$$

Cohesive tractions under progressive fracture (i.e., for increasing δ_{max}) or along a linear path back to the origin in case of unloading, are defined as follows:

$$\begin{Bmatrix} p_n \\ p_{ti} \end{Bmatrix} = \begin{Bmatrix} \partial\phi/\partial w_n \\ \partial\phi/\partial w_{ti} \end{Bmatrix} = e\frac{p_c}{\delta_c}\exp\left(-\frac{\delta}{\delta_c}\right)\begin{Bmatrix} w_n \\ \beta^2 w_{ti} \end{Bmatrix} \qquad \text{if } \delta = \delta_{\max} \text{ and } \dot{\delta} \geq 0$$

$$(i = 1, 2) \tag{10}$$

$$\begin{Bmatrix} p_n \\ p_{ti} \end{Bmatrix} = e\frac{p_c}{\delta_c}\exp\left(-\frac{\delta_{\max}}{\delta_c}\right)\begin{Bmatrix} w_n \\ \beta^2 w_{ti} \end{Bmatrix} \qquad \text{if } \delta = \delta_{\max} \text{ or } \dot{\delta} < 0$$

In mode I fracture, see Figure 10a, under progressive opening ($\delta = w_n > 0$; $\dot{\delta} = \dot{w}_n > 0$), the response of the interface is characterised by a tensile stress which first increases up to its maximum value p_c and then decreases asymptotically to zero in a softening regime. Non-holonomic behaviour and damage are exhibited when unloading occurs ($\dot{w}_n < 0$) starting from $w_n = \delta_{\max}$. Figure 10b visualises the modelled interface behaviour in the case of mode II fracture.

The parameters to identify in the interface model, Equations (8)–(10), are p_c, δ_c and β. In the FE discretisation adopted for the present illustrative examples (see Figure 11) the above interface model is employed where fracture is expected, namely either along the surface between film and substrate (see Figure 11a) or over the planes which are orthogonal to the upper face of the specimen through the diagonals of the pyramidal (Vickers) indenter (Figure 11b).

Like in previous simulations of indentation tests (Bhattacharya and Nix, 1998; Jayaraman et al., 1998; Cheng and Cheng, 1999), the commercial code Abaqus (1998) is employed with its large plastic strain capability. The contact interface between the indenter and the specimen is characterised by Coulomb friction without dilatancy, i.e., by a non-associative rigid-plastic model.

Boundary conditions are consistent with geometrical and physical symmetries: two-dimensional axial-symmetry for conical indentations of isotropic solids; symmetries in Vickers indentation of isotropic materials allow to reduce the domain, e.g.,

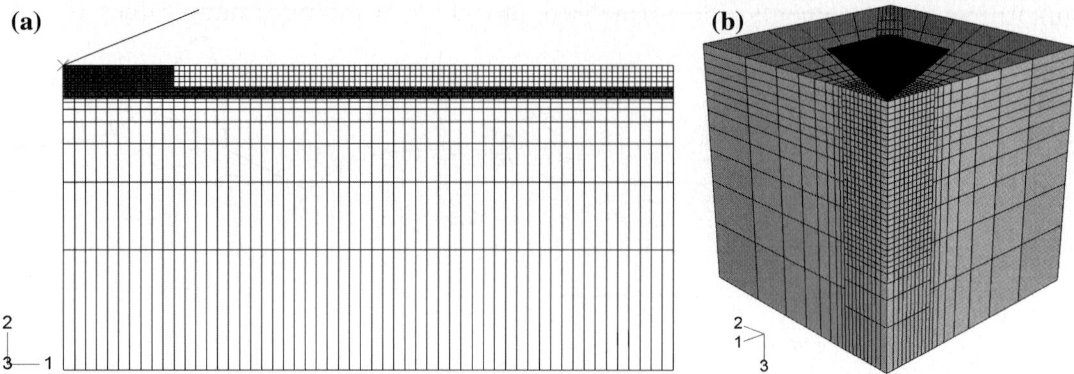

Figure 11. Finite element meshes adopted for simulation of: (a) Rockwell indentation of isotropic coating and substrate; (b) Vickers indentation of brittle bulk specimen.

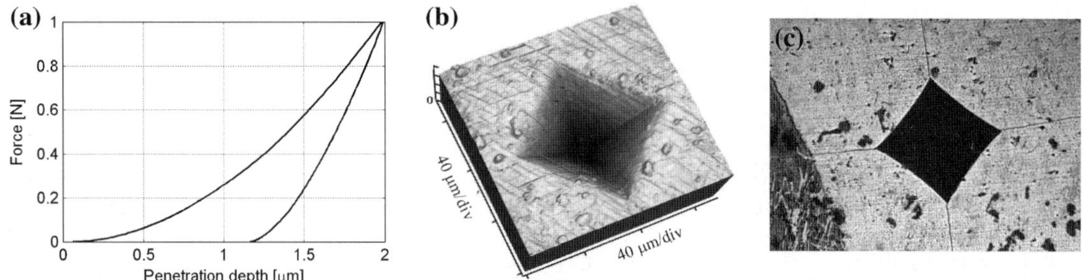

Figure 12. Experimental results on Zirconia specimen: (a) indentation curve; (b) imprint mapping by atomic force microscope and (c) crack visualization by optical microscopy.

to one quarter (Figure 11). The computing effort is further considerably reduced by preliminarily performing the condensation of the degrees-of-freedom belonging to the specimen portion exterior to the domain where all inelastic phenomena are *a priori* reasonably expected to be confined.

In both examples (a) and (b) in Figure 11, the indentation process is subdivided into M steps, run by index $i = 1 \ldots M$, and at each step the imposed force F_{mi} and the corresponding penetration depth of the indenter u_{mi}^{cur} are recorded, see the indentation curves in Figure 12a. Imprint displacement u_{mj}^{imp} normal to the indented surface (see Figure 12b) are measured in N selected points, run by index j. Typical instruments suitable to these measurements are AFMs at the micro-scale, laser profilometers at the macro-scale. In the case of Vickers indentation of brittle materials, Figures 11b and 12c, the crack length l_m^{cra} as well can be measured on the specimen surface.

The discrepancy function to be minimised encompasses, in the more general case, three kinds of contributions, namely:

$$\omega(\mathbf{x}) = \frac{1}{M} \sum_{i=1}^{M} \left(\frac{u_{ci}^{cur}(\mathbf{x}) - u_{mi}^{cur}}{u_{\max}^{cur}} \right)^2 + \frac{1}{N} \sum_{j=1}^{N} \left(\frac{u_{cj}^{imp}(\mathbf{x}) - u_{mj}^{imp}}{u_{\max}^{imp}} \right)^2 + \left(\frac{l_c^{cra}(\mathbf{x}) - l_m^{cra}}{l_{\max}^{cra}} \right)^2$$

$$(11)$$

where vector $\mathbf{x}$ collects the parameters p_c, δ_c and β to identify; subscript m marks experimentally measured quantities; subscript c indicates computed measurable quantities. Displacements are normalised by the relevant maximum values (i.e., by u_{max}^{cur}, u_{max}^{imp} and d_{max}^{cra}, respectively).

Equal 'weights' for all the contributions to the discrepancy norm (11) are assumed in the present illustrative exercises, carried out using computer-generated pseudo-experimental data; weights related to the measurement errors should be introduced when real experimental data are employed.

The dependence of the computed quantities on the material parameters is implicit in the FE model. Due to both geometrical (finite strains) and material nonlinearities allowed for, the objective function ω is non-explicit and expected to be generally non-convex. Like in Section 2, a conventional deterministic, batch (non-sequential) approach is adopted here for parameter identification, and the Trust Region algorithm (TRA) is employed once again as a tool for minimum search.

4.2. INVERSE ANALYSIS OF BRITTLE MATERIALS FRACTURE

Fracture processes induced by Vickers indentation in brittle materials are being studied, both from an experimental and numerical point of view, on Zirconia specimens. Micro indentation tests up to 1N load (see Figures 12a and b) and macro indentation tests up to 1kN load (see Figure 12c) have been performed. Figures 12b and c visualise typical images of the imprint after indentation; in the latter cracks are seen to develop in correspondence of the four corners of the Vickers indenter. Cracks turn out to open during the unloading indentation phase.

The test simulation has been carried out by employing associative Drucker–Prager model with linear isotropic hardening, and the above specified interface model.

Figure 13a shows the contour map of the computed displacements, orthogonal to the plane along which interface elements have been inserted, i.e., it visualises a 'half-penny crack' shape, which resembles that obtained experimentally.

Preliminary inverse analysis exercises, based on indentation of brittle materials, have been carried out by exploiting pseudo-experimental information in terms of indentation curve, imprint geometry and crack length, in order to identify the following material parameters: Young modulus E, compressive yield stress p_{cc}, hardening modulus H_{pl} (relating linearly the instantaneous elastic limit to the equivalent

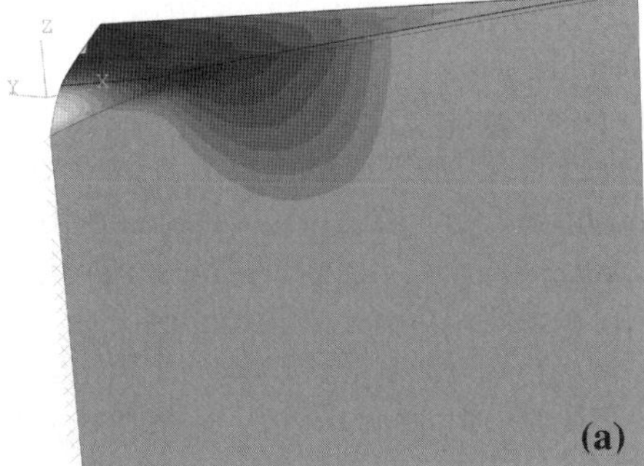
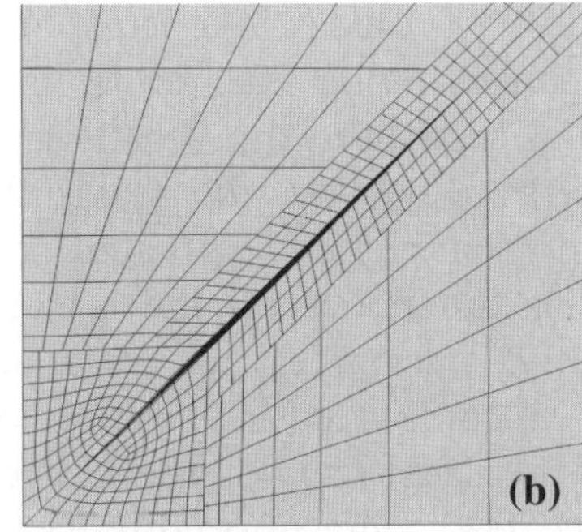

Figure 13. Half-penny shape crack in the computed imprint due to Vickers indentation: (a) contour map of displacements orthogonal to the plane along which interface elements have been inserted; (b) deformed mesh.

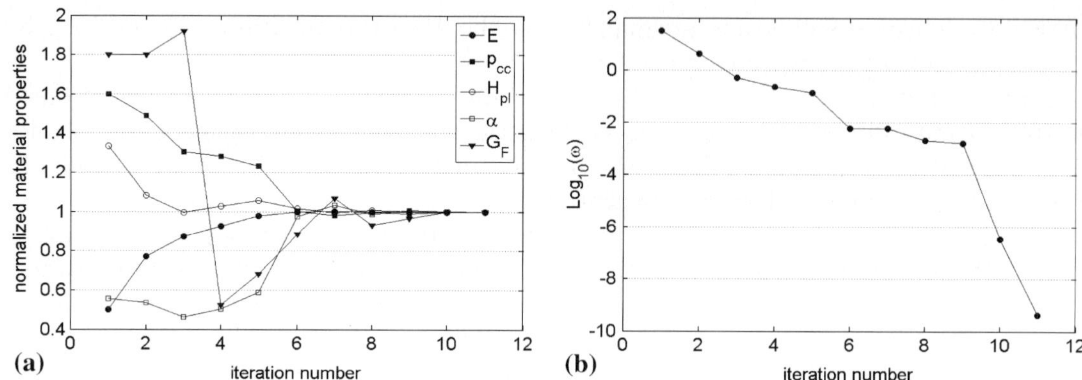

Figure 14. Convergence behaviour in the identification procedure based on indentation curve, imprint geometry and crack length data: (a) material parameters estimates; (b) progressive reduction of the discrepancy function ω.

plastic strain), friction angle α and critical displacement δ_c. Pseudo-experimental data have been generated assuming as material properties: $E = 80\,\text{GPa}$; Poisson ratio $\nu = 0.25$; $p_{cc} = 250\,\text{MPa}$; $H_{pl} = 149.75\,\text{GPa}$ and $\alpha = 27°$. In this case the tensile strength is a function of friction angle and compressive yield stress, namely:

$$p_c = p_{cc}\left(3 - \tan(\alpha)\right) / \left(3 + \tan(\alpha)\right) \tag{12}$$

The only parameter which characterises the interface is then the fracture energy or, equivalently, the critical displacement δ_c in Equation (10), which has been assumed equal to $0.207\,\mu\text{m}$. Parameter β has no relevance in this situation since pure mode I crack propagation is expected from symmetry considerations.

The sequence of parameter estimates, generated by the Trust Region algorithm, converges to the exact solution, as visualised in Figure 14a; Figure 14b shows the progressive reduction of the discrepancy function defined by Equation (11).

In a comparative inverse analysis exercise, pseudo-experimental information have been exploited in terms of indentation curve only, i.e., without information about imprint geometry and crack length. The behaviour of the inverse problem solution algorithm, in the absence of input disturbances, is visualised in Figure 15a: the resulting estimation of the fracture energy G_F is affected by a large error, but the algorithm returns poor estimates also of the Drucker–Prager model parameters p_{cc} and α. Figure 15b visualises the corresponding reduction of the non-dimensional discrepancy function ω, along the iteration sequence.

Figure 16 visualises the sensitivity of the vertical displacements in the residual imprint, with respect to the fracture energy G_F. The map is generated by interpolations of sensitivities evaluated in a number of locations corresponding to the FE discretization nodes on the upper surface of the indented specimen, for the set of constitutive parameters assumed above. The maximum value of the above sensitivity field is compared, in Table 1, to the maximum sensitivity of the indentation curve and the sensitivity of the crack length, both computed with respect the fracture energy after being normalised as follows:

$$L_{G_F} = \frac{\partial y}{\partial G_F}\frac{G_F}{y_{max}} \tag{13}$$

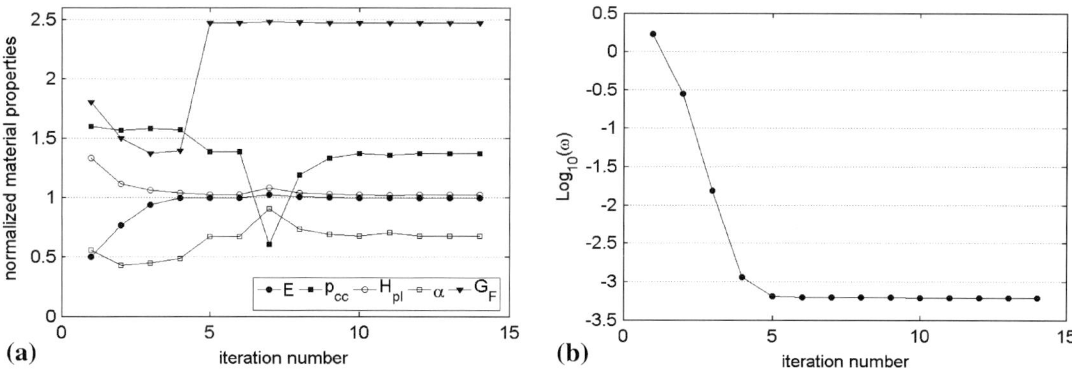

(a)   (b)

Figure 15. Convergence behaviour in the identification procedure based on indentation curve only: (a) material parameters estimates; (b) progressive reduction of the discrepancy function ω.

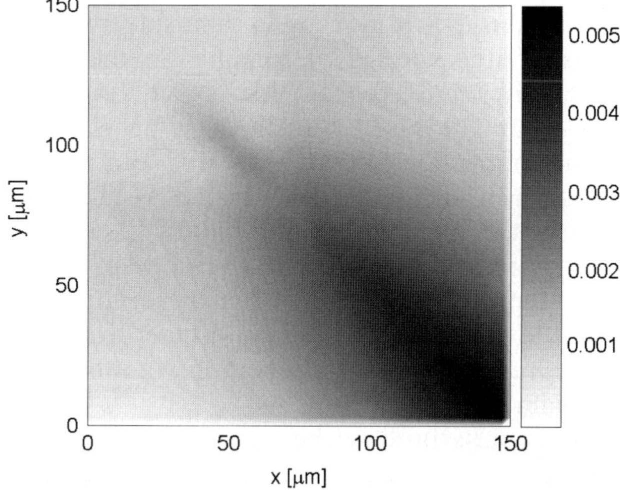

Figure 16. Sensitivity of the imprint geometry with respect to the fracture energy parameter G_F.

Table 1. Comparison among the maximum sensitivities of indentation curve, imprint geometry and crack length, respectively, with respect to fracture energy parameter G_F.

y	$max\left[L_{G_F}\right]$
u_{cj}^{cur}	**0.0019**
u_{cj}^{imp}	**0.0054**
l_c^{cra}	**0.025**

The derivatives with respect to G_F are evaluated taking as their argument the 'exact' parameters, originally adopted for computing the pseudo-experimental data. In the numerical computations, derivatives have been approximated by 1st-order forward finite-differences with 0.01% increment. Among the three kinds of available experimental data, namely indentation curve, imprint geometry and crack length, the largest sensitivity with respect to the fracture energy parameter (account taken of the different scales for the three variables) is exhibited by the crack length. Therefore, as qualitatively expected, crack length represents a crucial experimental information to be exploited in order to identify the material parameter G_F.

The above results on sensitivities are consistent with the already observed convergence difficulties in the identification process of the fracture energy and of other parameters, see Figure 15, when the only experimental data inserted into the inverse analysis procedure is the indentation curve. The main conclusion arising from the numerical tests on sensitivity, outlined in what precedes, is a confirmation that measurements on the residual deformed configuration of the specimen (namely imprint geometry and crack length) are generally bound to enhance the identification of material parameters through indentation tests.

4.3. Inverse analysis of coating-substrate delamination

When Rockwell indention is performed on a film-substrate system, two different delamination mechanisms are expected, depending on the ratio between the yield strengths of coating and substrate (see, e.g.: Abdul-Baqi and Van der Giessen, 2001; Li and Siegmund, 2004). In the case of ductile film on elastic substrate (e.g., gold on silicium), delamination phenomena are observed during the loading phase of the indentation process. Such phenomena materialise as mode II dominant fracture, caused by plastic flow in the film: the coating material spreads in horizontal direction due to the elastic substrate, see Figure 17a. The main effect on the indentation curves is a decrease of the applied force at equal penetration depth. As for a hard film on a ductile substrate, delamination usually occurs in mode I during unloading, due to the different recovery of elastic deformations between the hard film and the ductile substrate, see Figure 17b.

The inverse analysis exercise outlined below concerns a ductile film, modelled as elastic perfectly-plastic Huber-Mises material, on an elastic substrate. The parameters to identify in the adopted interface model, Equations (8) (10), are p_c, δ_c and β. Pseudo-experimental information are exploited, generated through direct analysis by assuming the following parameter set (subscript f and s indicate film and

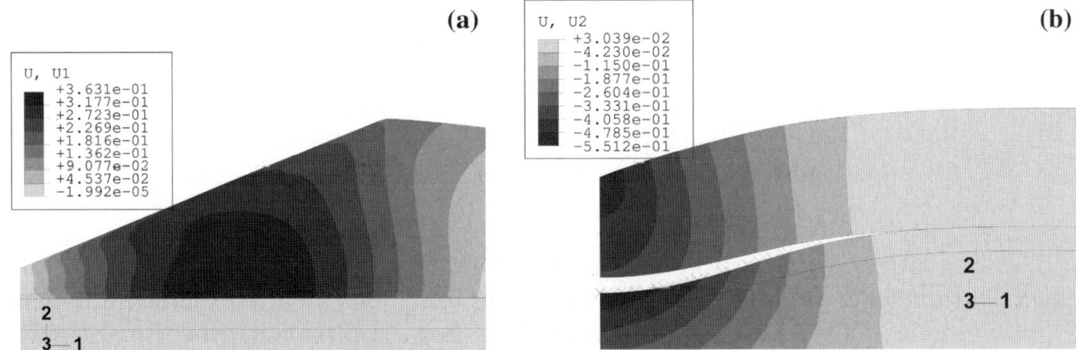

Figure 17. Delamination displacement fields due to indentation: (a) horizontal displacements in a case of ductile film on elastic substrate; (b) vertical displacements in a case of hard film on ductile substrate.

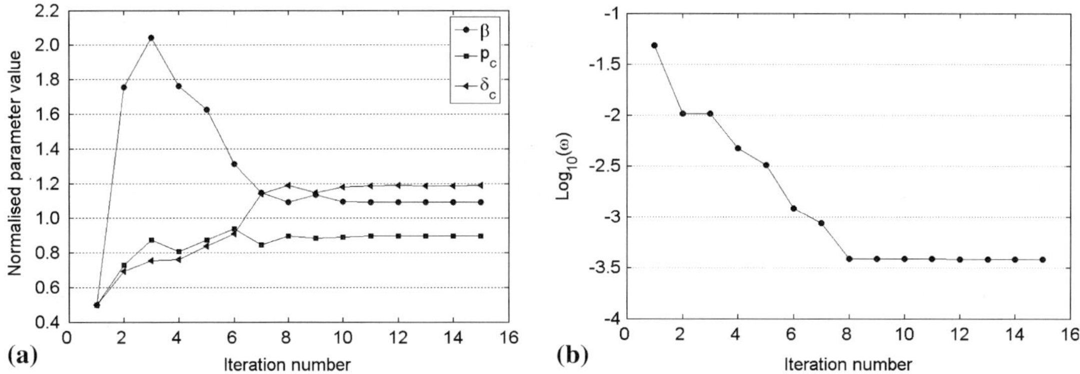

Figure 18. Convergence of Trust Region algorithm in the identification procedure based on both indentation curves and imprint geometry data: (a) normalised estimates of the parameters governing the delamination of a ductile film from a substrate; (b) progressive reduction of the discrepancy function ω.

substrate, respectively): $E_f = 80\,\text{GPa}$, $v_f = 0.4$, $\sigma_{yf} = 200\,\text{MPa}$, $E_s = 200\,\text{GPa}$, $v_s = 0.3$; $p_c = 100\,\text{MPa}$, $\delta_c = 0.002\mu\text{m}$, $\beta = 0.8$.

The solution process of the inverse problem implies once again the minimisation of the discrepancy function $\omega(\mathbf{x})$ according to Equation (11), but without the last addend. The Trust region algorithm converges toward the optimal estimates as shown in Figure 18a. In the absence of input disturbances, the maximum error in the identification of δ_c amounts to 19% difference with respect to the exact value. The progressive reduction of the non-dimensional discrepancy function ω along the iteration sequence and its stabilisation around a minimum value is visualised in Figure 18b. Significant residual errors in the parameter estimates are noticed as expected, since the vertical displacements of material points are little affected by the delamination process, i.e., are not quite sensitive to the sought parameters. If the input of the inverse analysis procedure could include horizontal residual displacements, the sought parameters (and those in the interface model among them) would probably become easily and efficiently identifiable by the methodology presented in this paper. But such a prospect requires future developments of the experimental mechanics implications of the present methodology.

5. Conclusions

In what precedes inverse analyses in fracture mechanics have been reviewed and briefly discussed with reference to specific but representative topics investigated by the authors' team, and with the following further limitations: quasi-brittle fracture of cementitious materials and delamination along coating-substrate interfaces; quasi-static (not dynamic) external actions; emphasis on recent or current methodological developments and their motivation and potential usefulness, even if not all corroborated by practical applications; mainly deterministic batch identification techniques.

In general terms, it can be said nowadays that the EKF methodology and its recent developments, if combined with a suitable mathematical model of the experimental tests, represent powerful and reliable tools in different application fields, which not only identify meaningful material parameters but also provide quantitative assessment of their uncertainties to structural analyses purposes in engineering practice, see e.g., Bittanti et al. (1984), Bolzon et al. (2002), Nakamura et al. (2003), Corigliano and Mariani (2004). However, even traditional least-square deterministic approaches to calibration of simple cohesive crack models lead to still challenging mathematical and numerical problems such as MPEC, briefly considered in what precedes.

The specific issues considered in this paper require and deserve further interdisciplinary investigations from various standpoints, particularly as for computational problems and experimental equipment. The consequences of systematic (non stochastic) modeling errors in test simulations on the resulting estimates of material parameters, should attract more and more attention in future researches. Desirable are future extensions of the inverse analysis methodology to mixed-mode fracture models and to recently developed methods of modeling smooth transitions from diffused damage to strain localization and to crack as displacement discontinuity. Of course, other areas of fracture mechanics, in primis ductile and dynamic fracture, give rise to interesting inverse analysis problems.

Despite the above limitations, even a small sample of engineering-motivated recent research results of fracture parameter identifications can evidence the remarkable potentialities of inverse analysis of fracture processes, combined with experimentation, in diverse mechanical technologies.

Acknowledgements

The research results referred to in this paper have been achieved in two projects (on concrete dams and on inverse analyses) co-financed by the Italian University Ministry (MIUR). Thanks are expressed to former student M. Lettieri for his contribution to Section 3.

References

ABAQUS/Standard (1998). *Theory and User's Manuals*. Release 6.2-1, Pawtucket, RI (USA).

Abdul-Baqi, A. and Van der Giessen, E. (2001). Indentation-induced interface delamination of a strong film on a ductile substrate. *Thin Solid Films* **381**, 143–154.

Abdul-Baqi, A. and Van der Giessen, E. (2002). Numerical analysis of indentation-induced cracking of brittle coatings on ductile substrates. *International Journal of Solids and Structures* **39**, 1427–1442.

Alvaredo, A.M. and Torrent, R.J. (1987). The effect of the shape of the strain-softening diagram on the bearing capacity of concrete beams. *Materials and Structures* **20**, 448–454.

Ardito, R., Bartolotta, L., Ceriani, L. and Maier, G. (2004). Diagnostic inverse analysis of concrete dams with static excitation. *Journal of the Mechanical Behavior of Materials* **15**(6), 381–389.

Barenblatt, G.I. (1962). The mathematical theory of equilibrium cracks in brittle fracture. *Advances in Applied Mechanics* **7**, 55–129.

Bažant, Z.P. and Planas, J. (1998). *Fracture and Size Effect in Concrete and Other Quasi-brittle Materials*, CRC Press: Boca Raton, FL (USA).

Bhattacharya, A.K. and Nix W.D. (1998). Finite element simulation of indentation experiments. *International Journal of Solids and Structures* **24**, 881–891.

Bittanti, S., Maier, G. and Nappi, A. (1984). Inverse problems in structural elasto-plasticity: a Kalman filter approach. In: A. Sawczuck and G. Bianchi (eds.) *Plasticity Today* Elsevier Applied Science, London, pp. 311–329.

Bocciarelli, M., Bolzon, G. and Maier, G. (2005). Parameter identification in anisotropic elastoplasticity by indentation and imprint mapping. *Mechanics of Materials* **37**, 855–868.

Bolzon, G. and Cocchetti, G. (2003). Direct assessment of structural resistance against pressurized fracture. *International Journal for Numerical and Analytical Methods in Geomechanics* **27**, 353–378.

Bolzon, G., Cocchetti, G., Maier, G., Novati, G. and Giuseppetti, G. (1994a). Boundary element and finite element fracture analysis of dams by the cohesive crack model: a comparative study. In: E. Bourdarot, J. Mazars and V. Sauma (eds.) *Dam Fracture and Damage* Balkema, Rotterdam, pp. 69–78.

Bolzon, G. and Corigliano, A. (1997a). A discrete formulation for elastic solids with damaging interfaces. *Computer Methods in Applied Mechanics and Engineering* **140**, 329–359.

Bolzon, G., Fedele, R. and Maier, G. (2002). Identification of cohesive crack models by Kalman filter. *Computer Methods in Applied Mechanics and Engineering* **191**, 2847–2871.

Bolzon, G., Ghilotti, D. and Maier, G. (1997b). Parameter identification of the cohesive crack model, In: H. Sol and C.W.J. Oomens (eds.) *Material Identification Using Mixed Numerical Experimental Methods* Kluwer Academic Publisher, Dordrecht, pp. 213–222.

Bolzon, G. (1996). Hybrid finite element approach to quasi-brittle fracture. *Computer and Structures* **60**, 733–741.

Bolzon, G. and Maier, G. (1998). Identification of cohesive crack models for concrete by three-point bending tests, In: R. de Borst, N. Bicanic, H. Mang and G. Meschke (eds.) *Computational Modelling of Concrete Structures* Balkema Publishers, Rotterdam, pp. 301–310.

Bolzon, G., Maier, G. and Novati, G. (1994b). Some aspects of quasi-brittle fracture analysis as a linear complementarity problem, In: Z.P. Bažant, Z. Bittnar, M. Jirásek and J. Mazars (eds.) *Fracture and Damage in Quasi-brittle Structures* E and FN Spon, London, pp. 159–174.

Bolzon, G., Maier, G. and Panico, M. (2004). Material model calibration by indentation, imprint mapping and inverse analysis. *International Journal of Solids and Structures* **41**, 2957–2975.

Bolzon, G., Maier, G. and Tin-Loi, F. (1995). Holonomic and non-holonomic simulations of quasi-brittle fracture:a comparative study of mathematical programming approaches, In: F.H. Wittman (ed.) *Fracture Mechanics of Concrete Structures* Aedificatio Publishers, Freiburg, pp. 885–898.

Bolzon, G., Maier, G. and Tin-Loi, F. (1997c). On multiplicity of solutions in quasi-brittle fracture computations. *Computational Mechanics* **19**, 511–516.

Brühwiler, E. and Wittmann, F.H. (1990). Wedge-splitting test, a new method of performing stable fracture mechanics tests. *Engineering Fracture Mechanics* **35**, 117–125.

Bui, H.D. (1994). *Inverse Problems in the Mechanics of Materials: an Introduction*, CRC Press, Boca Raton, FL (USA).

Camacho, G.T. and Ortiz, M. (1996). Computational modelling of impact damage in brittle materials. *International Journal of Solids and Structures* **33**, 2899–2938.

Carpinteri, A. (1989). Softening and snap-back instability of cohesive solids. *International Journal for Numerical Methods in Engineering* **29**, 1521–1538.

Carpinteri, A., Chiaia, B. and Invernizzi, S. (2001). Numerical analysis of indentation fracture in quasi-brittle materials. *Engineering Fracture Mechanics* **71**, 567–577.

CEB-FIP Model Code 1990 (1993). Design Code, Thomas Telford Ltd., London.

Cen, Z. and Maier, G. (1992). Bifurcations and instabilities in fracture of cohesive-softening structures: a boundary element analysis. *Fatigue and Fracture of Engineering Materials and Structures* **15**, 911–928.

Cetinel, H., Uyulgan, B., Tekmen, C., Ozdemir, I. and Celik, E. (2003). Wear properties of functionally gradient layers on stainless steel substrates for high temperature applications. *Surface and Coatings Engineering* **174–175**, 1089–1094.

Cheng, Y.T. and Cheng, C.M. (1999). Scaling relationships in conical indentation of elastic perfectly-plastic solids. *International Journal of Solids and Structures* **36**, 1231–1243.

Cocchetti, G. and Maier, G. (2003). Elastic-plastic and limit-state analyses of frames with softening plastic-hinge models by mathematical programming. *International Journal of Solids and Structures* **40**, 7219–7244.

Cocchetti, G., Maier, G. and Shen, X. (2002). Piecewise-linear models for interfaces and mixed mode cohesive cracks. *Computer Methods in Engineering and Science* **3**, 279–298.

Coleman, T.F. and Li, Y. (1996). An interior trust region approach for nonlinear minimisation subject to bounds. *SIAM Journal on Optimization* **6**, 418–445.

Comi, C., Maier, G. and Perego, U. (1992). Generalized variable finite element modeling and extremum theorems in stepwise holonomic elastoplasticity with internal variables. *Computer Methods in Applied Mechanics and Engineering* **96**, 213–237.

Corigliano, A. and Mariani, S. (2004). Parameter identification in explicit structural dynamics: performance of the extended Kalman filter. *Computer Methods in Applied Mechanics and Engineering* **193**, 3807–3835.

Cottle, R.W., Pang, J.S. and Stone, R.E. (1992). *The Linear Complementarity Problem*, Academic Press, Boston.

Denariè, E., Saouma, V., Iocco, A. and Varelas, D. (2001). Concrete fracture process zone characterization with fiber optics. *ASCE Journal of Engineering Mechanics* **127**, 494–502.

Dirkse, S.P. and Ferris, M.C. (1995). The PATH solver: a non-monotone stabilization scheme for mixed complementarity problems. *Optimization Methods and Software* **5**, 123–156.

Elices, M., Guinea, G.V. and Planas, J. (1992). Measurement of the fracture energy using three-point bend tests: Part 3 – Influence of cutting the P-δ tail. *Materials and Structures* **25**, 327–334.

Facchinei, F. and Pang, J.S. (2003). *Finite-Dimensional Variational Inequalities and Complementarity Problems*, Vol. II, Springer Series in Operations Research and Financial Engineering, Berlin.

Fedele, R., Maier, G. and Miller, B. (2005). Identification of elastic stiffness and local stresses in concrete dams by in situ tests and neural networks. *Structure and Infrastructure Engineering* **1**(3), 165–180.

Fiacco, A.V., McCormick and G.P. (1968). *Nonlinear Programming: Sequential Unconstrained Minimization Techniques*, Wiley, New York (reprinted by SIAM Publications, Philadelphia, 1990).

Frangi, A. and Maier, G. (2002). The symmetric Galerkin boundary element method in linear and nonlinear fracture mechanics, In: D. Beskos and G. Maier (eds.) *Boundary Element Advances in Solid Mechanics* CISM Lecture Notes, Vol. 440, Springer Verlag, Wien.

Gioda, G. and Maier, G. (1980). Direct search solution of an inverse problem in elastoplasticity: identification of cohesion, friction angle and in situ stress by pressure tunnel tests. *International Journal for Numerical Methods in Engineering* **15**, 1823–1848.

Goodman, R.E. (1980). *Introduction to Rock Mechanics*, John Wiley and Sons, New York.

Guinea, G.V., Planas, J. and Elices, M. (1992). Measurement of the fracture energy using three-point bend tests: Part 1 – Influence of experimental procedures. *Materials and Structures* **25**, 212–218.

Guinea, G.V., Planas, J. and Elices, M. (1994). A general bilinear fit for the softening curve of concrete. *Materials and Structures* **27**, 99–105.

Haykin, S. (1999). *Neural Networks*. A Comprehensive Foundation, 2nd Edition, Prentice Hall, Upper Saddle River, NJ (USA).

Hillerborg, A. (1991). Application of the fictitious crack model to different types of materials. *International Journal of Fracture* **51**, 95–102.

Jayaraman, S., Hahn, G.T., Oliver, W.C., Rubin, C.A. and Bastias, P.C. (1998). Determination of monotonic stress-strain curve of hard materials from ultra-low-load indentation tests. *International Journal of Solids and Structures* **35**, 365–381.

Jirasek, M. and Bažant, Z.P. (2001). *Inelastic Analysis of Structures*, John Wiley and Sons, New York.

Karihaloo, B. (1995). *Fracture Mechanics and Structural Concrete*, Longman Scientific and Technical, Burnt Mill, Harlow.

Kleiber, M., Antunez, H., Hien, T.D. and Kowalczyk, P. (1997). *Parameter Sensitivity in Nonlinear Mechanics. Theory and Finite Element Computations*, John Wiley and Sons, New York.

Li, W. and Siegmund, T. (2004). An analysis of the indentation test to determine the interface toughness in a weakly bonded thin film coating-substrate system. *Acta Materialia* **52**, 2989–2999.

Luo, Z.Q., Pang, J.S. and Ralph, D. (1996). *Mathematical Programs with Equilibrium Constraints*, Cambridge University Press, Cambridge (UK).

Maier, G. (1968). On softening flexural behaviour of beams in elasto-plasticity. *Rendiconti dell'Istituto Lombardo di Scienze e Lettere* **102**, 648–677, (in Italian).

Maier, G. (1970). A matrix structural theory of piecewise-linear plasticity with interacting yield planes. *Meccanica* **5**, 55–66.

Maier, G., Ardito, R. and Fedele, R. (2004). Inverse analysis problems in structural engineering of concrete dams. In: Z.H. Yao, M.W. Yuan e and W.X. Zhong (eds.) *Computational Mechanics* Tsinghua University Press & Springer-Verlag, Beijing, pp. 97–107.

Maier, G. and Comi, C. (2000). Energy properties of solutions to quasi-brittle fracture mechanics problems with piecewise linear cohesive crack models, In: A. Benallal (ed.) *Continuous Damage and Fracture* Elsevier, Amsterdam, pp. 197–205.

Maier, G., Giannessi, F. and Nappi, A. (1982). Identification of yield limits by mathematical programming. *Engineering Structures* **4**, 86–98.

Maier, G., Novati, G. and Cen, Z. (1993). Symmetric Galerkin boundary element method for quasi-brittle fracture and frictional contact problems. *Computational Mechanics* **13**, 74–89.

Mróz, Z. and Stavroulakis, G.E. (eds.) (2005). *Identification of Materials and Structures*, CISM Lecture Notes, Vol. 469, Springer-Verlag, Wien.

Nakamura, T., Yu, G., Prchlik, L., Sampath, S. and Wallace, J. (2003). Micro-indentation and inverse analysis to characterize elastic-plastic graded materials. *Materials Science and Engineering A* **345**, 223–233.

Nash, J.F. (1951). Non-cooperative games. *Annals of Mathematics* **54**, 286–295.

Norgaard, M., Pulsen, N. and Ravn, O. (2000). New developments in state estimation for nonlinear systems. *Automatica* **36**, 1627–1638.

Ohring, M. (1991). *The Materials Science of Thin Films*, Academic Press, New York.

Ortiz, M. and Pandolfi, A. (1999). Finite-deformation irreversible cohesive elements for three-dimensional crack-propagation analysis. *International Journal for Numerical Methods in Engineering* **44**, 1267–1282.

Pang, J.S. and Tin-Loi, F. (2001). A penalty interior point algorithm for a parameter identification problem in elasto-plasticity. *Mechanics of Structures and Machines* **29**, 85–99.

Planas, J., Elices, M. and Guinea, G.V. (1992). Measurement of the fracture energy using three-point bend tests: Part 2-Influence of bulk energy dissipation. *Materials and Structures* **25**, 305–312.

RILEM Draft Recommendation (1985). Determination of the fracture energy of mortar and concrete by means of three-point bend tests on notched beams. *Materials and Structures* **18**, 285–290.

Rose, J.H., Ferrante, J. and Smith, J.R. (1981). Universal binding energy curves for metals and bimetallic interfaces. *Physical Review* **9**, 675–678.

Spearing, S.M. (2000). Materials issues in microelectromechanical systems (MEMS). *Acta Materialia* **48**, 179–196.

Stavroulakis, G.E. (2000). *Inverse and Crack Identification Problems in Engineering Mechanics*, Kluwer Academic Publishers, Durtrecht.

Stavroulakis, G.E., Bolzon, G., Waszczyszyn, Z. and Ziemanski, L. (2003). Inverse Analysis, In: B. Karihaloo, R.O. Ritchie and I. Milne (eds.) *Comprehensive Structural Integrity* Vol. 3, Elsevier Science, Oxford.

Tin-Loi, F. and Que, N.S. (2001a). Parameter identification of quasi-brittle materials as a mathematical program with equilibrium constraints. *Computer Methods in Applied Mechanics and Engineering* **190**, 5819–5836.

Tin-Loi, F. and Que, N.S. (2002a). Identification of cohesive crack fracture parameters by evolutionary search. *Computer Methods in Applied Mechanics and Engineering* **191**, 5741–5760.

Tin-Loi, F. and Que, N.S. (2002b). Nonlinear programming approaches for an inverse problem in quasi-brittle fracture. *International Journal of Mechanical Sciences* **44**, 843–858.

Tin-Loi, F. and Xia, S.H. (2001b). Non-holonomic elastoplastic analysis involving unilateral frictionless contact as a mixed complementarity problem. *Computer Methods in Applied Mechanics and Engineering* **190**, 4551–4568.

Volinsky, A.A., Moody N.R. and Gerberich, W.W. (2002). Interfacial toughness measurement for thin film on substrates. *Acta Materialia* **50**, 441–466.

Waszczyszyn, Z. (ed.) (1999). *Neural Networks in the Analysis and Design of Structures*, CISM Lectures Notes, Vol. 404, Springer Verlag, Wien.

Wittmann, F.H. and Hu, X. (1991). Fracture process zone in cementitious materials. *International Journal of Fracture* **51**, 3–18.

Xiaodong, L., Dongfeng, D. and Bhushan, B. (1997). Fracture mechanisms of thin amorphous carbon films in nano-indentation. *Acta Materialia* **45**, 4453–4461.

Xu, X.P. and Needleman, A. (1994). Numerical simulation of fast crack growth in brittle solids. *Journal of the Mechanics and Physics of Solids* **42**, 1397–1434.

Zhi-He, J., Glaucio, H., Paulino, R.H. and Dodds, J. (2003). Cohesive fracture modeling of elastic–plastic crack growth in functionally graded materials. *Engineering Fracture Mechanics* **70**, 1885–1912.

International Journal of Fracture (2006) 138:75–100
DOI 10.1007/s10704-006-7155-5

© Springer 2006

Nanoprobing fracture length scales

W.W. GERBERICH[1], W.M. MOOK[1], M.J. CORDILL[1], J.M. JUNGK[1],
B. BOYCE[2], T. FRIEDMANN[2], N.R. MOODY[3] and D. YANG[4,*]
[1]*Department of Chemical Engineering and Materials Science, University of Minnesota, Minneapolis,
MN 55455, USA*
[2]*Sandia National Laboratories, Albuquerque, NM 87185, USA*
[3]*Sandia National Laboratories, Livermore, CA 94550, USA*
[4]*Hysiton, Inc., Minneapolis, MN 55404, USA*
**Author for correspondence (E-mail: wgerb@tc.umn.edu)*

Received 1 March 2005; accepted 1 December 2005

Abstract. Historically fracture behavior has been measured and modeled from the largest structures of earthquakes and ships to the smallest components of semiconductor chips and magnetic recording media. Accompanying this is an evolutionary interest in scale effects partially due to advances in instrumentation and partially to expanded supercomputer simulations. We emphasize the former in this study using atomic force microscopy, nanoindentation and acoustic emission to probe volumes small in one, two and three dimensions. Predominant interest is on relatively ductile Cu and Au films and semi-brittle, silicon nanoparticles. Measured elastic and plastic properties in volumes having at least one dimension on the order of $10 - 1000\,\text{nm}$, are shown to be state of stress and length scale dependent. These in turn are shown to affect fracture properties. All properties can vary by a factor of three dependent upon scale. Analysis of fracture behavior with dislocation-based, crack-tip shielding is shown to model both scale and stress magnitude effects.

1. Introduction

Historically interest in fracture took a bounce in the 1920's after ships sank, e.g. the Titanic (Garzke et al., 1997) and Griffith (1921), working in the steel industry, formulated his seminal contribution. Later, in the 1940s, an additional bounce occurred again after Liberty ships sank (Parker, 1957). Here, the brittle-ductile transition concept started to give ground to early quantitative concepts (Irwin, 1948; Orowan, 1950). Repeating the theme in the 1960's after missiles "sank" on the test stand, produced the surge in the fracture mechanics field that spawned linear-elastic fracture mechanics, LEFM initiated by Irwin (1960), and elastic–plastic fracture mechanics, EPFM by Rice (1969) and Hutchinson (1968). The 1980's broke the pattern probably due to no major wars or disasters and the perception that the major problems had been solved. Interest in the field of fracture abated with the drives in funding at major institutions being high tech information-based industries. In retrospect this turned out to be a major benefactor to the fracture field since this resulted in two parallel but seemingly disconnected developments. One was the atomic force microscope (AFM) developed by Binnig, Quate, and Gerber (1987), while the other involved the huge gains in the computational power of supercomputers. The first led to stand-alone and AFM-based nanoindenters, the second to multi-scale modeling capabilities. In terms of scale, most are familiar with Moore's law and the doubling phenomena of information storage and computation requiring ultra-fine films and lines. These have the required measures of microstructure and property integrity, often involving AFM and nanoindentation instruments. From a mechanical property perspective, the 1980s and 1990s have been consumed with processing, imaging microstructure, and measuring modulus, hardness and strength properties at the nanoscale. More recently this has changed as large-scale manufacturing of MicroElectroMechanical, MagnetoOptical and Magnetically-Coupled-Shape Memory Systems (MEMS, MOPS, and MACS) have been envisioned. One might say that the "sputnik" driving force for research and development of the 2000s is nanotechnology.

But what is the impact of this upon the field of fracture? Until recently there have been relatively few major studies involving either AFM equipped nanoindenters or multi-scale modeling. For example, how many fracture toughness measurements exist for LIGA nickel, a major material proposed for MEMS? Moreover, if the difficult and interesting problems were mostly solved by the end of the 1980's, this will neither attract major participation in nor more funding of basic and applied research. Let us emphasize our perception that the fracture field is currently robust and will continue to grow in prominence. First and foremost there will be technological developments driving the need. Second, there are many unresolved problems due to scale effects. Consider a length scale, ℓ, as might be associated with a film thickness, nanowire width, or a nanodot radius small in one, two or three dimensions. For a nanocrystalline solid this may also be a grain size. A schematic in Figure 1 illustrates the research opportunities for understanding and measuring the common mechanical properties in such small units. In Figure 1(a) it is suggested that modulus may either increase or decrease at length scales much less than 100 nm. Furthermore, it may have different values at a given length scale. We hasten to add that this is usually not a true length scale effect but one brought about indirectly due to very large hardnesses or flow stresses. Additionally, the sign of the deviation is associated with the sign of the stress taking compression to be positive. Next, the commonly referred to indentation size effect (ISE) is seen in Figure 1(b). Increases in hardnesses by factors of two, three or more are now commonly observed, e.g. Nix and Gao (1998), and Corcoran et al. (1997). Largely unresolved is what happens below 20 or 30 nm penetration depths, using this as a length scale. Does the hardness continue to increase, plateau or actually decrease? Since yield strength is proportional to hardness ($\sigma_{ys} \sim H/3$), one can look at the similarity of Figure 1(c) and think of the abscissa as the inverse square root of that in Figure 1(b). Now, however, the length scale is the grain size and the popular interpretations are quite different, being dislocation-based, strain-gradient plasticity for the ISE (Nix and Gao, 1998; Gerberich et al., 2002) and the Hall-Petch grain-size dependence (Hall, 1951; Petch, 1953) for nanocrystalline solids.

An ongoing controversy continues in materials science both computationally and experimentally associated with very small grain sizes. This involves whether or not grain boundary processes eventually cause a decrease in strength at grain sizes on the order of 10 nm. Finally the relatively unexplored area of fracture toughness, K_{IC}, as a function of length scale, offers a rich set of opportunities. We now know that for adhered films small in one-dimension there tends to be a delamination toughness plateau at thicknesses below some value on the order of 100 nm (Volinsky et al., 2002). This is indicated as the Griffith line (Griffith, 1925) in Figure 1(d). At larger

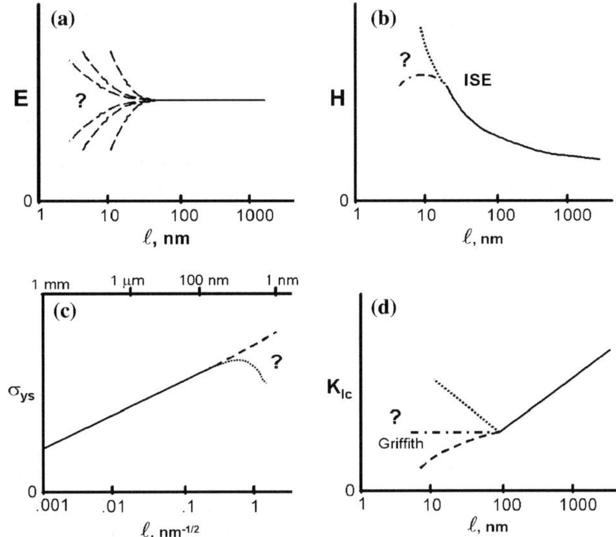

Figure 1. Research opportunities for understanding and measuring common mechanical properties, such as, (a) deviation of elastic modulus, (b) indentation size effect, (c) Hall–Petch behavior, and (d) fracture toughness at small size scales.

thicknesses in ductile films, plastic energy dissipation can account for the increase. But is the increase due to an increase in volume of material available for plastic flow or to a decrease in yield stress due to the grain size being larger for thicker films? Additionally, at very small scales there is the question of what happens to nanowires and nanodots that may support very large stresses resulting in the modulus changes exhibited in Figure 1(a). Since $K_{IC} \sim [2E\gamma_s]^{1/2}$ in this regime, one might expect the upper branch for nanodots in compression and the lower branch for nanowires in tension (Figure 1(d)). Admittedly, it is conceptually difficult to grasp a 1 nm defect compromising the strength of anything. However, physicists routinely worry about sub-nanometer defects in carbon nanotubes (Yu et al., 2000; Belytschko et al., 2004). To illustrate the problem, a 1 nm defect loaded to 10 GPa represents a stress intensity factor of 0.56 MPa m$^{1/2}$ and at fracture would be a strain-energy release rate of 3 J/m^2 for a 100 GPa modulus material. This is about equal to twice the surface energy for a host of materials.

As fracture resistance relates to both elastic and elastic–plastic behavior, consider briefly some scale dependencies of the properties schematically shown in Figure 1. Prior to that it is appropriate to give an overview of a few of the techniques used in measuring elastic, plastic, and fracture properties at the nanoscale.

2. Background

2.1. EXPERIMENTAL PROBES

Here, we will generally list several references that have used both nanoindentation and microtensile instruments to measure elasticity and yield behavior. The exceptions are where we have used nanoindenters to determine the compressive flow and fracture properties of nanoparticles. Additionally, a new acoustic probe which shows promise for investigating flow and fracture events will be discussed.

2.1.1. *Modulus hardness and yield strength from nanoindentation*

While there may be many "home-built" nanoindentation systems, they all most likely borrow their transducer design from existing commercial systems. Of these, the most common commercial nanoindentation platforms are Hysitron (with the Tribo-Indenter® and TriboScope®) and MTS (with the NanoIndenter XP® and DCM). The Hysitron transducers are capable of both indentation tests and scanning probe microscopy, where the same transducer and tip are used for both functions. Consequently Hysitron transducers employ a capacitive loading technique that limits the maximum load to approximately 10 mN while maintaining a low thermal drift with high mechanical stability. Unlike the Hysitron instruments, MTS transducers apply and regulate the load by manipulating a current that passes through a solenoid. This allows for much higher peak loads, but tends to increase thermal drift and decrease instrument response time.

The major deviation from most of the literature involves evaluation of nanoparticles. For modulus and yield strength both upper and lower bound estimates have been made. The upper bound uses the mean contact pressure, appropriate to small displacements while the lower bound uses the approximate deformed shape of a right cylinder at large displacements. These are explained in more detail in Section 2.2 for modulus.

2.1.2. *Modulus and yield strength from microtensile tests*

Traditional tensile tests have also been scaled down to the micro- and nano-regime by Read (1998) using MEMS-based transducers or other specialized load cell configurations by Huang and Spaepen (2000). These tests have had moderate success determining elastic modulus and yield strengths with freestanding thin films, but require substantial experimental expertise to operate. Atomistic simulations by Gall et al. (2005) of nanowire tensile tests have also given significant insight into length scale behavior.

2.1.3. *Fracture properties via nanoindentation*

A relatively large body of literature is beginning to accumulate using nanoprobes to examine channel cracking (through thickness) or blistering (delamination) associated with thin films (Becit, 1979; Begley and Ambriso, 2003; Cordill et al., 2004). A new development is using a nanoindenter to nucleate cracks in nanostructures, in particular spherical nanoparticles (Gerberich et al., 2003; Mook et al., 2005). As the indenter tip radius of curvature that is used is large compared to the nanoparticle this is essentially a compression test of the sphere between two flat plates. By imaging a nanosphere with an AFM-based nanoindenter, Mook et al. (2005) measured the dimensions of a nanosphere and repeatedly compressed it to higher values until it fractured. The fracture instability and subsequent analysis of the fracture toughness is discussed in a following section.

2.2. MODULUS OF ELASTICITY

While we just consider the upturn in modulus at small length scales as depicted in Figure 1(a), the downturn is also expected. The upturn was found by Gerberich et al. (2003) and Mook et al. (2005) during compression testing of silicon and titanium nanospheres. For the series of silicon and titanium nanospheres evaluated to date we have determined modulus as function of both mean contact pressure as an upper bound and average stress in the whole sphere as a lower bound. We could compare the results to those theoretical models and experimental determinations of modulus under high pressure. Experimentally, we have found moduli from an elastoplastic unloading method (Gerberich et al., 1993). This uses the total displacement of a spherical contact into a planar surface and the residual displacement on unloading. Utilizing the cumulative total displacement minus the residual displacement for each subsequent indent gives us a smaller average unloading stiffness and effectively is ultraconservative. As explained by Mook et al. (2005) this gave the contact radius and with the load, P, allowed us to calculate the reduced modulus, E^*, from an analysis by Johnson (1985) giving

$$E^*_{\text{U.B.}} = \frac{32P}{3\pi^2 a \delta^u_E} \tag{1}$$

where δ^u_E is the maximum displacement minus the residual displacement. This "rigid punch" unloading displacement will overestimate the stiffness of the entire sphere and therefore is taken as an upper-bound modulus. By using conservation of volume, the average cross-section of an equivalent right cylinder was used to calculate the average stress and strain. This gave a lower bound elastic modulus of

$$E^*_{\text{L.B.}} = \frac{3P\,(2r - \delta_{\text{cum}})^2}{4\pi r^3 \delta^u_E} \tag{2}$$

where $2r$ is the nanosphere diameter and δ_{cum} is the cumulative plastic displacement. We attempted to use the Oliver–Pharr (1992) method but this gives us unrealistically large moduli, possibly due to creep-in effects as discussed elsewhere by Mook et al. (2005). These ultraconservative modulus measurements of the bounds were compared to both theory and experiment largely generated by the geophysical community's interest in high pressures of over 300 GPa at the earth's center (Murnaghan, 1967; Christensen et al., 1995).

With respect to theory, a considerable body of relatively empirical literature has been generated by geologists and recently more precisely by geophysicists, materials scientists, and physicists using quantum mechanical approaches, e.g. by Moriarty et al., 2002, Van Vliet et al. (2003) and Gall et al. (2004). For example, with the linear-muffin-tin orbital (LMTO) method followed up by full potential (FP) LMTO and later extended to a model generalized pseudo potential theory (MGPT), Moriarty et al. 2002 has studied the effects of pressure on both phase transitions and elastic constants in transition metals. Experimentally, Christensen et al. (1995) have conducted extensive diamond-anvil studies to examine high-pressure effects on both modulus and yield strength. As one might expect, there are sufficient similarities between the diamond anvil experiments squeezing a powder sample between two flats (e.g., 23 μm diamond flats) and squeezing a single sphere between two platens (Shipway and Hutchings, 1993; Majzoub and Chaudhri, 2000).

Given this at least qualitative similarity, we compared our modulus vs. hardness data to Murnaghan's equation of state (Murnaghan, 1967). To first order this reduces to a pressure enhanced bulk modulus given by

$$K_{(p)} = K_o + K'_o p \tag{3a}$$

where K_o is the pressure-free bulk modulus given by

$$K_o = \frac{3\lambda + 2\mu}{3} \tag{3b}$$

with λ the Lame's constant and μ the shear modulus. If we take some license with Equation (3) and assume the same general proportionality applies to Young's modulus and further take Poisson's ratio to be independent of pressure, we find

$$E_{(p)} = E_o + 12\,(1 - 2v)\,p. \tag{4}$$

Knowing full well that Poisson's ratio is mildly pressure dependent, we nevertheless take $v = 0.218$ for silicon. With this, the proportionality constant in Equation (4), really dE/dp, becomes 6.77 for silicon. Experimental values determined from Equation (4) fit midway between the lower bound and upper bound estimates of Equations (1) and (2) (see Figure 2). It is satisfying that no upper bound data point fell below the theoretical estimate and no lower bound estimate fell above. To first order, this relationship was verified for five silicon and four titanium nanospheres of different diameters. The results can be understood in terms of a high compressive stress decreasing the atomic spacing, r_o. Since modulus is roughly affected by atomic radius spacing according to $E \sim 1/r_o^4$, one might expect a 40 GPa stress representing a strain of 20 percent to increase the modulus by a factor of more than two as shown in Figure 2. Again, we emphasize this is an indirect length scale effect brought on by the stress changing the lattice spacing, in effect making a new material. The reason the effect is greater at smaller dimensions is due to the increased mean pressure or stresses sustainable. The opposite trend in modulus could occur in the tension of nanowires as shown by Gall et al. (2004) due to an increase in the lattice constants as suggested in Figure 1(a). As this indirect effect is caused by high mean pressure or stresses, it is useful to next examine the indentation size effect (ISE) depicted in Figure 1(b).

2.3. HARDNESS

First it is emphasized that the increase in hardness or mean contact stress associated with the ISE does not normally translate into any observable modulus increase, particularly in single crystals. This is for two reasons. First and foremost the stresses in our and others studies of Au and Cu were only 3 GPa (Corcoran et al., 1997; McElhaney et al., 1998). At this magnitude one might expect no greater than about a 14 percent change. Second, the normal stresses reduce deeper into a single crystal making elastic displacements preferred further below the contact area where stresses and hence moduli are lower. This may not be the case for very thin films. The ISE is seen for both indentation into single crystal Fe-3%Si and thin-film, nanocrystalline Au in Figure 3. The theoretical fits are from recent paper addressing surface energy, γ_s, effects where hardness is given by (Tymiak et al., 2001; Gerberich et al., 2005)

$$H = \frac{27\alpha\gamma_s}{(2\delta R)^{1/2}} \tag{5}$$

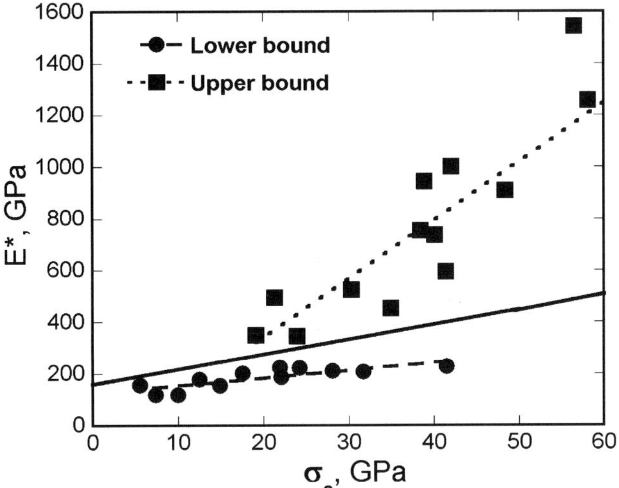

Figure 2. Experimental values for elastic modulus calculated from Equation (4) (solid line) fit between estimates for elastic modulus from Equations (1) and (2). The relationships have been applied to silicon and titanium nanospheres of different diameters (Mook et al., 2005).

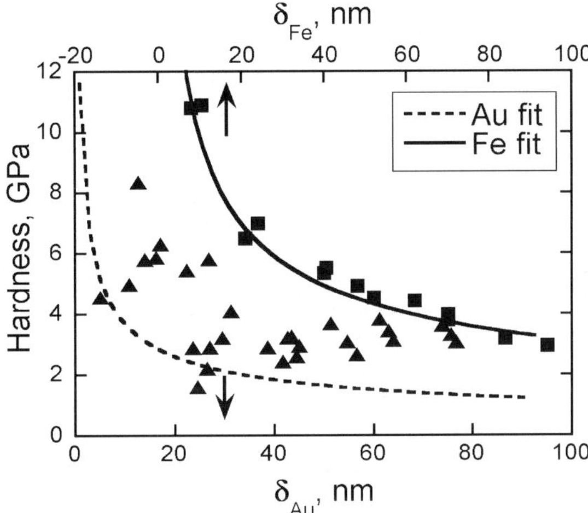

Figure 3. Indentation size effect length scale for Fe-3% Si and Au with theoretical fits modeled with Equation (5). For Au, $R = 150\,\text{nm}$ and for Fe, $R = 80\,\text{nm}$.

where R is the diamond tip radius and α is taken to be 7.5 for Fe and 5.0 for Au. The appropriate surface energies are 1.95 and 1.49 J/m^2, respectively. Regarding effects on modulus, a modulus measurement into Fe-3%Si was not accurate at 10 nm depths due to surface roughness. For 7 GPa pressures the most change one could expect would be 15 percent but no effect was detected. On the other hand, large effects were measured on very thin Au films but it is not clear at this time how much of that may be attributed to large pressures and how much to pile-up and creep-in artifacts.

As to the ISE effect, the length scale represented in Figure 3 is the displacement that is an extrinsic variable. One can calculate the intrinsic length scale from strain-gradient plasticity theory as Nix and Gao (1998) have done which is then a constant, δ^*.

Alternatively, Horstemeyer and Baskes (1991) and later Gerberich et al. (2002) have suggested that the volume-to-surface ratio is a length scale. Here volume, V, was associated with the plastic zone size created by the indentation and area was the surface under the indenter contact. This led to a constant length scale, V/S, for the first several hundred nanometers of penetration. One could just as easily hypothesize an evolutionary length scale during penetration because of both the surface energetics changing with contact area and the dislocation density increasing. Using a from the contact area as the evolutionary length scale, the hardness of the Fe-3%Si vs. a^{-1} gives an extrapolated value of about 750 MPa in Figure 4. This is similar to three times the flow stress for the bulk single crystal as one might expect when converting hardness to flow strength. Note that a is the same length scale as in Equation (5) if one uses the geometric contact radius for a spherical diamond tip. Considering that hardness as a function of an inverse length scale is qualitatively similar to a Hall–Petch plot, we next consider the yield strength.

2.4. YIELD STRENGTH

The time-honored Hall–Petch relationship (Hall, 1951; Petch, 1953) for yield strength dependence is an inverse square-root grain size relation as indicated in Figure 1(c). Some actual data for nanocrystalline nickel, is shown in Figure 5, recently obtained at Sandia National Laboratories by Hansen (2004). This exhibits no drop-off in the flow stress at very small grain sizes. Also interesting is that nanocrystalline Ni which is cold-rolled has a much higher slope than extrapolated polycrystalline nickel. As pointed out by Hansen (2004) and Hughes (2004) there is also evidence that it is dislocation plasticity that controls strength and not some grain boundary diffusion or sliding mechanism. Already, in constrained systems such as thin films (Mook et al., 2004), nanoparticles (Gerberich et al., 2003), and nanoboxes (Mook et al., 2004), it is known that two or three length scales are needed due to the evolutionary nature of the microstructure during severe deformation.

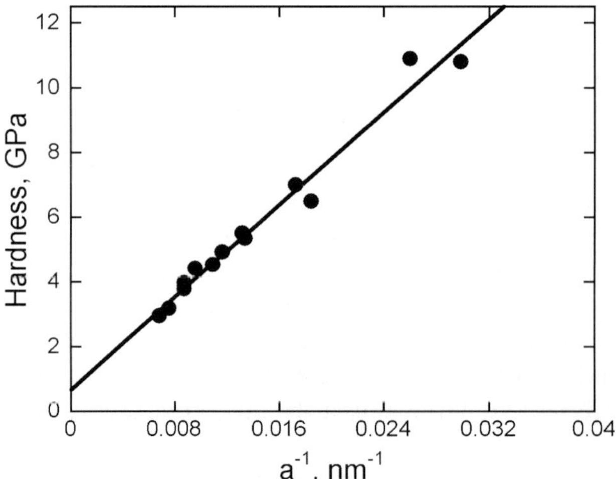

Figure 4. Using a V/S length scale relationship, a Hall–Petch type relationship is found for the Fe-3% Si.

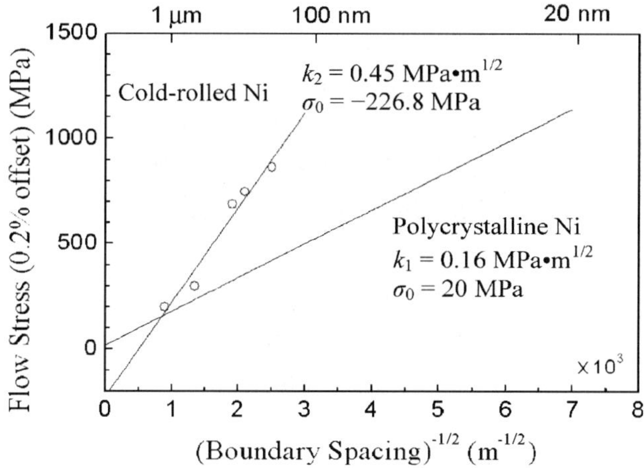

Figure 5. The flow stress–boundary spacing relationship at room temperature for polycrystalline Ni and for cold-rolled Ni (Hansen, 2004).

One case in point is the compression of silicon nanospheres where it was shown that the mean contact stress, σ_c, for a given load, was strongly dependent upon the length scale (Gerberich et al., 2003; Mook et al., 2004). In this case the particle diameter was used as the length scale. These nanospheres were nearly perfect single crystals, an example of one being shown in Figure 6. There were no dislocations and no discontinuities in the lattice except for vacancies. Taking a cross-section of the data loaded to $35\,\mu$N for different particle diameters (Gerberich et al., 2003) produces Figure 7(a). It is seen that this is considerably different from a Hall–Petch plot in terms of its extrapolation to large scales. One reason for that is these spheres were sequentially loaded a number of times. For one nanoparticle then the plastic strain history could be quite different from another. A better way to illustrate any length scale effect is to show all data for the sequential runs with stresses determined at a number of positions for each run. This is shown for the 50 and 93 nm diameter spheres in Figure 7(b). Five sequential runs are shown for the 50 nm sphere and seven for the 93 nm diameter sphere. Cumulative displacement is used since a residual plastic displacement occurred after each run. This was determined from the height of the nanoparticle after each run. From the minimum in the data that occurs at about half the radius for the two spheres, there is nearly a 20 GPa difference in mean contact stress similar to what is indicated in Figure 7(a). More importantly the mean contact flow stress starts out high, goes through a minimum and then rapidly increases. This can be interpreted in terms of a surface dominant regime at small displacements and a work-hardening dominant regime at larger displacements due to constrained plasticity. As explained in detail by Gerberich et al. (2005), once a sufficient number of dislocation loops have been emitted and trapped in the silicon between the diamond tip and the sapphire substrate, considerable back stresses evolve. Due to this evolutionary dislocation structure the length scale varies with cumulative displacement. Gerberich et al. (2005) has proposed a transition model for contact stress moving from the surface-dominant regime to one controlled by dislocation hardening. This is given by

$$\sigma_c = \frac{4\alpha\gamma_s r^2}{\alpha^2 (r - \beta a)} \tag{6}$$

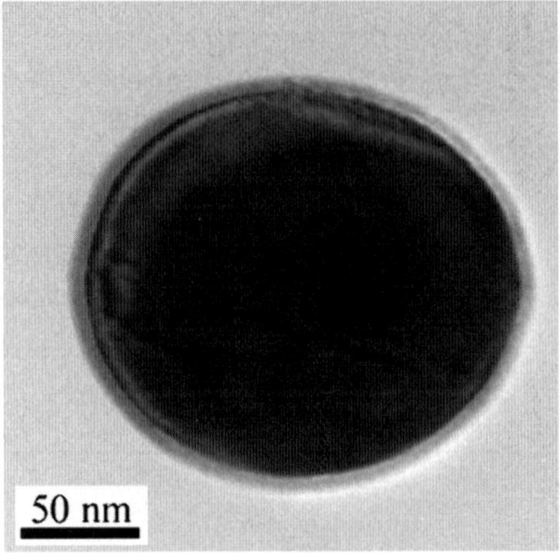

Figure 6. Bright-field TEM image of a single-crystal, defect free Si nanoparticle (Philips CM30, 300 kV). The light outer ring is an oxide film. Image courtesy of C.R. Perrey and C.B. Carter, University of Minnesota.

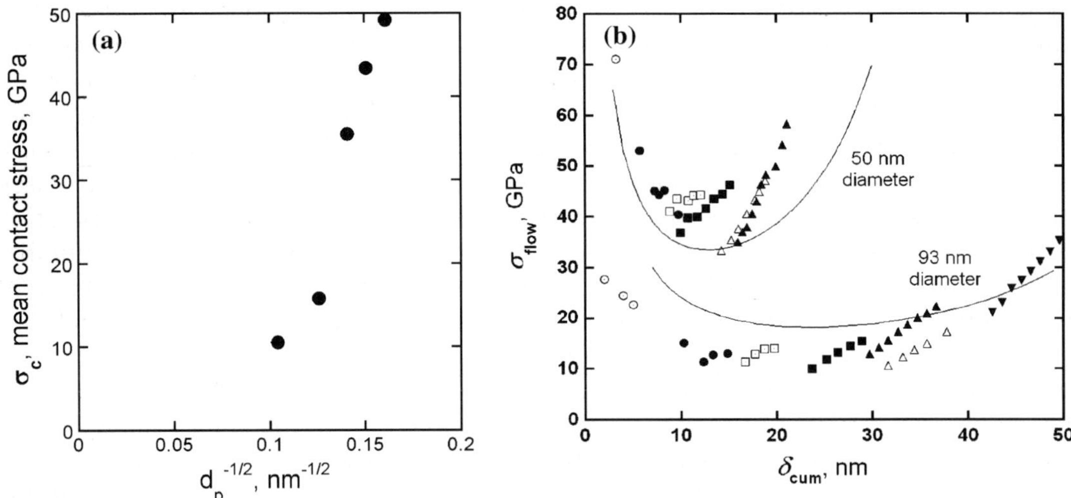

Figure 7. (a) A portion of contact stress data for Si nanospheres of different diameters. The data is considerably different from a Hall–Petch plot when extrapolated to large scales. (b) Length scale effect for two Si nanospheres, 50 and 93 nm in diameter. Stresses were determined at a number of increasing displacement magnitudes for each indentation run.

where the symbols are the same as in Equation (5) except r is the sphere radius, α is 20 for silicon, $\gamma_s = 1.56\,\mathrm{J/m^2}$ and a value of β equal to unity was used. Equation (6), representing the solid curves is shown for both diameter spheres in Figure 7(b). Since $a = (\delta r)^{1/2}$ for small displacements, we see that σ_c for a nanosphere initially varies as r/δ while the indentation hardness varies a $1/a$. While both of these give an indentation size effect, the constrained flow behavior of the nanosphere gives considerable hardening as $\beta a \to r$ in the bracketed term of the denominator of Equation (6).

What we see then is that depending on the length scale and/or confinement in structures small in one or several dimensions, the flow stress can vary in a complex fashion. The degree of complexity supercedes what was implied by the simple sketch of Figure 1(c). Given that modulus effects are fairly straightforward, hardness still not well understood and yield or flow stress now showing great complexity, we next turn to the most difficult subject, that of fracture.

2.5. FRACTURE

Here, we only present in a preliminary way what most agree is the mechanism for interfacial delamination toughness of ductile metallic films. The reader is referred to the extensive papers and reviews by Volinsky et al. (2002) and Cordill (2004) that have been written to describe the test techniques using embedded films for 4-point bending, superlayers for debonding films or indentation-induced delamination among others. The strain energy release rate appears to be controlled by the Griffith criterion below a critical length scale, here found to be a film thickness somewhat less than a 100 nm. In an extensive series of studies, Volinsky et al. (2002) and Lane et al. (2000) have shown that for films larger than the critical length scale, the plastic volume dissipates more energy to increase fracture resistance. For two material combinations on Si, this is seen in Figure 8. Lane et al. (2000) find a similar effect using approaches pioneered by Suo et al. (1993), and Wei and Hutchinson (1997) coming from the continuum side. From the dislocation side we have modeled this with the original Lin and Thompson (1986) model to account for the increases shown in Figure 8. What we agree on is that there is a leveraging of the thermodynamic work of adhesion such that a power law as used by Mook et al. (2004) or exponential dependence used by Volinsky et al. (2002) and Lin and Thomson (1986) on the adhesion energy occurs in thicker films. We consider this to be affected by dislocation shielding and return to this subject in the last section.

One other aspect of interest is that an R-curve dependence has been observed for increasing delamination crack sizes (Volinsky et al., 2002a; Cordill et al., 2005; Gerberich et al., 2003a). This was dramatically shown in microscratch induced delaminations of Au films from SiO_2 by Gerberich et al. (2003a). The abrupt load

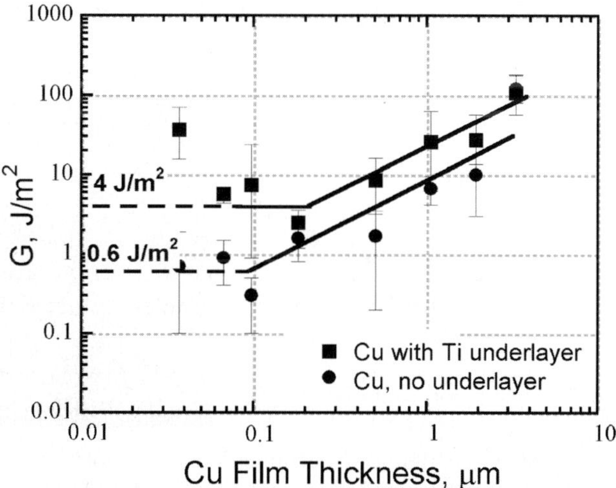

Figure 8. Interfacial fracture energy increases with increasing thickness as shown for a Cu/Ti film system and a Cu system both deposited on silicon substrates. Below a critical thickness the energy plateaus.

drops in the lateral force along with the AFM imaged delamination arrest regions were used to determine the fracture resistance at arrest. The load drops and corresponding G_R values at arrest are shown in Figures 9(a) and (b). At the time we did not carefully strip off the film and examine the surface for fiducial marks that are often left behind by microscopic debris left at arrest point. The latter has been described in detail by Volinsky et al. (2002a). More recently, a tungsten (W) film was peeled off of a Cu layer bonded to a silicon wafer. The W was utilized as a super-layer to provide sufficient residual stress to debond the interface in the shape of telephone cord buckles. This is discussed more fully in a companion paper by Cordill et al. (2005). When removing the W, the weakest interface being W/Cu, fiducial marks shown in Figure 9(c) represented crack arrest. One can see from the parabolic shape of the marks that the stress intensity is larger at the leading edge. We have modeled such stepwise slow crack growth as an R-curve effect with a volume to surface length scale derived from nanoindentation induced plastic zones. As discussed by Gerberich et al. (2003a), this gave a strain-energy release rate

$$G_R = \frac{12\sigma_{ys}^2 h}{E}\left(\frac{\Delta b}{b_o}\right)^{1/2} \tag{7}$$

where b_o, the initial interfacial crack radius was on the order of the film thickness, h. Here σ_{ys} is the yield strength, E is the modulus, and Δb is the incremental increase in crack growth. The calculated value at arrest can be seen to give a reasonable representation of the R-curve in Figure 9(b). We will return to the length scale issue associated with time-dependent cracking in Section 3.

2.6. ACOUSTIC EMISSION

For probing fracture length scales, one time-honored technique is acoustic emission (AE). From earthquakes to submarines to microcracks in rocket motor cases, acoustic emission has been used for either remote or on-device sensing. Contact probes, of course, are most sensitive since signal strength increases as the probe is moved closer to the epicenter. A quarter of a century ago, Fleischmann et al. (1977), considered

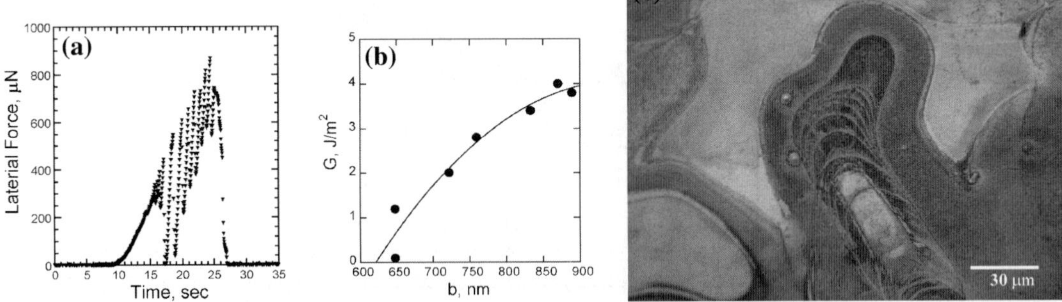

Figure 9. (a) Microscratch data for an induced delamination on a gold film. The load drops in the lateral force and AFM of crack arrest regions were used to calculate the interfacial fracture resistance shown in (b). The theoretical solid curve is Equation (7). (c) Optical micrograph of Cu surface after the removal of a buckled tungsten film. Fiducial marks indicate crack growth and arrest.

the analogy between a moving line dislocation and a moving sound wave to find the AE intensity of an acoustic wave from

$$|\mathrm{AE}_\perp(\omega, r_\mathrm{o})| = \frac{\rho_\mathrm{o} u_\mathrm{o} L v_\mathrm{o}}{\pi r_\mathrm{o}} \left| \sin \frac{\pi \omega \lambda}{v_\mathrm{o}} \right| \tag{8}$$

where ρ_o is a relative density, u_o is the source displacement, L is the dislocation segment length, v_o is the dislocation velocity, r_o is the distance of the sensor from the source, ω is the frequency and λ is the dislocation glide distance. One can draw the parallel to the crack problem where the AE intensity would be proportional to the stress intensity just as the local displacement at the crack tip would be. That is for $\sin(x) \sim x$ in the argument of Equation (8), one finds an AE amplitude associated with a crack advance of area ΔA giving

$$|\mathrm{AE}_\mathrm{c}(\omega, r_\mathrm{o})| \simeq \kappa(\rho_\mathrm{o}, \omega) r_\mathrm{o}^{1/2} \frac{K_\mathrm{I}}{r_\mathrm{o} E} L\lambda \simeq \frac{\kappa(\rho_\mathrm{o}, \omega)}{r_\mathrm{o}^{1/2}} \frac{K_\mathrm{I}}{E} \Delta A \tag{9}$$

where $\kappa(\rho_\mathrm{o}, \omega)$ depends on the density and frequency of the wave, r_o is the sensor distance from the crack tip, K_I is the applied stress intensity, E is the material modulus and ΔA is the area swept out by the crack advance. The $r_\mathrm{o}^{1/2}$ results due to the dependence of displacement on $r^{1/2}$ behind the crack tip. Some time ago Gerberich and Jatavallabhula (1980) had proposed this proportionality between local stresses or stress intensity times the dislocation or crack area swept out and AE intensity. We will explore this in the last section as a potential probe of plasticity and fracture at the nanoscale.

To summarize this background, we have briefly reviewed a number of experimental probes for examining the mechanical behavior of small volumes. We have further suggested that atypical length scale and stress effects make the understanding of even simple properties like modulus and strength difficult. We next turn to fracture with the prospect that complexity is magnified by its dependence on both the modulus and strength as well as an increasing set of defects.

3. Fracture

Aside from thin film studies, there is very limited information on fracture at the nanoscale. The development of crack-tip shielding arguments, first in the bulk and second at thin-film interfaces will be given. This will then be followed by application to quasi-brittle nanospheres of silicon (Mook et al., 2005). We will discuss the often-used classroom schematic of toughness vs. yield strength as illustrated in Figure 10. This diagram, along with data like it appear no less than a dozen times in the recent ASM Handbook on Fatigue and Fracture (1996). The question will be how length scale might shift the solid curve. It is well known that smaller grain size improves both fracture toughness and yield strength giving a shift upwards and to the right. As shown by Bazant (2004), very large volumes can decrease yield strength slightly and fracture toughness sometimes greatly as has been treated by the Weibull statistics of defect populations. But can other length scale or state-of-stress effects cause similar shifts? If we consider strain energy release rate as constant for a given material, then Figure 1(a) might suggest yes since $K_\mathrm{c} = [E G_\mathrm{c}]^{1/2}$. This will be considered later in discussing the fracture of 20–100 nm spheres. Finally, we will return to

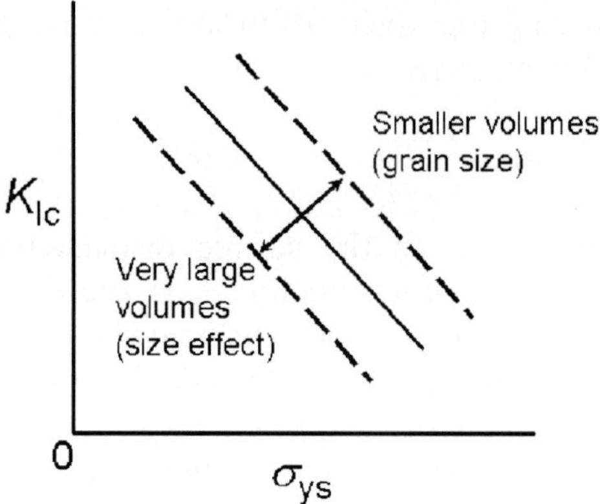

Figure 10. Classic illustration of toughness vs. yield strength.

the acoustic emission probe that has some promise in assisting the investigation of both dislocation and fracture instabilities in small volumes.

3.1. CRACK-TIP SHIELDING IN BULK

The Rice–Thomson model (Rice and Thomson, 1974) originally focusing on dislocation emission vs. fracture has led to many crack-tip shielding models. Many of these have been recently reviewed by Gerberich et al. (2003b). Computationally there is currently a resurgence in activity due to the large-scale atomistic and discretized dislocation simulations possible (Van Swygenhoven and Spaczer, 1989; Curtin and Miller, 2003). A couple of analytical/computational approaches are mentioned here due to their applicability to nanocrystalline structures, nanowires or nanoparticles. A seminal series of papers by Anderson and Rice (1986), Thomson (1986) and Li (1986) addressed various issues of dislocation emission and crack-tip shielding. Li (1986) allowed Huang and Gerberich (1992) to examine the blockage of dislocation emission by a grain boundary. This dislocation-shielding model described the distance, c, between the crack tip and the nearest dislocation as affected by the grain boundary blockage of the dislocation pile-up. With d the distance between the crack tip and the grain boundary, the important parameter was

$$\xi^2 = \frac{d-c}{d}. \tag{10}$$

Given S_1 and S_2 functions of stress intensity and ξ more completely described elsewhere by Anderson and Rice (1986), Thomson (1986) and Li (1986), one can find the number of dislocations in equilibrium with a crack tip embedded in a crystal of size, d. This is given by

$$N = \frac{2S_1}{\pi}\sqrt{\frac{d}{b}}\left[E(\xi) - \left(1-\xi^2\right)H(\xi)\right] + \frac{AS_2^2}{2\tau_f}\left[1 - \left(1-\xi^2\right)\left\{\frac{H(\xi)}{E(\xi)}\right\}^2\right] \tag{11}$$

where $A = \frac{\mu}{2\pi(1-v)}$, τ_f is a friction stress for dislocation motion and $E(\xi), H(\xi)$ are complete elliptical integrals. An asymptotic solution was obtained demonstrating that $H(\xi)\sqrt{1-\xi^2}$ tended to zero as $\xi \to 1^-$. With $E(1)$ equal to 1, Equation (11) became

$$N \simeq \frac{4\,(1-v)}{\sqrt{\pi}\,\mu}\,\frac{\sqrt{d}}{b}\left[K - \tau_f\sqrt{\frac{d}{\pi}}\right] \qquad (12)$$

with K the applied stress intensity and v, μ and b the Poisson's ratio, shear modulus and Burgers vector for the material of interest. Chen et al. (1981) applied this to several fine-grained ferrite structures by assuming that about 10^4 dislocations could be emitted prior to brittle fracture at $100\,\mathrm{K}$. This number of dislocations was consistent with computer simulations of a super/discrete array of dislocations at a crack tip in single crystal Fe-3% Si studies by Huang and Gerberich (1992) and Li et al. (1990). Using this for N, knowing that $4(1-v)/\sqrt{\pi} = 1.63$ and $\mu = 77\,\mathrm{GPa}$ for ferrite, the missing parameter was τ_f which was taken as $\sigma_{ys}/2$, half the yield strength at $100\,\mathrm{K}$. From Chen et al. (1981), this can be given by

$$\sigma_{ys}\big|_{100\mathrm{K}} = \sigma_o + k_y d^{-1/2} \qquad (13)$$

with $\sigma_o = 356\,\mathrm{MPa}$ and $k_y = 0.73\,\mathrm{MPa}\ \mathrm{m}^{1/2}$. Even though the fit in Figure 11 approaches the single crystal case, it falls well below the other data. This is as it should be since out of plane crack-tip deflection and bridging in a polycrystalline array would be additive effects to just grain boundary blockage of crack-tip emitted dislocations. In addition, one might expect these additive effects to be larger at increasing grain size as seen in Figure 11. To assess how realistic 10^4 shielding dislocations are, one can use a crack tip shear strain of Nb/d to show this implies a crack-tip strain of 0.25 for a $10\,\mu\mathrm{m}$ grain but only 0.0025 for a $1\,\mathrm{mm}$ grain. This is consistent with expectations as one would expect much larger crack-tip strains for the tougher fine-grained ferrite. From a fracture mechanics viewpoint, one can back-calculate the number of dislocations from the observations assuming all are crack-tip emitted giving rise to crack-tip displacement, $\delta_t \sim K_{IC}^2/2\sigma_{ys}E$. This gives

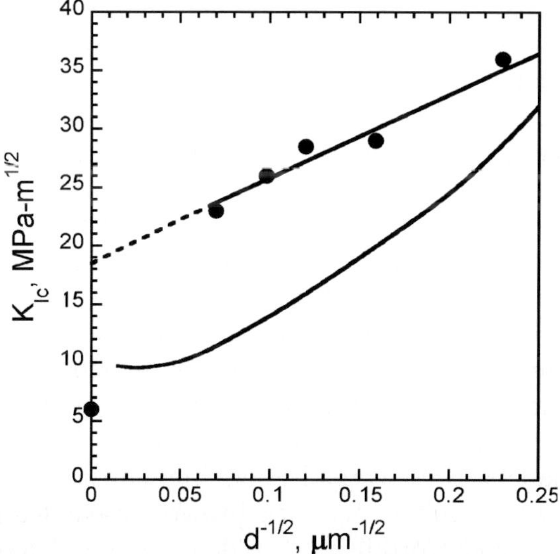

Figure 11. A Hall-Petch type relationship for the fracture toughness dependence of ferrite on grain size at $100\,\mathrm{K}$ (after Katz et al., 1993); the solid curve at the bottom is the theoretical prediction from the blocked slip band shielding model, Equation (12).

$$N = \frac{\delta_t}{b} = \frac{K_{IC}^2}{2\sigma_{ys}Eb}. \tag{14}$$

Using the observed data in Figure 11 and Equation (13) for the yield strength one finds $N = 27,000$ and $10,000$ at $10\mu m$ and $1000\mu m$, respectively. If we would have used 27,000, the fit would have been better for the coarsest grains. With this first order confirmation, Katz et al. (1993) used this and the previous shielding estimates by Anderson and Rice (1986), Thomson (1986) and Li (1986) to examine fracture in smaller volumes.

3.2. CRACK-TIP SHIELDING IN THIN FILMS

From a crack-tip shielding context, one can understand the R-curve behavior in terms of the smallest length scale that governs the local crack-tip stress field. This is more local than the length scale based upon the plastic zone size and film thickness used to determine Equation (7) by Gerberich et al. (2003a). We will not consider crack-tip shielding in detail since there are a number of papers, an overview (Volinsky et al., 2002) and a detailed encyclopedia (see Gerberich et al., 2003b) account of how crack-tip emitted dislocations can shield the tip-stress from that created by the applied stress intensity. Here, it is the distance, c, of the last emitted dislocation from the crack-tip, as constrained closer to the tip by the previous ones that is the critical length size. This has been given by Volinsky et al. (2002) and Gerberich et al. (2000) as

$$G_I = \frac{\sigma_{ys}^2 c}{5.96\,E} \exp\left\{\frac{k_{IG}}{0.76\sigma_{ys}c^{1/2}}\right\} \tag{15}$$

where $k_{IG} = [EW_d]^{1/2}$ is the local Griffith value based upon the thermodynamic adhesion energy, W_d. Since both G_R and G_I of Equations (7) and (15) can be considered as crack growth resistance, these may be set equal to solve for crack-tip shielding in the Au film crack arrest data of Figure 9(b). Knowing that $h = 250$ nm, $\sigma_{ys} = 500$ MPa, $k_{IG} = 0.3$ MPam$^{1/2}$ and values of $\Delta b/b_o$ cited in Gerberich et al. (2003a), this allowed calculation of c by setting Equation (7) equal to (15). From this we calculated that c decreased from 19.2 to 15.4 nm as the crack grew from 65 to 890 nm during stable slow crack growth. This implies that as the crack grew more dislocations are emitted driving the nearest dislocation closer to the crack tip. This greater shielding allowed a greater driving force. Of course, as the blister size increases, the stress intensity decreases if the load is constant. Thus, it requires an increasing penetration load to provide a series of fracture instabilities as observed in indentation-induced blisters of Volinsky et al. (2002a). For spontaneous telephone cord buckles, the residual stress is constant. Nevertheless intermittent slow crack growth occurs as indicated by the arrest marks in Figure 9(c).

For sometime we had been concerned about the considerable "scatter" associated with superlayer indentation studies of crack growth resistance present in the studies of Volinsky et al. (2002) and Cordill et al. (2005). It now becomes clear that this is not scatter but an R-curve effect that depends upon the indentation load. For example, in Figure 12 we determined G_I for a series of increasing loads that gave the "error" bars around each data point. However, with a constant value of k_{IG} close

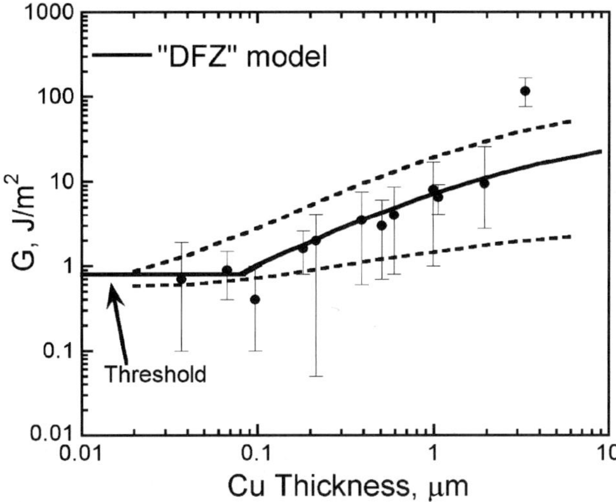

Figure 12. Interfacial fracture energy of Cu films calculated using indentation blisters and linear elastic fracture mechanics. Dashed lines model the upper and lower bounds for crack tip shielding given an R-curve behavior. The solid line indicates the dislocation free zone (DFZ) model, Equation (15).

to that for dislocation emission, one finds that G increases from about 0.6–1.35 J/m^2 during a small amount of subcritical crack growth in the thinnest film. For the thickest film, an increase from about 2–38 J/m^2 is calculated. The limits were fit by picking two values of c, i.e., the so-called dislocation-free zone. However, these are extremely small being 12 nm for the upper bound data and 40 nm for the lower bound. These lower and upper limits found for all thicknesses are the dashed lines in Figure 12 and correspond reasonably well to the range of values calculated from many experiments at each thickness representing a range of loads. We want to make it clear here (Cordill et al., 2004) that the data points and "error bars" are from calculations of the blister sizes using linear elastic fracture mechanics for the driving force while the dashed boundaries and the solid curve are determined from the crack-tip shielding model. To summarize then for these thin films small in one dimension, crack-tip shielding by emitted dislocations are proposed to play a dominant role. This, however, is not a certainty as many external sources from grown-in dislocations during sputtering could have provided shielding as well.

3.3. CRACK-TIP SHIELDING IN NANOSPHERES

On much firmer ground are experiments where dislocation sources are sparse or non-existent. We were fortunate in the first experiments conducted to have evaluated silicon nanospheres such as the one in Figure 6 which were nearly perfectly round, in general defect free, and work hardened by the multiplication of dislocations (Gerberich et al., 2003). We were further fortunate to have experienced fracture in these nanospheres that could be identified by the fragmentation of the nanoparticles at a critical load. We were able to calculate an upper bound contact stress based upon the geometric contact area of a sphere squeezed between two flat platens. The shape of the deformed sphere was followed by atomic force microscopy as detailed elsewhere in Gerberich et al. (2003) and Mook et al. (2005). When the sphere fractured the shape completely changed as shown by one example in Figure 13. Since we were imaging the spheres with a 1μm radius diamond tip, the only way the tip can dip in between spheres of 40–100 nm in diameter is if they fragment. Having the critical load and contact area allowed a determination of the contact stress at fracture. The only assumption we made was that a crack was nucleated in the oxide films on the order of 1.5–2.0 nm thick at the edge of the contact where the stresses were highest. With a crack jumping into the more brittle oxide film, it could go critical where the driving force exceeded the resistance. Mook et al. (2005) loaded the spheres to ever increasing loads until fracture was detected. Experimentally, then, Mook et al. (2005) proposed that the fracture toughness could be determined from the crack size in the oxide film and the upper bound contact stress. This is shown in Figure 14(a) as a function of nanosphere diameter. The critical strain energy release rate was determined by using the average modulus as the stresses associated with Equation (4) vary as a function of position in the sphere. As seen in Figure 14(a) and (b), we now see that K_{IC} approximately varies as $1/d$ and G_{IC} as $1/d^2$ where d is the particle diameter. This strain energy release rate magnitude and trend agreed quite well with the work per unit fracture area based upon the elastic strain energy density and the volume to fracture area of each sphere. This result of increasing fracture resistance of smaller volumes is considerably at odds with the thin films. For the latter fracture resistance increases with larger dimensions of the small volume, in this case the thickness as in Figures 8 and 12.

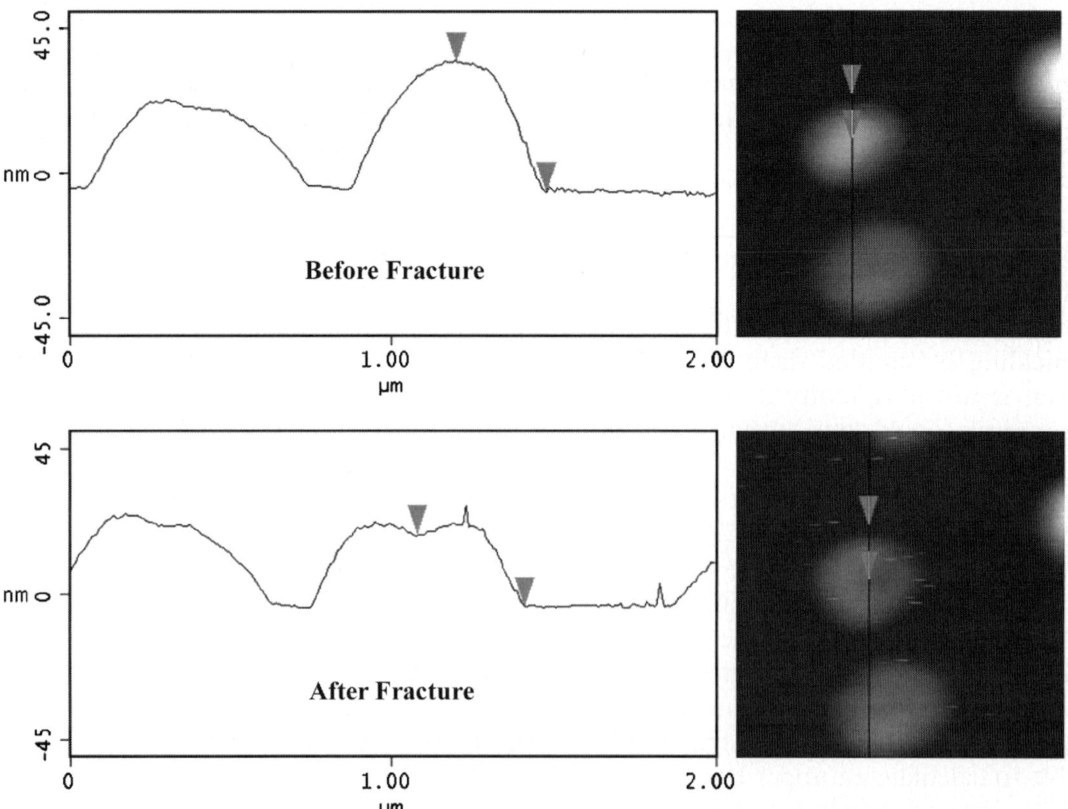

Figure 13. Scanning probe microscopy images and cross-sections before and after a 44 nm particle was fractured showing a dip of 3 nm in between fragments.

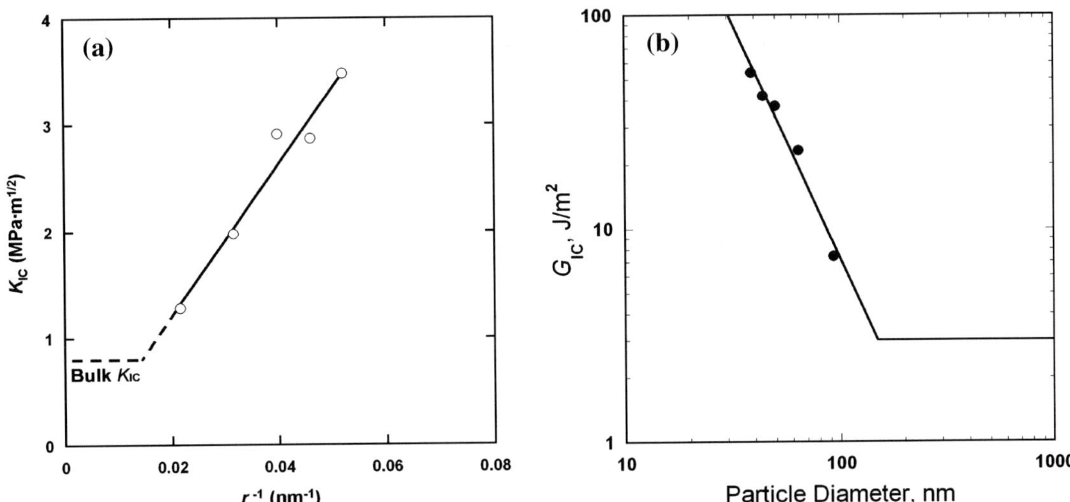

Figure 14. (a) Loading a single nanosphere until a fracture event occurs lead to the proposed fracture toughness model from a crack in the oxide film surrounding the particle (b) The critical strain energy release rate, G_{Ic}, was also calculated using the average elastic modulus associated with Equation (4) and the high contact pressures.

Theoretically, we first attempted to use the dislocation-shielding model of Equation (15) along with the value of c estimated from a dislocation pileup to model toughness in the spheres. Such an approach had been previously used by Michot and George (1982) and Gerberich (1985) to determine the shielding effect in bulk silicon. The number of dislocations could be taken as the appropriate displacement divided by the Burgers vector. This resulted in extremely small stand-off distances from the crack-tip for c but more importantly the opposite trend for toughness compared to the data in Figure 14(b). We quickly realized that dislocation arrays at a crack tip confined in a volume small in three-dimensions could be very different compared to a thin-film small in one dimension. The latter was modeled with crack-tip shielding that had also successfully been used to model single crystal behavior where the plasticity is unimpeded. Similarly, for a thin film one can propose dislocation activity at a small angle to the crack tip or along an interface where flow is relatively unimpeded in two dimensions. This is not possible for these nanoparticles surrounded by an oxide film acting as a barrier to dislocation egress just as a grain boundary might. For that reason we used the slip blockage model of Equation (12) as had been verified to first order for polycrystalline ferrite. The necessary values for Equation (12) are given in Table 1. The friction stress was taken as one-quarter of the mean contact compressive stress. Since stresses away from the contact decrease to less than half the contact stress and the shear stress is half the compression, this is a good estimate. The shear modulus was taken as a constant value of 66 GPa recognizing that on average throughout the sphere this would vary as discussed above. For Si, $4(1-v)/\sqrt{\pi}$ is 1.77. The remaining parameter is the number of shielding dislocations, N. We assumed this to be proportional to the number of dislocations emitted from the diamond-tip contact edge where stresses were highest. This gave

$$N = \alpha_\perp \delta_{cum}/2b \tag{16}$$

Table 1. Parameters for analysis of silicon nanosphere fracture.

Sphere diameter	(a)	(b)	(c)			MPa m$^{1/2}$		
d, nm	d^*, nm	δ_{cum}, nm	N	τ_{f}, GPa		k_{IG}	$K_{\text{IC}}^{\text{shield}}$	$K_{\text{IC}}^{\text{obs}}$
93	67.5	25.5	27.0	4.86		0.79	1.50	1.27
63.5	37.4	20.9	22.1	8.46		0.87	2.00	1.94
50.2	30.9	22.6	23.9	14.1		0.96	2.65	2.88
44	26.2	24.3	22.1	13.9		0.95	2.63	2.84
39	22.5	23.5	27.1	18.9		1.03	3.06	3.5
$\sim 20^{\text{(d)}}$	~ 10	~ 10	~ 25	~ 33.3		~ 1.2	3.88	–

(a) d^* = height at the end of the run where it fractured; (b) δ_{cum} = cumulative displacement at fracture; (c) N from Equation (16) using $\alpha_\perp \sim 0.5$ (d) last row is extrapolated.

at the top and bottom of the sphere where δ_{cum} is the cumulative plastic displacement at the point of fracture. This was determined by measuring the height of the deformed sphere prior to the loading run that produced failure and adding to it the loading displacement of the current run producing failure. As explained in Gerberich et al. (2005a), this gives a good measure of the cumulative plastic strain. One further point is that the several runs prior to failure foreshortened the sphere so that the deformed height, d^*, is smaller than the original height or particle diameter, d. As this is the correct dimension for slip blockage in this model, d^* is used as shown in Table 1. With a $b = 0.236$ nm for Si, Equations (14) and (16) along with the data in Table 1 give values of fracture toughness, K_{IC}. At larger volumes, the data shown in Figures 14(a) and (15) are seen to decrease toward the accepted value of about 0.8 MPa m$^{1/2}$ for bulk single crystal silicon. Theoretically, the fracture toughness calculated follows the data trend and would predict that the bulk value would be reached at a sphere diameter of about 195 nm. This is found by assuming that N ~ 25 persists to larger sphere diameters at fracture. If this is constant than differentiating Equation (12) gives a minimum at $d = 195$ nm. Placed back into Equation (12) this gives the K_{IC} at the minimum to be 0.99 MPa m$^{1/2}$ reasonably close to the accepted value. One other pair of calculations were conducted, the first was to evaluate the minimum G_{I} if only the surface energy were involved on the resistance side. We calculated this as k_{IG} from $[2E\gamma_{\text{s}}]^{1/2}$ with E now being considered dependent on the large compressive stresses and γ_{s} equal to 1.56 J/m^2 for silicon. Also determined was a continuum estimate using the stress intensity solution for the plastic zone diameter being equal to the distorted sphere height, d^*, giving

$$K_{\text{IC}} = \sigma_{\text{ys}} \sqrt{\pi d^*}. \tag{17}$$

Here, σ_{ys} was taken as $2\tau_{\text{f}}$ used in Table 1. These three estimates give the same decreasing trend in fracture resistance with increasing length scale (see Figure 15). For the good fit of the shielding model we did use an adjustable parameter of $\alpha_\perp = 0.5$ which implies either that all dislocations emitted do not act as shielding dislocations and/or the slip plane is not ideally oriented for maximum shielding. We point out while the models have assumed prismatic punching it is even more likely that shear loops are involved and hence $\alpha_\perp < 1$.

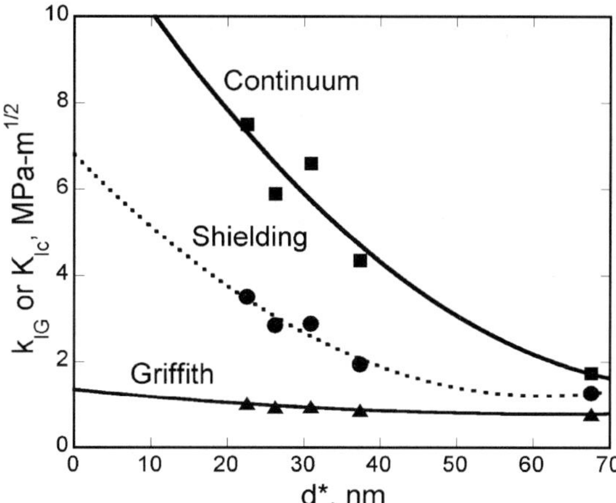

Figure 15. Stress intensity using continuum, dislocation shielding, and Griffith models for increasing nanosphere diameters. The models have assumed prismatic punching of dislocations but it is even more likely that shear loops are also included.

This discussion brings us back to Figure 1(d) where we now see that at length scales greater than about 100 nm, toughness can increase in volumes small in one dimension as seen in Figures 8 and 12. Additionally, we see that there are good reasons both elastically and based upon crack-tip shielding for fracture resistance to increase below 100 nm for volumes small in three dimensions as seen in Figure 15. We hasten to add that much of this is tenuous, and is based upon too few experimental results conducted in a scale regime not easily assessed. Still, the results are intriguing. To obtain a better measure of when fractures are occurring and perhaps even assessing the character of fracture dynamics, we have started to conduct experiments using acoustic emission. This will be briefly discussed in the next section.

3.4. ACOUSTIC EMISSION PROBES

In conjunction with colleagues at Sandia National Laboratories (Jungk et al. 2005) we have been conducting thin film studies of fracture in thin diamond films. In preparation for that, we calibrated a piezoelectric sensor attached to the diamond probe of a nanoindenter. This was accomplished by producing yield excursions into MgO and Al_2O_3 single crystals. The result in Figure 16 is in terms of the acoustic energy in arbitrary units as integrated under the waveform shown in Figure 16(b). The incremental energy of the transients $Pd\delta$ in Figure 16(a) gave the nearly proportional relationship shown in Figure 16(c). The best fit was given by

$$AE_{ind}^{au} = 2 \times 10^{16} \, (Pd\delta)^{4/3} \, J^{-4/3} \tag{18}$$

where $Pd\delta$ is the external work in joules done to produce the instabilities in Figure 16(a) and AE_{ind}^{au} is the acoustic energy produced under the indenter in arbitrary units. A similar set of experiments was conducted for indentation produced cracks in 110 nm thick diamond films. If the same correlation is assumed as in Equation (18), then all that is needed is a measure of work done in producing cracks at the indentation site. After acoustic emission was detected during indentation of the diamond

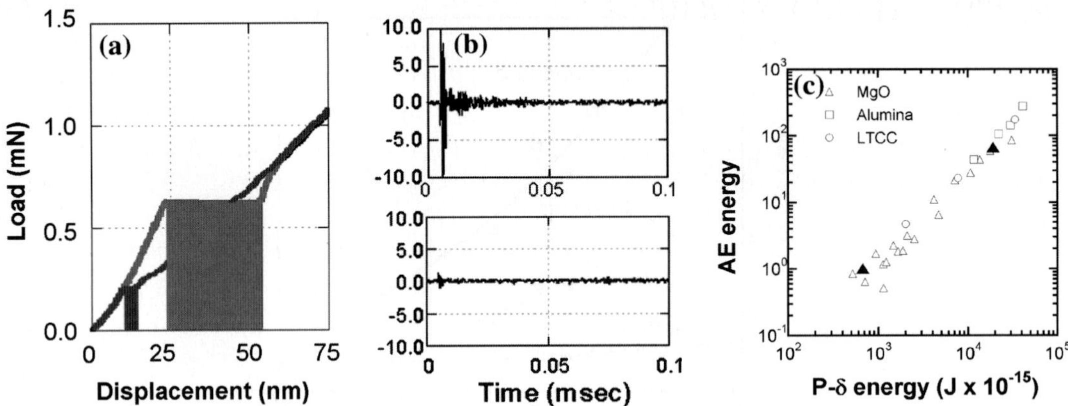

Figure 16. Loading and acoustic data for two acoustically monitored nanoindentation experiments into (100) MgO. As shown (a), excursions were observed to occur over a range of loads, likely depending on indenter tip proximity to crystal defects. These displacement excursions coincided with generation of acoustic signals (b) that when integrated, could be directly linked with the acoustic energy. A linear relationship (c) was observed in the correlation between released load-displacement energy and the integrated acoustic energy for three different ceramics over nearly four orders of magnitude.

film, the samples were unloaded and crack lengths were determined by microscopy. The result is shown in Figure 17. An analytical first-order estimate of the work done in advancing the crack was accomplished in terms of the stress intensity solution for indentation-induced cracks in thin films. As presented in detail by Jungk et al. (2005), this gives

$$P\mathrm{d}\delta = \frac{\alpha K_I^2}{E}\left[\frac{c^3 h(1+v)}{2\pi(1-v)}\right]^{1/2} \tag{19}$$

with α a proportionality factor, c the crack length produced and h the film thickness. It is seen that this has work units. Given that the AE energy is quite large even for zero channel crack lengths, we modified Equation (18) to

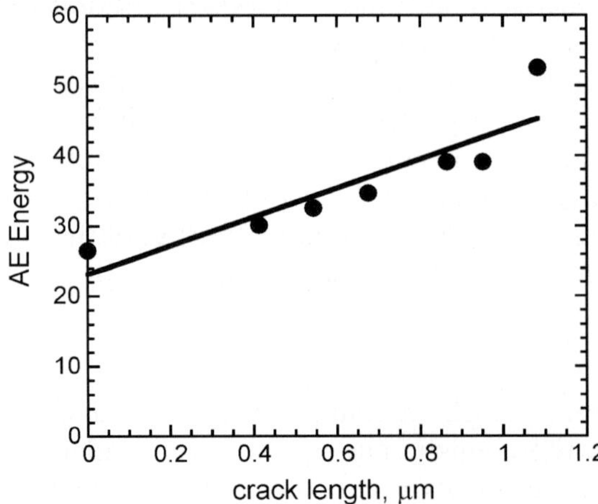

Figure 17. The observed linear relationship between the integrated acoustic energy and the measured radial (or channel) crack length in a 110 nm diamond film. The nonzero intercept likely corresponds to released acoustic energy from local film delamination and/or surface fracture along the indenter face.

$$AE_{crack}^{au} = 23 + \left(2 \cdot 10^{16}\right) (Pd\delta)^{4/3} \, J^{-4/3} \tag{20}$$

where now $Pd\delta$ is taken from Equation (19). With $\alpha = 1.62 \times 10^{10} \mathrm{m}^{3/2}$, this is seen to reproduce the data well in Figure 17. We propose that the AE energy is non-zero at "zero" crack length due to either delamination or circumferential cracking at the indenter edge not easily detected by scanning electron microscopy. This is currently under investigation.

For crack detection in a 100 nm Si nanosphere, we can consider $Pd\delta$ as a work per unit fracture area times the fracture area of the sphere, A_s. This gives $Pd\delta$ to be $G_{Ic}A_s$. Taking G_{IC} to be 8 J/m^2 from Figure 14(b) for the 100 nm sphere, this gives 6.3×10^{-14} J of work in fracturing the sphere. In a previous work, Tymiak et al. (2004) showed that on sapphire using the same detection system, measurable emission for displacement jumps due to yielding or twinning in sapphire represented $Pd\delta$ values on the order of 10^{-11} J, much larger than those shown in Figure 16(a). Also, if we examine our original calibration in Figure 16(c), it is suggested that we would need an order of magnitude more sensitivity to detect the 100 nm sphere fracture. That may not be the case since the crack tip source for the emission is directly under the transducer. Since the source to transducer distance in Equations (8) and (9) are critical, this may make such detection feasible.

4. Summary

Evidence for large mean contact stress directly and length scales indirectly affecting elasticity has been shown. Beyond that large local stresses, surface energy and length scale confinement in one, two and three dimensions lead to complex flow stress behavior at the nanoscale. Some of these phenomena are reviewed briefly with the intent of demonstrating how such effects impact fracture at length scales of 40–4000 nm for thin films and 20–200 nm for nanoparticles. Increases of about a factor of three in modulus of elasticity, hardness and strength are shown to be dependent on several factors including high mean contact pressures and confinement in small volumes. To probe how such equation of state effects impact on fracture, we have used microscopic, atomic force and acoustic sensor probes to examine fracture length scales. In thin films of copper and gold, elastic and plastic effects coupled with dislocation shielding give increasing delamination energies with increasing film thickness. For such adhered films small in one-dimension, R-curve effects are also measured and modeled. In spherical nanoparticles an opposite effect is seen in that increased toughness occurs with decreasing diameters of silicon nanospheres. It is proposed that this change is due to the difference in combined hardening, state of stress, and length scale effects that occur in volumes small in one-dimension compared to those small in three dimensions. Such behavior is also analyzed with a blocked slip band model appropriate to crack-tip shielding. For semi-brittle silicon, fracture toughness increases from about 0.8 MPa-m$^{1/2}$ for bulk silicon in tension to over 3 MPa-m$^{1/2}$ for nanospheres in compression.

Acknowledgements

The authors wish to thank T. Buchheit of Sandia National Laboratories (Albuquerque, NM) and R. Nay (Hysitron Inc, Minneapolis, MN) for assistance with the acoustic emission and C.B. Carter and C.R. Perrey (University of Minnesota) for the electron microscopy. This work was supported by the National Science Foundation under grants DMI 0103169, CMS-0322436, an NSF-IGERT program through grant DGE-0114372 and the United States Department of Energy Office of Science, DE-AC04-94AL85000.

References

Anderson, P.M. and Rice, J.R. (1986). Dislocation emission from cracks in crystals or along crystal interfaces. *Scripta Metallurgica*, **20**, 1467–1472.

ASM Handbook on Fatigue and Fracture (1996). ASM International Materials, 19, Park, OH.

Bazant, Z.P. (2004). Scaling theory for quasibrittle structural failure, *PHAS*, **101**, National Academy of Sciences, September.

Becit, M.R. (1979). Fracture of a surface layer bonded to a half space. *International Journal of Engineering and Science* 17, 287–295.

Begley, M.R. and Ambriso, J.M. (2003). Channel cracking during thermal cycling of thin film multilayers. *International Journal of Fracture* **119/120**, 325–338.

Belytschko, T., Xiao, S.P., Schatz, G.C. and Ruoff, R.S. (2004). *Atomistic Simulations of Nanotube Fracture,* Dept. of Mechanical Engineering Northwestern Unversity, www.tam.northwestern.edu/tb/nano/tubefrac.

Binnig, G., Quate, C.F. and Gerber, C. (1987). Atomic force microscope. *Physical Review Letters* **56**, 930–933.

Chen, Y.T., Atteridge, D.G. and Gerberich, W.W. (1981). Dislocation dynamics of Fe-binary alloys: I. Low temperature plastic flow. *Acta Metallurgica* **29**, 1171–1185.

Christensen, N.E., Ruoff, A.L. and Rodriguez, C.O. (1995). Pressure strengthening: a way to multimegabar static pressures. *Physical Review B* **52**, 9121–9124.

Corcoran, S.G., Colton, R.J., Lilleodden, E.T. and Gerberich, W.W. (1997). Anamolous plastic deformation of surfaces: nanoindenation of gold single crystals. *Physical Review B* **55**, 16057–16060.

Cordill, M.J., Bahr, D.F., Moody, N.R. and Gerberich, W.W. (2004). Recent developments in thin film adhesion measurement. *IEEE Transactions on Device Manufacturing and Reliability* **4**, 163–168.

Cordill, M.J., Moody, N.R. and Bahr, D.F. (2005). The effects of plasticity on adhesion of hard films on ductile interlayers, *Acta Materialia*, accepted.

Curtin, W.A. and Miller, R.E. (2003). Atomistic/continuum coupling in computational materials science. *Modelling and Simulation in Materials Science and Engineering* **11**, R33–R68.

Fleischmann, P., Lakestani, F., Baboux, J.C. and Rouby, D. (1977). Spectral and energy analysis of a moving ultrasonic source-application of acoustic emission to aluminum during plastic deformation. *Materials Science and Engineering* **29**, 205–212.

Gall, K., Diao, J. and Dunn, M. (2004). Strength of gold nanowires. *Nanoletters* **4**, 2431–2436.

Gall, K., Diao, J., Dunn, M.L., Haftel, M., Bernstein, N. and Mehl, M.J. (2005). Tetragonal Phase Transformation in Gold Nanowires, *Journal of Engineering Materials and Technology*, submitted.

Garzke, Jr., W.H, Brown, D.K., Matthias, P.K., Cullimore, R., Wood, D., Livingston, D., Leighty, H.P., Foecke, T. and Sandiford, A. (1997). *Titanic, the Anatomy of a Disaster*, Report from the Marine /Forensic Panel (SD-7), Soc. Of Naval Architects and Marine Engineers, 1.1–1.47.

Gerberich, W.W. and Jatavallabhula, K. (1980). A review of acoustic emission from source controlled by grain size and particle fracture, in *Nondestructive Evaluation*, (edited by Buck, O. and Wolf, S.M.) TMS, Warrendale, PA, 319–348.

Gerberich, W.W. (1985). Interaction of microstructure and mechanism in defining K_{Ic}, K_{Iscc}, or ΔK_{th} values In: *Fracture: Interactions of Microstructure, Mechanisms, and Mechanics*, (edited by Wells, J.M. and Landes, J.D.,) TMS, Warrendale, PA, 49.

Gerberich, W.W., Yu, W., Kramer, D., Strojny, A., Bahr, D.F, Lilleodden, E.T. and Nelson, J. (1993). Elastic loading and elastoplastic unloading from nanometer level indentations for modulus determinations. *Journal of Materials Research* **13**, 421–439.

Gerberich, W.W., Volinsky, A.A. and Tymiak, N.I. (2000). A brittle to ductile transition in adhered thin films. *Materials Research Society Symposium* **594**, 51–363.

Gerberich, W.W., Tymiak, N.I., Grunlan, J.C., Horstemeyer, M.F. and Baskes, M.I. (2002). Interpretations of indentation size effects. *Journal of Applied Mechanics* **69**, 433–442.

Gerberich, W.W., Mook, W.M, Perrey, C.R., Carter, C.B., Baskes, M.I., Mukherjee, R., Gidwani, A., Heberlein, J., McMurry, P.H. and Girshick, J.L. (2003). Superhard silicon nanospheres. *Journal on the Mechanics and Physics of Solids* **51**, 979–992.

Gerberich, W.W., Jungk, J.M., Li, M., Volinsky, A.A., Hoehn, J.W. and Yoder, K. (2003a). Length scales for the fracture of nanostructures. *International Journal of Fracture* **119/120**, 387–405.

Gerberich, W.W., Jungk, J.M. and Mook, W.M. (2003b). Crack-dislocation interactions. in *Comprehensive Structural Integrity: Interfacial and Nanoscale Failure*, (edited by Gerberich, W. and Yang, W.), ch. 10, 357–382.

Gerberich, W.W., Cordill, M.J, Mook, W.M., Moody, N.R., Perrey, C.R., Carter, C.B., Mukherjee, R. and Girshick, S.L. (2005). A boundary constraint energy balance criterion for small volume deformation. *Acta Materialia*, accepted.

Gerberich, W.W., Mook, W.M.,Cordill, M.J., Carter, C.B., Perrey, C.R., Heberlein, J. and Girshick, S.L. (2005a). Reverse plasticity in single crystal silicon nanospheres. *International Journal of Plasticity*, accepted.

Griffith, A.A. (1921). The phenomena of rupture and flow in solids. *Philosophical Transactions of the Royal Society of London* **A221**, 163–198.

Griffith, A.A. (1925). The theory of rupture. *Proceedings of the 1st Congress on Applied Mechanics*, Delft, 55–63.

Hall, E.O. (1951). The deformation and ageing of mild steel III. Discussion of results. *Proceedings of Physical Science B* **64**, 747–753.

Hansen, N. (2004). Hall-Petch relation and boundary strengthening. *Scripta Metallurgica* **51**, 801–806.

Horstemeyer, M.F. and Baskes, M.I. (1991). Atomistic finite deformation simulations: a discussion on length scale effects in relation to mechanical stresses. *Transactions of the ASME. Journal of Engineering Materials and Technology* **121**, 114–119.

Huang, H. and Gerberich, W.W. (1992). Crack-tip dislocation emission arrangements for equilibrium – II. Comparisons to analytical and computer simulation models. *Acta Metallurgica et Materialia* **40**, 2873.

Huang, H. and Spaepen, F. (2000). Tensile testing of free-standing Cu, Ag and Al thin films and Ag/Cu multilayers. *Acta Materialia* **48**, 3261–3269.

Hughes, D.A. (2004). Sandia National Laboratories, Livermore, CA, private communication.

Hutchinson, J.W. (1968). Singular behaviour at the end of a tensile crack in a hardening material. *Journal of the Mechanics and Physics of Solids* **16**, 3–31.

Irwin, G.R. (1948). Fracture dynamics. In: *Fracturing of Metals*, Am. Soc. For Metals, Cleveland, 147–166.

Irwin, G.R. (1960). *ASTM Bulletin*, Jan., 29.

Johnson, K. (1985). *Contact Mechanics*. Cambridge Univ. Press, U.K., 57.

Jungk, J.M., Boyce, B.L., Buchheit, T.E, Friedmann, T.A., Yang, D. and Gerberich, W.W. (2005). Indentation fracture toughness and acoustic energy release in diamond films, in preparation.

Katz, Y., Keller, R.R., Huang, H. and Gerberich, W.W. (1993). A dislocation shielding model for the fracture of semibrittle crystals. *Metallurgical Transactions A* **24A**, 343–350.

Lane, M., Dauskardt, R.H., Krishna, N. and Hashim, I. (2000). Adhesion and reliability of copper interconnects with Ta and TaN barrier layers. *Journal of Materials Research* **15**, 203–211.

Li, J.C.M. (1986). *Scripta Metallurgica* **20**, 1477.

Li, M., Chen, X.-F., Katz, Y. and Gerberich, W.W. (1990). Dislocation modeling and acoustic emission observation of alternating ductile/brittle events in Fe-3wt.%Si crystals. *Acta Metallurgica et Materialia* **38**, 2435–2453.

Lin, J.H. and Thomson, R. (1986). Cleavage, dislocation emission, and shielding for cracks under general loading. *Acta Metallurgica* **34**, 187–206.

Majzoub, R. and Chaudhri, M. (2000). High-speed photography of low-velocity impact cracking of solid spheres. *Philosophical Magazine Letters* **80**, 387.

McElhaney, K.W., Vlassak, J.J. and Nix, W.D. (1998). Determination of indenter tip geometry and indentation contact area for depth-sensing indentation experiments. *Journal of Materials Research* **13**, 1300–1306.

Michot, G. and George, A. (1982) In situ observation by x-ray synchrotron topography of the growth of plasticity deformed regions around crack tips in silicon under creep conditions. *Scripta Metallurgica* **16**, 519–524.

Mook, W.M., Jungk, J.M., Cordill, M.J., Moody, N.R., Sun, Y., Xia, Y. and Gerberich, W.W. (2004). Geometry and surface state effects on the mechanical response of Au nanostructures. *Zeischrift fur Metallkunde* **95**, 416–424.

Mook, W.M., Perrey, C.R. Carter, C.B., Mukherjee, R., Girschick, S.L., McMurry, P.H. and Gerberich, W.W. (2005). Scale effects on nanoparticle modulus and fracture, *Physical Review B*, submitted.

Moriarty, J.A., Belak, J.F., Rudd, R.E., Soderlind, P., Streitz, F.H. and Yang, L.H. (2002). Quantum-based atomistic simulation of materials properties in transition metals. *Journal of Physics: Condensed Matter* **14**, 2825–2857.

Murnaghan, F. (1967). *Finite Deformation in an Elastic Solid*. Dover Publ., New York.

Nix W.D., Gao H (1998). Indentation size effects in crystalline materials: a law for strain gradient plasticity. *Journal of the Mechanics and Physics of Solids* **46**, 411–425.

Oliver, W.C. and Pharr, G.M. (1992). An improved technique for determining hardness and elastic modulus using load and displacement sensing indentation experiments. *Journal of Materials Research* **1**, 1564–1583.

Orowan, E. (1950). *Fatigue and Fracture of Metals*. MIT Press, Cambridge, MA, 139.

Parker, E.R. (1957). *Brittle Behavior of Engineering Structures*. National Academy of Sciences, National Research Council, J. Wiley, New York.

Petch, N.J. (1953). *Journal of the Iron and Steel Institute* **173**, 25.

Poirier, J.-P. (2000). Murnaghan's integrated linear equation of state. In: *Introduction to the Physics of the Earth's Interior, 2nd Ed.*, Cambridge University Press, 65.

Read, D.T. (1998). Piezo-actuated microtensile test apparatus. *Journal of Testing & Evaluation* **26**, 255–259.

Rice, J.R. (1969). A path independent integral and the approximate analysis of strain concentration by notches and cracks. *Journal of Applied. Mechanics* **35**, 379–386.

Rice, J.R. and Thomson, R. (1974). Ductile versus brittle behaviour of crystals. *Philosophical Magazine* **29**, 74–97.

Shipway, P.H. and Hutchings, I.M. (1993). Fracture of brittle spheres under compression and impact loading. I. Elastic stress distributions. *Philosophical Magazine* **A67**, 1389–1404.

Suo, Z., Shih, F. and Varias, A. (1993). A theory for cleavage cracking in the presence of plastic flow. *Acta Metallurgica et Materialia* **41**, 551–557.

Thomson, R. (1986). Dislocation emission from cracks in crystals or along crystal interfaces. *Scripta Metallurgica* **20**, 1473.

Tymiak, N.I., Kramer, D.E., Bahr, D.F., Wyrobek, T.J. and Gerberich, W.W. (2001). Plastic strain and strain gradients at very small indentation depths. *Acta Materialia* **49**, 1021–1034.

Tymiak, N.I., Daugela, A., Wyrobek, T.J. and Warren, O.L. (2004). Acoustic emission monitoring of the earliest stages of contact-induced plasticity in sapphire. *Acta Materialia* **52**, 553–563.

Van Swygenhoven, H. and Spaczer, M. (1989). Competing plastic deformation mechanisms in nanophase metals. *Physical Review B* **60**, 22–25.

Van Vliet, K., Li, J., Zhu, T., Yip, S. amd Suresh, S. (2003). Quantifying the early stages of plasticity through nanoscale experiments and simulations. *Physical Review B* **67**, 104–105.

Vlassak, J.J. (2003). Channel cracking in thin films on substrate of finite thickness. *International Journal of Fracture* **119/120**, 299–323.

Volinsky, A.A., Moody, N.R. and Gerberich, W.W. (2002). Interfacial toughness measurements for thin films on substrates. *Acta Materialia* **50**, 441–466.

Volinsky, A.A., Moody, N.R., Kottke, M.L. and Gerberich, W.W. (2002a). Fiducial mark and nanocrack zone formation during thin-film delaminating. *Philosophical Magazine A* **82**, 3383–3391.

Wei, Y. and Hutchinson, J.W. (1997). Nonlinear delamination mechanics for thin films. *Journal of the Mechanics and Physics of Solids* **45**, 1137–1159.

Yu, M.F., Lourie, O., Dyer, M.J., Moloni, K., Kelly, T.F. and Ruoff, R.S. (2000). Strength and breaking mechanism of multiwalled carbon nanotubes under tensile load. *Science* **287**, 637–640.

International Journal of Fracture (2006) 138:101–137
DOI 10.1007/s10704-006-7156-4

© Springer 2006

Application of fracture mechanics concepts to hierarchical biomechanics of bone and bone-like materials

HUAJIAN GAO[1,2]
[1]*Max Planck Institute for Metals Research, Heisenbergstrasse 3, D-70569 Stuttgart, Germany (E-mail: hjgao@mf.mpg.de)*
[2]*Present address: Division of Engineering, Brown University, Providence, RI 02912, USA*

Received 1 March 2005; accepted 1 December 2005

Abstract. Fracture mechanics concepts are applied to gain some understanding of the hierarchical nanocomposite structures of hard biological tissues such as bone, tooth and shells. At the most elementary level of structural hierarchy, bone and bone-like materials exhibit a generic structure on the nanometer length scale consisting of hard mineral platelets arranged in a parallel staggered pattern in a soft protein matrix. The discussions in this paper are organized around the following questions: (1) The length scale question: why is nanoscale important to biological materials? (2) The stiffness question: how does nature create a stiff composite containing a high volume fraction of a soft material? (3) The toughness question: how does nature build a tough composite containing a high volume fraction of a brittle material? (4) The strength question: how does nature balance the widely different strengths of protein and mineral? (5) The optimization question: Can the generic nanostructure of bone and bone-like materials be understood from a structural optimization point of view? If so, what is being optimized? What is the objective function? (6) The buckling question: how does nature prevent the slender mineral platelets in bone from buckling under compression? (7) The hierarchy question: why does nature always design hierarchical structures? What is the role of structural hierarchy? A complete analysis of these questions taking into account the full biological complexities is far beyond the scope of this paper. The intention here is only to illustrate some of the basic mechanical design principles of bone-like materials using simple analytical and numerical models. With this objective in mind, the length scale question is addressed based on the principle of flaw tolerance which, in analogy with related concepts in fracture mechanics, indicates that the nanometer size makes the normally brittle mineral crystals insensitive to cracks-like flaws. Below a critical size on the nanometer length scale, the mineral crystals fail no longer by propagation of pre-existing cracks, but by uniform rupture near their limiting strength. The robust design of bone-like materials against brittle fracture provides an interesting analogy between Darwinian competition for survivability and engineering design for notch insensitivity. The follow-up analysis with respect to the questions on stiffness, strength, toughness, stability and optimization of the biological nanostructure provides further insights into the basic design principles of bone and bone-like materials. The staggered nanostructure is shown to be an optimized structure with the hard mineral crystals providing structural rigidity and the soft protein matrix dissipating fracture energy. Finally, the question on structural hierarchy is discussed via a model hierarchical material consisting of multiple levels of self-similar composite structures mimicking the nanostructure of bone. We show that the resulting "fractal bone", a model hierarchical material with different properties at different length scales, can be designed to tolerate crack-like flaws of multiple length scales.

Key words: Biological materials, bone, buckling, flaw tolerance, fracture, hierarchical materials, nacre, size effects, stiffness, strength, structural optimization, toughness.

1. Introduction

Biological materials are bottom-up designed systems formed from billions of years of natural evolution. In the long course of Darwinian competition for survival, nature has evolved a huge variety of hierarchical and multifunctional systems from nucleic acids, proteins, cells, tissues, organs, organisms, animal communities to ecological systems. Multilevel hierarchy a rule of nature. The complexities of biology provide an opportunity to study the basic principles of hierarchical and multifunctional systems design, a subject of potential interest not only to biomedical and life sciences, but also to nanosciences and nanotechnology. Systematic studies of how hierarchical structures in biology are related to their functions and properties can lead to better understanding of the effects of aging, diseases and drugs on tissues and organs, and may help developing a scientific basis for tissue engineering to improve the standard of living. At the same time, such studies may also provide guidance on the development of novel nanostructured hierarchical materials via a bottom-up approach, i.e. by tailor-designing materials from atomic scale and up. Currently we barely have any theoretical basis on how to design a hierarchical material to achieve a particular set of macroscopic properties. The new effort aiming to understand the relationships between hierarchical structures in biology and their mechanical as well as other functions and properties may provide challenging and rewarding opportunities for mechanics in the 21st century.

With the above objective in mind, we have studied the nanostructural mechanical properties of bone-like materials such as bone, tooth and shells (Gao et al., 2003, 2004; Ji and Gao, 2004a,b, 2006; Ji et al., 2004a; Guo and Gao, 2005; Liu et al., 2006) as well as bio-inspired materials such as biomorphous metal–matrix composites (Ji et al., 2004b) and superhard nanocrystalline coating (Kaufmann et al., 2005). Bone-like materials exhibit complex hierarchical structures over many length scales. For example, sea shells have 2 to 3 levels of lamellar structure (Currey, 1977; Jackson et al., 1988; Menig et al., 2000, 2001), while vertebral bone has 7 levels of structural hierarchy (Currey, 1984; Landis, 1995; Rho et al., 1998; Weiner and Wagner, 1998). Although the higher level structures of bone and bone-like materials show great complexity and variations, they exhibit a generic nanostructure at the most elementary level of structural hierarchy (Figure 1) consisting of nanometer sized hard mineral crystals arranged in a parallel staggered pattern in a soft protein matrix (Jaeger and Fratzl, 2000; Gao et al., 2003; Fratzl et al., 2004a). For example, the nanostructure of tooth enamel shows needle-like (15–20 nm thick and 1000 nm long) crystals embedded in a relatively small volume fraction of a soft protein matrix (Warshawsky, 1989; Tesch et al., 2001; Jiang et al., 2005). The nanostructure of dentin and bone consists of plate-like (2–4 nm thick and up to 100 nm long) crystals embedded in a collagen-rich protein matrix (Landis, 1995; Landis et al., 1996; Roschger et al., 2001), with the volume ratio of mineral to matrix on the order of 1 to 2. Nacre is made of very high volume fraction of plate-like crystals (200–500 nm thick and a few micrometers long) with a small amount of soft matrix in between (Currey, 1977; Jackson et al., 1988; Kamat et al., 2000; Menig et al., 2000; Wang et al., 2001). Figure 1 illustrates that bone and nacre are constructed with basically the same type of nanostructure made of staggered hard plate-like inclusions in a soft matrix. This staggered nanostructure is primarily subjected to uniaxial loading, as shown schematically

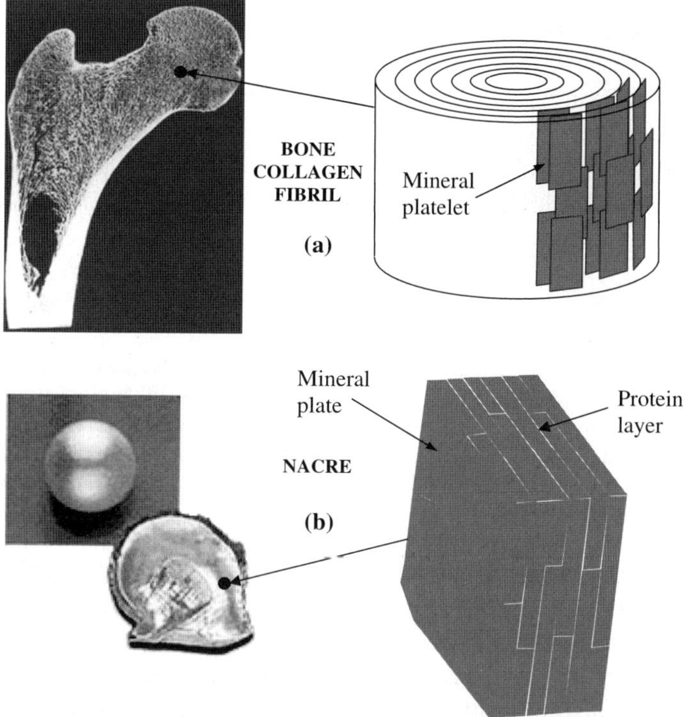

Figure 1. Nanostructures of nacre and bone. (a) The nanostructure of bone consists of plate-like mineral crystals 2–4 nm in thickness and up to 100 nm in length embedded in a collagen-rich protein matrix. (b) The elementary structure of nacre consists of plate-like mineral crystals 200–500 nm in thickness and a few micrometers in length with a very small amount of soft matrix in between.

in Figure 2(a). Under uniaxial tension, the path of load transfer in the staggered nanostructure follows a tension-shear chain with mineral platelets under tension and the soft matrix under shear (Figure 2b). Research has shown that tendons and wood also deform by shearing of a soft matrix between stiff fibres (Fratzl et al., 2004b). In wood cell walls, stiff cellulose fibrils are aligned in a soft hemicellulose–lignin matrix (Brett and Waldron, 1981; Fengel and Wegener, 1984). The toughness and other properties of bone-like materials have been investigated from various points of view including their hierarchical structures (Kessler et al., 1996; Menig et al., 2000, 2001; Kamat et al., 2000), the mechanical properties of protein on dissipating fracture energy (Smith et al., 1999; Thompson et al., 2001; Hassenkam et al., 2004; Fantner et al., 2004), the surface asperities of mineral plates (Wang et al., 2001), the mineral bridges in nacre (Song et al., 2003) and the reduction of stress concentration at a crack tip (Okumura and de Gennes, 2001). The importance of organic matrix on the properties of biocomposites has been demonstrated by testing under various wet, dry, baked and boiled conditions (Fantner et al., 2004; Neves and Mano, 2005). Fracture mechanics concepts have been applied to address the question why the elementary structure of biocomposites is generally designed at the nanometer length scale (Gao et al., 2003). Recent research has also shown that the nanometer sized crystallites in human tooth enamel exhibit remarkable resistance against chemical dissolution (Tang et al., 2004; Wang et al., 2005).

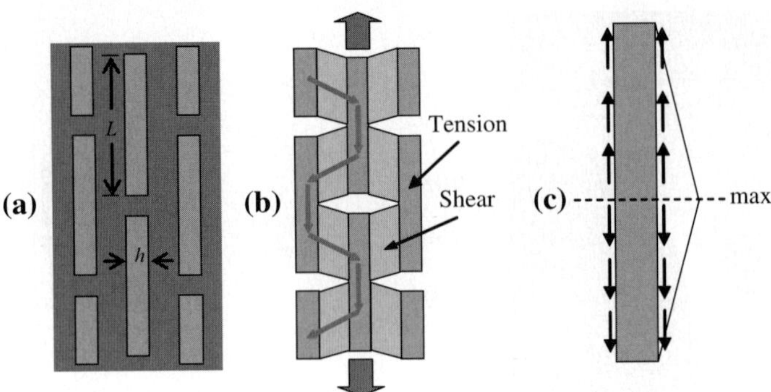

Figure 2. (a) The nanostructure of staggered hard plates in a soft matrix represents a convergent design of natural evolution. (b) The primary load bearing zones of biological nanostructure showing mineral crystals primarily in tension and the soft matrix primarily in shear. (c) The free body diagram displaying the forces acting on a single mineral plate. At the nanostructure level, the load is mainly uniaxial and is transferred along a tension-shear chain as illustrated in (b).

A central hypothesis adopted in our studies (Gao et al., 2003, 2004; Gao and Ji, 2003; Ji and Gao, 2004a,b) is that bone and bone-like materials have been evolved to tolerate crack-like flaws at multiple size scales. As brittle bone would severely diminish an animal's chance to survive, from a Darwinian point of view, evolution will tend to select those design strategies that tend to suppress brittle crack propagation. In addition, the self-sensing, self-adapting and self-repairing capabilities of bone require constant removal and replacement of old and damaged materials with fresh and healthy materials. The fact that these renewal processes should occur at the same time while an animal is conducting its normal activities also suggests that bone must be designed to tolerate crack-like flaws at all relevant sizes. In the presence of crack-like flaws, the optimal state of a material which induces the maximum strength corresponds to a uniform distribution of stress in the still uncracked material, which then fails by uniform rupture, rather than by crack propagation. This optimal state, referred to as the flaw tolerance solution, can be achieved simply by size reduction (Gao et al., 2003; Gao and Chen, 2005). Below a critical structural size, the material fails no longer by propagation of a pre-existing crack, but by uniform rupture at the limiting strength of the material. This concept has been applied to understand not only the staggered nanostructure of bone (Gao et al., 2003), but also the fibrillar nanostructure of gecko (Gao and Yao, 2004; Gao et al., 2005; Tang et al., 2005). Yao and Gao (2006) have further shown that flaw tolerance is an important principle in the bottom-up designed hierarchical structures of gecko for robust and releasable adhesion. In these biological systems, it has been shown that the flaw tolerance solution emerges as soon as the characteristic size of the critical structural link is reduced to a critical size. Gao and Chen (2005) considered flaw tolerance solutions in a thin strip containing interior or edge cracks under uniaxial tension and showed that below a critical size the strip has the intrinsic capability to tolerate cracks of all sizes. The concept of flaw tolerance provides an important analogy between Darwinian selection of robust nanostructures in biology and fracture mechanics concepts of notch insensitivity, fracture size effects and large scale yielding or bridging.

This paper will attempt to address the following questions with respect to the hierarchical nanocomposite structures of bone-like materials. (1) The smallest building blocks of biological materials are generally designed at the nanoscale. Why is nanoscale so important to biological materials? (2) Bone contains a high volume fraction of a protein-rich soft material. How does nature create a stiff composite in spite of the soft phase? (4) The mineral is usually brittle and has low fracture energy. How does nature create a tough composite in spite of high volume fraction of a brittle phase? (3) The protein and mineral have widely different strengths. How does nature balance materials with widely different strength levels? (5) Can the nanostructure of bone be understood from a structural optimization point of view? If so, what is the objective function? (6) The long and slender mineral particles in bone are susceptible to buckling under compressive loading. How does nature solve the buckling problem? (7) While sea shells exhibit 2–3 levels of structural hierarchy, bone has 7 levels of hierarchy. What is the role of structural hierarchy? Most of these questions have already been discussed in our recent publications. Here we summarize the key ideas and results. The reader is encouraged to consult various references given in the paper for more details.

2. The length scale question: why is nanoscale important for biological materials?

Bone and bone-like materials have adopted a generic elementary structure with characteristic length scale in the nanometer regime (Figure 2a). Why is nanoscale so important for biological materials? This question has been addressed by Gao et al. (2003) using fracture mechanics concepts. Under uniaxial tensile stress, the mineral crystals are primarily under tension while the soft matrix transfers load between neighboring crystals via shear (Figure 2b). Within a mineral crystal, the maximum tension occurs at the mid-section of the plate (Figure 2c). For a robust nanostructure, the mineral crystals should not be sensitive to crack-like flaws. Consider an edge cracked mineral crystal (Figure 3a, b). Under the constraint from the soft matrix, the stress field near the crack would be similar to that in a laterally constrained strip under uniaxial tension (Figure 3c). By symmetry, the edge cracked plate in Figure

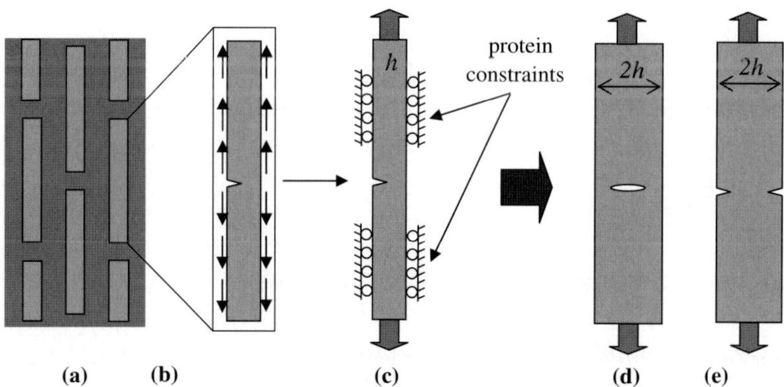

Figure 3. Flaw tolerance of the hard plates in the biological nanostructure. (a) The staggered hard–soft structure. (b) A virtual edge crack is assumed to exist in a hard plate. (c) The hard plate is primarily subject to uniaxial tension under constraint from the surrounding matrix. This problem is converted to two approximately equivalent problems: (d) a center crack and (e) a double edge crack in a strip twice as wide.

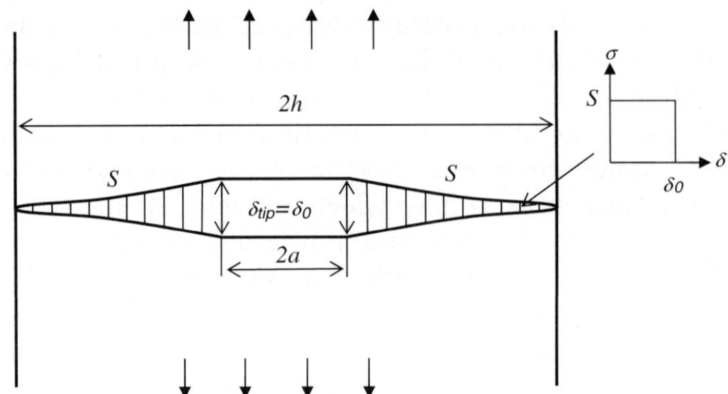

Figure 4. The flaw tolerance solution of a center cracked strip based on the Dugdale model. A Dugdale interaction law is assumed in the plane of the crack. The condition of flaw tolerance is equivalent to requiring δ_{tip} not to exceed δ_0 for any crack size a. In this case, the opening displacement in the plane of the crack outside the crack region should lie within the range of cohesive interaction δ_0.

3(c) with thickness h should be approximately equivalent to the problem of a strip of width $2h$ containing a center interior crack (Figure 3d) or two symmetrically placed edge cracks (Figure 3e). The flaw tolerance solutions for the strip problems in Figure 3(d) and Figure 3(e) have been studied by Gao and Chen (2005) using Griffith and Dugdale models.[1] Here we briefly summarize the calculations based on the Dugdale model with the following traction-separation law (Dugdale, 1960)

$$\sigma(\delta) = \begin{cases} S & \delta \leqslant \delta_0, \\ 0 & \delta > \delta_0, \end{cases} \tag{1}$$

where $\sigma = \sigma(\delta)$ is the normal traction, S is the strength of the material, δ_0 is the range of cohesive interaction and δ is the normal separation in the plane of the crack (Figure 4). In the Dugdale model, the flaw tolerance condition that cracks of any size in the range $0 \leqslant a < h$ do not grow is

$$\delta_{\text{tip}}(0 \leqslant a/h < 1) \leqslant \delta_0, \tag{2}$$

where δ_{tip} denotes the separation at the crack–tip.

Figure 4 depicts the flaw tolerance solution of a center cracked strip with a uniform distribution of normal stress S outside the crack region, regardless of the crack size. The crack–tip separation can be calculated from linear elasticity as

$$\delta_{\text{tip}} = \frac{4(1 - \nu^2)Sh}{\pi E} f(a/h), \tag{3}$$

where E is Young's modulus and $f(a/h)$ is a dimensionless function which can be numerically calculated. An approximate solution of $f(a/h)$ is obtained by Gao and Chen (2005) based on a periodic crack solution as

---

[1]This problem has been previously considered by Carpinteri (1982, 1997) with respect to a tension collapse that precedes brittle crack propagation in a strip. Carpinteri (1982) introduced a "brittleness number" as $s = K_{IC}/\sigma_u \sqrt{h}$ where K_{IC} denotes the fracture toughness, σ_u the strength and h the size of the strip. Carpinteri showed that if this "brittleness number" is larger than a critical value around 0.54, tension collapse occurs before brittle crack propagation for any crack length.

$$f\left(a/h\right)=\int_{a/h}^{1}\ln\frac{\sin\pi\left(\xi+a/h\right)/2}{\sin\pi\left(\xi-a/h\right)/2}\mathrm{d}\xi. \tag{4}$$

The flaw tolerance condition (2) can then be cast in the form

$$h\leqslant h_{\mathrm{cr}}=\min_{0\leqslant a/h<1}h^{*}\left(a/h\right),\quad h^{*}\left(a/h\right)=\frac{\pi}{4\left(1-v^{2}\right)f\left(a/h\right)}\ell_{\mathrm{ft}}, \tag{5}$$

where ℓ_{ft} is defined as

$$\ell_{\mathrm{ft}}=\frac{\delta_{0}E}{S}=\frac{\Gamma E}{S^{2}}. \tag{6}$$

In writing the second equation above, we have used the relation $\Gamma=S\delta_{0}$ between the fracture energy Γ, the cohesive strength S and the interaction range δ_{0} in the Dugdale model.

Assuming $v=0.3$, the function $h^{*}\left(a/h\right)/\ell_{\mathrm{ft}}$ is plotted in Figure 5 using the approximate solution for $f\left(a/h\right)$ in Equation (4) and numerical calculations from a finite element (FEM) analysis. Figure 5 indicates that, for both types (center and double edge) of cracks, $h^{*}\left(a/h\right)/\ell_{\mathrm{ft}}$ has a minimum value around $0.32\pi\approx1$ for a center crack with crack size close to one half of the strip width. Therefore, the Dugdale model predicts that the condition for a strip to achieve flaw tolerance is

$$h\leqslant h_{\mathrm{cr}}=\ell_{\mathrm{ft}}=\frac{\Gamma E}{S^{2}}=\frac{\delta_{0}E}{S}, \tag{7}$$

which can also be expressed in terms of a dimensionless number, referred to as the flaw tolerance number, as

$$\Lambda_{\mathrm{ft}}=\frac{\Gamma E}{S^{2}h}\geqslant1. \tag{8}$$

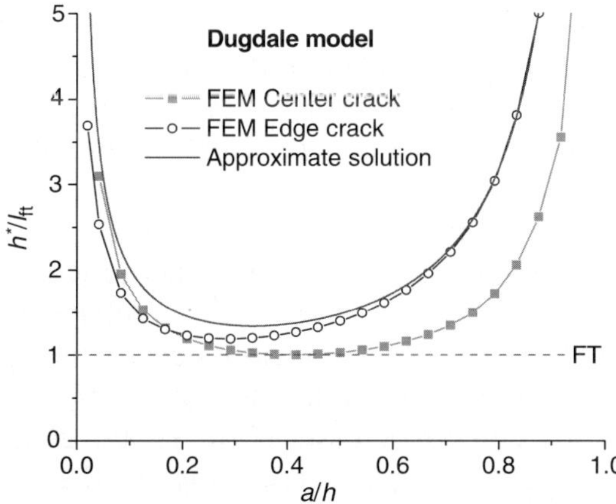

Figure 5. The normalized critical size for a strip of width *2h* to tolerate a center crack of length *2a* or two symmetric double edge cracks each of length *a*. The minimum value of such curves corresponds to the critical size for the strip to tolerate cracks of all sizes. The calculations are based on the Dugdale model (Gao and Chen, 2005).

Strips thinner than the critical thickness given in Equation (7) are predicted to tolerate center and edge cracks of all sizes. A mineral crystal with characteristic dimension satisfying Equation (8) has the intrinsic capability to tolerate crack-like flaws of all sizes. Taking the equality in Equation (8) yields the critical condition

$$\Lambda_{ft} = \frac{\Gamma E}{S^2 h} = 1,\tag{9}$$

which will be applied later to design a hierarchical material with multiscale flaw tolerance using a bottom-up approach.

We note that similar length scales with various physical interpretations to $\ell_{ft} = \Gamma E/S^2$ have appeared in the classical Dugdale–Barenblatt model (Dugdale, 1960; Barenblatt, 1985), and large scale yielding or bridging models applied to mechanics of earthquake rupture (Rice, 1980), notch insensitivity (Bilby et al., 1963; Suo et al., 1993), fracture size effects in concrete (Bazant, 1976; Hillerborg et al., 1976; Carpinteri, 1982, 1997; Bazant and Cedolin, 1983; Bazant and Planas, 1998) and in sea ice (Mulmule and Dempsy, 2000), fiber bridging in composites (Bao and Suo, 1992; Cox and Marshall, 1994; Massabo and Cox, 1999), as well as in computational cohesive models (Needleman, 1987; Tveergaard and Hutchinson, 1992; Xu and Needleman, 1994; Camacho and Ortiz, 1996). Such length scale also appeared in dynamic failure phenomena such as the minimum fragment sizes in comminuting of glassy materials (Kendall, 1978; Karihaloo, 1979) and in dynamic fragmentation (Drugan, 2001). The concept of flaw tolerance in biological systems can thus be closely related to various established phenomena and concepts in fracture mechanics. It is interesting to compare the concept of flaw tolerance to that of large scale yielding or bridging; in the latter case, one compares the size of a crack with that of the yield or bridging zone near the crack tip; if these sizes are comparable, the fracture strength is predicted to approach the yield strength of the material. In the flaw tolerance conditions expressed in Equations (7)–(9), it is not the size of a flaw but the size of a material which is linked to a critical length scale. The focus of flaw tolerance is thus on the intrinsic capability of a material to tolerate crack-like flaws below a critical structural size. The flaw tolerance number defined in Equation (9) corresponds to twice the square of Carpinteri's (1982) brittleness number. The flaw tolerance solution corresponds to optimizing the strength of a material taking into account the possible existence of crack-like flaws. From this point of view, Equations (7)–(9) indicate that strength optimization in the presence of crack-like flaws can be achieved simply by reducing the characteristic structural size to a critical value.

For brittle materials, the fracture energy is twice the surface energy ($\Gamma = 2\gamma$) and the limiting strength is sometimes also called the theoretical strength of the material ($S = \sigma_{th}$). For biominerals, taking $\gamma = 1\,\text{J/m}^2$, $E = 100\,\text{GPa}$ and $\sigma_{th} = E/30$ leads to an estimate of ℓ_{ft} around $18\,\text{nm}$.[2] This result has led to the postulate that the nanometer size of mineral crystals in bone is selected to ensure optimal fracture strength and maximum tolerance of crack-like flaws (Gao et al., 2003). The Young's modulus of various brittle materials can vary, depending on the atomic structure and the

---

[2] A previous estimate of the critical size for flaw tolerance h_{cr} based on a thumbnail crack with depth equal to half of the plate thickness is (Gao et al., 2003) $h_{cr} = \pi \gamma E/\sigma_{th}^2 = \pi \ell_{ft}/2$. This gives an estimate of h_{cr} around $30\,\text{nm}$.

purity of the material, in quite a wide range between a few gigapascals to around a thousand gigapascals. For example, the Young's modulus of $CaCO_3$ can be as low as 50 GPa and that of diamond can reach as high as 1200 GPa. Typical estimates of the theoretical strength can range between 1% and 10% of the Young's modulus. If we take $\gamma = 1 \text{ J/m}^2$, $E = 50 - 1000 \text{ GPa}$, and $\sigma_{th} = (1\% - 10\%)E$, the characteristic length ℓ_{ft} for flaw tolerance can be estimated to vary in the range

$$\ell_{ft} \approx 2\text{Å}-400 \text{ nm}. \tag{10}$$

Therefore, the critical length of flaw tolerance can vary from near atomic scale for materials like diamond to a few hundred nanometers for biominerals. Recent studies on mechanical strength of single wall carbon nanotubes and SiC and beta-Si_3N_4 nano-whiskers (Pugno and Ruoff, 2004) seem to show that nano-defects in these nearly perfect nanostructures can strongly affect their strength. A possible explanation is that the characteristic length of flaw tolerance for these materials is on the atomic scale so that the resulting nanostructures remain sensitive to the presence of nano-defects. For biological materials, the inorganic components in bone, teeth and sea shells have significantly lower Young's modulus and lower strength, hence become insensitive to crack-like flaws at relatively larger size scales (on the order of a few to a few hundred nanometers). In other words, materials with nearly perfect atomic structures and high Young's modulus may remain sensitive to crack-like flaws even down to atomic scales while materials with lower Young's modulus and less perfect lattice structures or less pure chemical constituents are likely to approach their intrinsic limiting strength at tens to hundreds of nanometers size scales.

In summary, a short answer to the length scale question is that the nanometer size makes normally brittle mineral crystals insensitive to crack-like flaws or flaw tolerant. The flaw tolerance corresponds to optimizing the strength of a material in the presence of crack-like flaws. This optimal state of material can be achieved simply by size reduction. The load transfer in the staggered nanostructure of bone-like materials follows a tension-shear chain, in which the tension loaded brittle mineral crystals are susceptible to brittle fracture. By designing the mineral crystals at the nanoscale, nature finds a way to render these nominally brittle materials insensitive to crack-like flaws. This design strategy, when combined with the extraordinary energy absorbing capabilities of the soft protein matrix, leads to robust biocomposites.

3. The Stiffness Question: How Does Nature Create a Stiff Composite in Spite of High Content of a Soft Matrix?

The soft matrix in bone has quite high volume fraction ($\sim 60\%$) and can be a few orders of magnitude softer than both the mineral crystals and bone. How does nature create a stiff composite in spite of such high content of a soft material? The stiffness of the staggered nanostructure of bone has been analyzed previously by Jaeger and Fratzl (2000) and Kotha et al. (2000). Here we address the stiffness question using the tension-shear chain model shown in Figure 2(b), following the analysis by Gao et al. (2003) and Ji and Gao (2004a).

Applying force equilibrium to an individual mineral crystal (Figure 2c) indicates that the tensile stress in mineral varies linearly with the distance x from the center of the platelet,

110 *H. Gao*

$$\tilde{\sigma}_{\mathrm{m}}(x) = \sigma_{\mathrm{m}}\frac{L-2|x|}{L}. \tag{11}$$

The maximum stress occurs near the center of the platelet and is related to the shear stress of protein as

$$\sigma_{\mathrm{m}} = \rho\tau_{\mathrm{p}}, \tag{12}$$

where $\rho = L/h$ is the aspect ratio of the mineral crystals. The average stress in the composite is thus

$$\sigma = \tfrac{1}{2}\Phi\sigma_m = \tfrac{1}{2}\rho\Phi\tau_{\mathrm{p}}, \tag{13}$$

where Φ denotes the volume fraction of mineral. The average strain in the composite structure obeys the kinematical relationship

$$\varepsilon = \frac{\Delta_{\mathrm{m}} + 2\varepsilon_{\mathrm{p}}h(1-\Phi)/\Phi}{L} \tag{14}$$

where Δ_{m} is the elongation of mineral crystals, ε_{p} is the shear strain in the soft matrix layers between the crystals, and h and L are the width and length of mineral crystals shown in Figure 2(a). Note that Equations (11) and (14) should be valid even under large nonlinear deformation as long as the shear strain in the soft matrix stays uniform.

The constitutive behaviors of biological materials, especially the soft matrix, are usually very complex. Within linear elastic approximations, we can write

$$\Delta_{\mathrm{m}} = \frac{\sigma_{\mathrm{m}}L}{2E_{\mathrm{m}}}, \quad \varepsilon_{\mathrm{p}} = \frac{\tau_{\mathrm{p}}}{G_{\mathrm{p}}}, \tag{15}$$

where E_{m} denotes the Young's modulus of mineral and G_{p} the shear modulus of protein. Inserting Equation (15) into Equation (14) while making use of Equation (13) yields an estimate for the composite stiffness (Gao et al., 2003) as

$$\frac{1}{E} = \frac{4(1-\Phi)}{G_{\mathrm{p}}\Phi^2\rho^2} + \frac{1}{\Phi E_{\mathrm{m}}}. \tag{16}$$

Normally, the modulus of protein G_{p} can be much smaller (up to 3 orders of magnitude) than that of mineral E_{m}. For moderate aspect ratios, the second term in Equation (16) can be dropped and we obtain a simple scaling law

$$E \propto G_{\mathrm{p}}\rho^2, \tag{17}$$

which immediately shows that, in spite of the high volume fraction of the soft matrix, the biocomposite can always be stiffened via a sufficiently large aspect ratio. These calculations indicate that the large aspect ratio of mineral crystals may have been evolved to compensate for the low modulus of the protein matrix since it is the combination $G_{\mathrm{p}}\rho^2$ which appears in the expression for the composite stiffness.

In summary, a short answer to the stiffness question is that the large aspect ratio of mineral crystals in bone can fully compensate the softness of the matrix. The simple scaling law of Equation (17) shows that an aspect ratio of 30–40 would provide a magnification of 3 orders of magnitude over the stiffness of protein and bring

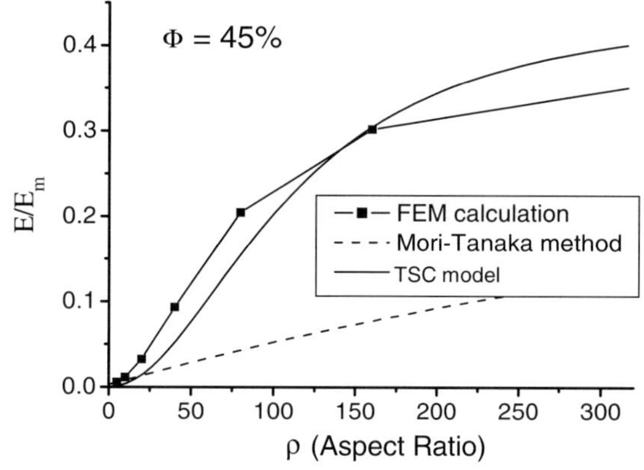

Figure 6. Comparison of the stiffness predicted by the tension-shear chain (TSC) model with finite element (FEM) calculations and the Mori–Tanaka method; E is the composite stiffness and E_m is the stiffness of mineral. The calculations were performed assuming $E_m/E_p = 1000$ and $\Phi = 45\%$ (Ji and Gao, 2004a).

the composite stiffness close to that of the mineral. Figure 6 shows that the stiffness estimate in Equation (16) based on the tension-shear chain (TSC) model agrees reasonably with results from a finite element analysis of the staggered nanostructure (Ji and Gao, 2004a). In contrast, the widely used Mori–Tanaka method (Mori and Tanaka, 1973) shows much poorer agreement with the FEM calculations, which may be attributed to the fact that the staggered alignment of crystals is not considered in the Mori–Tanaka method.

4. The toughness question: how does nature build a tough composite in spite of high content of a brittle phase

We have discussed that the staggered alignment and large aspect ratios of mineral crystals play an essential role in creating a stiff biocomposite in spite of a high volume fraction of the soft matrix. The mineral crystals thus provide the required structural rigidity for bone-like materials. However, a rigid structure by definition does not deform much and is usually brittle. How does nature build toughness into the structure?

To estimate the fracture energy of the staggered nanostructure of bone-like materials, consider a crack growing in an infinite medium made of the staggered biocomposite as shown in Figure 7. Assume that the dissipation of fracture energy during crack growth is confined to within a strip of localized deformation along the prospective crack path (Figure 7), reminiscent of the classical Dugdale model (Dugdale, 1962) of plastic yielding near a crack in a ductile sheet. In such a cohesive strip model, the fracture energy can be calculated from the integral

$$J_c = w \int \sigma(\varepsilon)\,d\varepsilon, \tag{18}$$

where w is the width of the localization strip and $\sigma(\varepsilon)$ is the stress–strain relation of the staggered biocomposite up to ultimate failure. We still use the tension-shear chain model of Figure 2(b) to evaluate the integral in Equation (18), assuming that

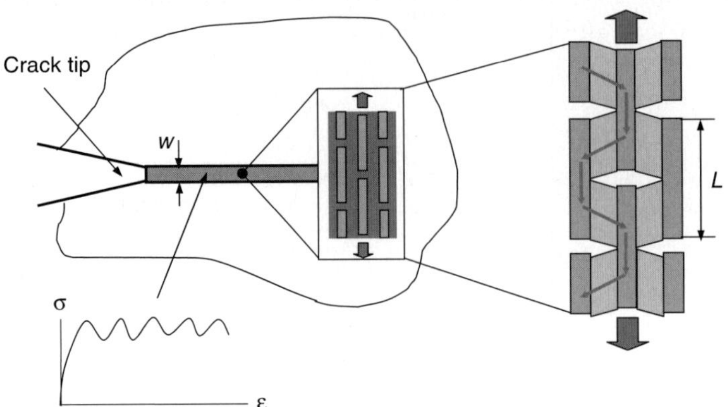

Figure 7. A Dugdale-type estimate for the fracture energy of biocomposites. The fracture energy dissipation is assumed to concentrate within a strip of localized deformation with width w. The tension–shear chain model is used to estimate the fracture energy in the localization strip. The stress–strain relation of material within the localization strip contributes to the fracture energy of the biocomposite. The hard plates are assumed to remain intact as the soft matrix undergoes large shear deformation. The width of the fracture process zone is assumed to be on the order of the length of the hard plates.

the deformation in the protein layers between the hard crystals remains uniform. At the composite level, crack propagation occurs by pulling the hard crystals out of the soft matrix. In principle, the width of the fracture localization zone w could be determined by considering the characteristic length scale associated with strain localization in the staggered biocomposite. However, such analysis would be complicated by the complex constitutive behaviour of protein. Assuming the mineral crystals are strong enough to remain intact during fracture, w should be larger than the length of the mineral crystals L,

$$w = \xi L, \quad \xi \geqslant 1. \tag{19}$$

For a conservative estimate, one can simply take $\xi = 1$ and $w = L$. The protein matrix is expected to play the dominant role in absorbing and dissipating fracture energy, in which case Equation (18) can be rewritten as (Ji and Gao, 2004a)

$$J_c = L \int \sigma d\varepsilon \approx (1 - \Phi) L \int \tau_p d\varepsilon_p = (1 - \Phi) L \Theta_p \min(S_p, S_{int}, S_m/\rho), \tag{20}$$

where

$$\int \tau_p d\varepsilon_p = \Theta_p \min(S_p, S_{int}, S_m/\rho) \tag{21}$$

denotes the deformation energy dissipated by the soft matrix per unit volume, τ_p being the shear stress in the plastically deforming soft matrix which is limited by the yield strength S_p of protein (corresponding to the stress required for domain unfolding), the protein–mineral interface strength S_{int} and the limiting strength of the mineral crystals S_m; Θ_p denotes the effective strain to which the soft matrix can deform before failure.

The fracture energy estimate in Equation (20) is simple and insightful. The suggested mechanisms of toughness enhancement are briefly discussed in the following.

It is quite obvious that the toughness of bio-composites should increase with the volume fraction of protein $(1 - \Phi)$: The more protein, the more material absorbs and dissipates fracture energy.

Baring mineral fracture, the length of the mineral crystals L sets a minimum length scale for strain localization near the crack tip: The longer the plates, the more delocalized the crack-tip deformation and the larger the fracture energy.

The effective strain Θ_p, as a measure of the deformation range of protein, is a key parameter for fracture energy of the biocomposite. The hierarchical structures of proteins are ideally suited for absorbing and dissipating fracture energy. Proteins in nacre can undergo large deformation by gradual unfolding of their domain structures (Smith et al., 1999). It can take a large amount of deformation before the primary structure of protein, the peptide backbone, is directly stretched. Thus the molecular design of proteins is ideally suited for absorbing fracture energy. If the mineral crystals are strong enough to remain intact during the deformation and $S_{\text{int}} = S_p$, then Θ_p should include not only domain unfolding of protein molecules but also slipping along the protein–mineral interface. Therefore, it will be advantageous to let the interface have the same strength as protein to maximize the deformation range of the soft matrix (plus slipping along the interface). Appropriately designed interface strength is known to play a key role in the toughness of engineering composites (Evans, 1990). In bone-like materials, nature seems to have found a clever strategy which not only leads to an optimal design of the protein–mineral interface strength for maximize deformability of the soft matrix, but also significantly raises the stress level which operates the deformation. This will be discussed further in the following.

In order to achieve maximum toughness, large deformation alone is not sufficient as it is the area under the stress–strain curve which defines the fracture energy. It is also important to enhance the stress level in the soft matrix that operates protein deformation. The nanosized super-strong mineral crystals allow the protein and the protein–mineral interface to have an opportunity to enhance their strengths without fracturing the crystals. In bone, this enhancement seems to have been enabled by the mechanism of sacrificial bonds in which Ca^{2+} ions cross-link protein peptides with negative electric charges, forming relatively strong bonds with strength up to about 30% of the covalent bonds of the peptide backbone (Thompson et al., 2001). The Ca^{2+} induced sacrificial bonds bind not only different peptide groups of protein but also protein peptides with functional groups on the surfaces of the mineral crystals, thus also providing a solution to the problem of $S_{\text{int}} = S_p$ by raising them to similar high levels. In other words, the Ca^{2+} induced sacrificial bonds not only build up a large operating stress in protein, converting entropic elasticity behaviors of biopolymers to one that resembles metal plasticity via cross-linking, but also allow protein deformation and interface slipping to occur simultaneously under similar stress levels, making it possible also to maximize the deformation range of the soft matrix. Note that optimizing fracture strength of the mineral crystals via size reduction is a prerequisite for implementing the sacrificial bonds strategy. Otherwise, brittle fracture in mineral crystals may occur before the stress in the soft matrix can be raised to the desired level.

Therefore, from the toughness point of view, the optimal design is to raise the strength of the protein–mineral interface, together with that of the soft matrix, to a highest possible level without breaking the mineral crystals. For mineral crystals to remain intact while protein deforms, one should have the condition

$$S_p = S_{int} \leqslant S_m/\rho. \tag{22}$$

Equation (22) also shows why it is necessary to optimize the strength of mineral crystals via size reduction. Since large aspect ratios $\rho = 30$–40 are needed to compensate for the softness of the matrix, the mineral strength must exceed

$$S_m \geqslant (30 \sim 40) S_p \tag{23}$$

in order for the crystals to stay intact during protein deformation. The organic matrix and the inorganic mineral crystals are locally polarized and the interface strength is dominated by electrostatic interactions. Taking S_p to be around $(20$–$50)$ MPa, we can immediately estimate from Equation (23) that the mineral strength S_m needs to be on GPa levels, which is near the theoretical strength of the mineral. The analysis explained, from a different perspective, why it is important to have the size of mineral platelets chosen at the nanoscale: High strength, flaw tolerant mineral crystals play an essential role in maintaining a significant stress in the soft matrix to operate energy dissipating deformation in the protein and along the protein–mineral interface.

We can also understand the toughening mechanism of biocomposites from the following point of view. On the composite level, the structure is stiff and deformation remains small. In order to fully utilize the large deformation capability of the soft matrix, a strain amplification mechanism is provided by the staggered biocomposite structure. To see this, we rewrite Equation (14) into the following form:

$$\varepsilon_p = \frac{\Phi}{2(1-\Phi)}\rho\,(\varepsilon - \Delta_m/L) \propto \rho\varepsilon, \tag{24}$$

which immediately shows that the strain in the soft matrix is magnified over the composite strain by the aspect ratio of the mineral crystals, allowing the protein to fully deform and dissipate energy at the microstructural level without inducing large deformation on the composite level.

In summary, the toughness question has been analyzed as follows. The hierarchical domain structures of protein molecules naturally endow the protein-rich soft matrix with an intrinsic capability to deform to very large strains.[3] The staggered biocomposite structure provides a strain amplification mechanism by which the protein-rich soft matrix undergoes large deformation while the composite only deforms at a small strain. The formation of relatively strong secondary bonds via Ca^{2+} ions in bone provides a strategy that not only leads to an optimal design of the protein–mineral interface strength to maximize the deformation range of the soft matrix, but also raises the stress level that operates protein deformation in the soft matrix, resulting in maximum toughness. A pre-requisite to these toughening mechanisms in the soft matrix is that the mineral crystals must be designed on the nanometer length scale to prevent brittle fracture of mineral.

---

[3]The specific structures of proteins also allow the soft matrix to recover from moderate deformation levels by reverting to their natural folding state under proper chemical conditions over a period of time, thus giving rise to a self-repairing mechanism for minor damages. For severe damages, regeneration cells (osteoclasts and osteoblasts) are recruited to replace damaged bone tissues with fresh and healthy materials.

5. The strength question: how does nature balance the widely different strength levels of protein and mineral?

The deformation mechanisms of protein and mineral are quite different. While the deformation of mineral crystals is mostly elastic until failure at their limiting strength, proteins are expected to undergo very large deformation via unfolding of their hierarchical molecular structures formed from secondary bonding such as van der Waals, hydrogen and ionic (e.g., Ca^{2+}) bonds. Therefore, the strength of protein is expected to be much smaller (but with a much larger range of deformation) compared to that of mineral. How does nature balance the large strength differences between mineral and protein? Figure 2(b) shows that the mineral crystals are primarily subjected to tension while the protein matrix mainly under shear. The strength of the staggered biocomposite structure thus depends on which phase, mineral or protein, fails first. It follows from Equation (13) that

$$S = \min\left(\frac{\Phi \rho S_{p}}{2}, \frac{\Phi S_{m}}{2}\right), \tag{25}$$

where S_{p} denotes the strength of the protein[4] (or the protein–mineral interface) and S_{m} the tensile strength of the mineral crystals. If the mineral crystals fail first, then $S = \Phi S_{m}/2$, corresponding to mineral plates reaching their limiting strength S_{m}. On the other hand, if the soft matrix fails first, then $S = \Phi \rho S_{p}/2$, corresponding to the soft matrix reaching its limiting strength S_{p}.

From the toughness point of view, the optimal design condition in Equation (22) indicates that the strengths of the protein and the protein–mineral interface should be raised to the same level bounded by the fracture strength of the mineral crystals divided by the aspect ratio. From the stiffness point of view, the aspect ratio ρ should be as large as possible. From the flaw tolerance point of view, the strength of the mineral crystals could be raised near the theoretical strength of mineral via size reduction. The optimal design exploring the limits of all these requirements is thus

$$S_{p} = S_{int} = S_{m}/\rho. \tag{26}$$

Therefore, nature uses the aspect ratio of mineral crystals to strike a balance between the widely different strength levels of protein and mineral. In this way, the softer and weaker protein matrix would be subjected to proportionally smaller stress. Complementary to the mechanism of strain amplification in protein discussed in the previous section, the stress in the mineral crystals can be said to have been magnified with respect to that in the soft matrix by the aspect ratio. By this design, an optimal balance is achieved for deformation in the organic (large deformation, small stress) and inorganic (small deformation, large stress) components of the biocomposite so that both are utilized to their maximum potential.

---

[4] For simplicity, in this paper we do not distinguish between the yield and fracture strengths of the soft matrix. In principle, they could have quite different values.

6. The Optimization Question: is Bone Nanostructure an Optimization?

Although the hierarchical structures of bone and bone-like materials are extremely complex, it is interesting that they have all adopted essentially the same staggered nanostructure at the most elementary level. Why does nature design the basic building blocks in this form? Is bone nanostructure a structural optimization? If so, what is the objective function? A first study of these questions has been conducted by Guo and Gao (2005). Here we briefly summarize the main result of this study and alert the reader to focused papers on this subject in the near future.

The problem considered by Guo and Gao (2005) is set up within a topology optimization framework as finding the optimal distribution of a hard and brittle inorganic phase relative to a soft and ductile organic matrix in a representative unit cell Y of the ultrastructure of bone (Figure 8a). The optimization problem is formulated based on the hypothesis that bone-like materials are designed with the objective to simultaneously optimize stiffness and toughness for optimal structural support and flaw tolerance. In accordance with this hypothesis, we consider topological optimization of the representative unit cell under uniaxial tension with the following objective function

$$f = -\left(\frac{E^{\text{eff}}}{E^0}\right)\left(\frac{\Gamma}{\Gamma^0}\right) \tag{27}$$

where E^{eff} is the composite stiffness in the direction of loading and Γ is the deformation energy absorbed by the soft phase before any material point in the unit cell reaches its ultimate failure strength; E^0 and Γ^0 are two normalization parameters for the objective function.

The material distribution (topology) in the unit cell is represented by an indicator function $\rho(x)$ which only takes the values of 0 or 1: $\rho(x) = 1$ indicates that the point x is occupied by the hard material and $\rho(x) = 0$ indicates occupation by the soft material. The optimization is conducted under fixed mineral volume fraction and the constraint that, during the process of deformation, the effective strain of any material point in the unit cell should not exceed a critical value set to be 1% for the hard phase and 150% for the soft phase. In the actual calculation, the unit cell is discretized into a number of blocks (Figure 8a) each of which is assumed to be occupied by either the hard or the soft material.

Formally, the optimization problem is formulated as follows:

Find $\rho(x) \in L^\infty(Y)$, $\alpha_{\bar{u}} \in R$

$$\min f = -\left(\frac{E^{\text{eff}}}{E^0}\frac{\Gamma}{\Gamma^0}\right), \tag{28}$$

S.t.

$$\iint_Y C(x)\varepsilon(u(x)) : \varepsilon(v(x)) \mathrm{d}Y = 0 \text{ in } Y \text{ for every } v \in U_{\text{ad}}, \tag{29}$$

$$\iint_Y \rho(x)\mathrm{d}Y = \overline{V} \quad \rho(x) \in \{0, 1\}, \tag{30}$$

$$u = \bar{u} = \alpha_{\bar{u}} u^0 \quad \text{on } \partial Y_{\bar{u}}, \tag{31}$$

$$\varepsilon^{\text{eq}}(x) = \sqrt{\varepsilon(u(x)) : \varepsilon(u(x))} \leqslant \rho(x)\varepsilon_{\text{f}}^{\text{m}} + [1 - \rho(x)]\varepsilon_{\text{f}}^{\text{p}} \quad \text{in } Y, \tag{32}$$

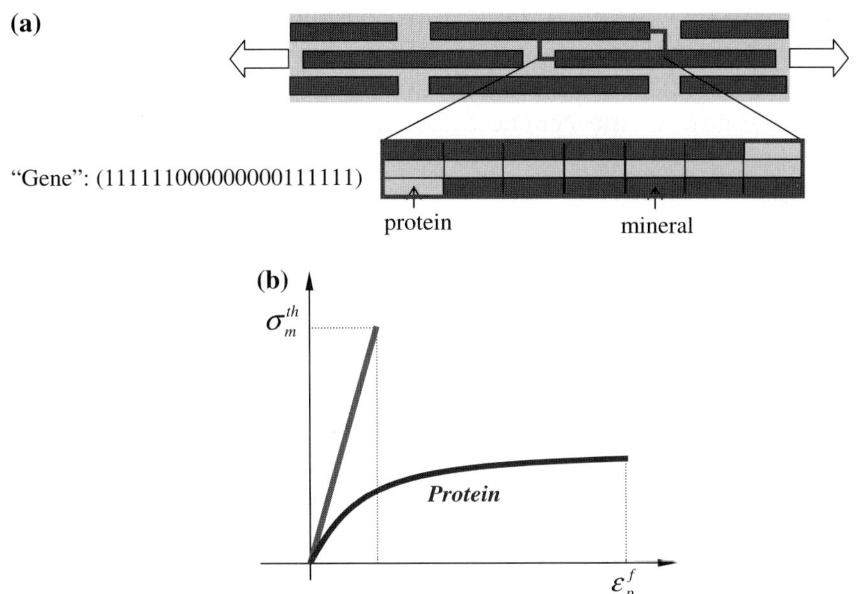

Figure 8. The nanostructure of bone as structural optimization. (a) Structural optimization in the representative unit cell of the nanostructure of bone. The unit cell is discretized into a number of material elements which is occupied by either the hard (red) or the soft (blue) material. The optimization is conducted under fixed volume fraction of hard (or soft) material and the constraint that the local stress or strain does not exceed the corresponding critical values in the hard or soft materials. In the genetic algorithm, the topology is represented by a binary "gene" with 1 indicating mineral and 0 indicating protein. (b) The simplified constitutive laws of hard and soft phases. The hard phase is modelled as linear elastic until breaking at its theoretical strength. The soft phase is modelled as a plastic material with no elastic rebound upon unloading.

where $C(x) = \rho(x)C^m + [1 - \rho(x)]C^P$ is the fourth order elasticity tensor at material point x, expressed as an interpolation between the elasticity tensors C^m of mineral and C^P of protein; $\varepsilon^{eq}(x)$ is the effective strain at x; ε_f^m and ε_f^P denote the failure strain of mineral and protein, respectively; $u^0 = (1, 0)^T$ is the displacement vector indicating that the structure is loaded in the x_1 direction; $\alpha_{\bar{u}}$ is the displacement magnitude of the applied loading; U_{ad} denotes the space of admissible test functions in the weak form of equilibrium equation (29); $\overline{V}$ is the given amount of hard material in the unit cell. For simplicity, the problem has so far been considered only in 2-D. Extension to 3-D is in principle straightforward.

The assumed constitutive behaviors are depicted in Figure 8(b). The hard material remains linear elastic until failure at 1% of elastic strain. The constitutive law of the soft phase is modeled as a plastic material with stress-strain law $\sigma_e = \sigma_f^P(1 - e^{-\gamma \varepsilon_e})$ where σ_f^P, σ_e and ε_e are yield stress, effective stress and effective strain, respectively. The soft phase is assumed to be a perfect energy absorber in the sense that all deformation energy is dissipated with no elastic rebound upon unloading. The constitutive law of the soft phase is $\sigma_{ij} = [f(\varepsilon_e)/\varepsilon_e](^4I_{ijkl})\varepsilon_{kl}$ where $^4I_{ijkl}$ is the fourth order identity tensor. Nonlinear finite element method is used to obtain the mechanical response of the structure.

The proposed optimization problem can be solved by different structural optimization methods. In the present study of the structure of bone-like materials, it seems also interesting to apply genetic algorithm (e.g., Goldberg, 1989) to solve the optimization problem. Genetic algorithm (GA) is an optimization method mimicking the genetic mechanisms of Darwinian selection and evolution. By repeatedly modifying a population of solutions, the genetic algorithm selects individuals from the current population to be parents and uses them to produce the offspring in the next generation. The population gradually "evolves" towards an optimal solution over a number of generations. The basic ingredients of GA can be briefly summarized as follows. First, an initial population is created with a number of randomly generated individuals. The current generation is then used to create the next generation by the following steps: (1) each member of the current population is ranked according to its fitness value; (2) parents are selected; (3) Children are created by making random changes to a single parent (mutation), by combining the genes of a pair of parents (crossover), and by allowing the best solutions in the current generation to automatically survive to the next generation (elite selection); (4) The current population is replaced with the next generation. The algorithm stops when one of the stopping criteria is met.

For our problem, the nanostructural optimization of bone is regarded as a "natural evolution" process driven by Darwin's selection principle. Every solution is coded by a binary "gene" which, as shown in Figure 8(a), represents a specific material distribution (1: mineral, 0: protein). A displacement loading factor is also coded in the gene (not shown in the example code in Figure 8a). In the present work, a binary code with $m+9$ bits is used where the first m bits represent material distribution in m discretized elements and the last 9 bits represent the value of the displacement loading factor whose value varies in [0.0, 5.1] with a resolution of 0.1. The constraints are enforced in a "hard kill" way, meaning that if a solution in the population violates the constraints, then its fitness value is set to minus infinity. The essential ingredients of adopted genetic algorithm are illustrated by a flow chart shown in Figure 9.

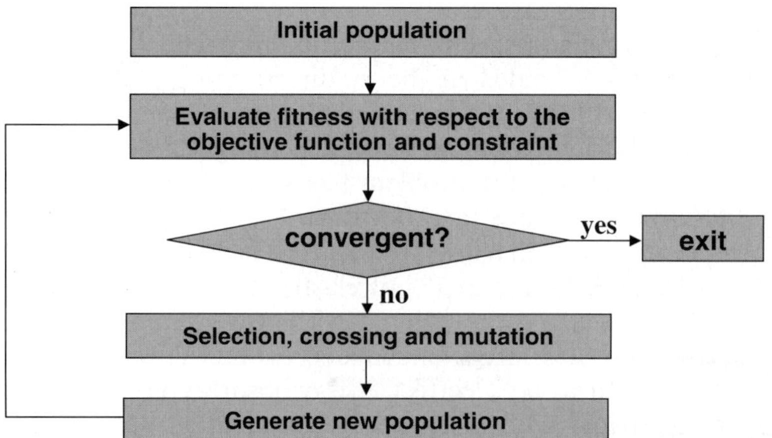

Figure 9. An illustrative flow chart of the genetic algorithm (GA) adopted in the study of bone nanostructure (Guo and Gao, 2005). GA is based on the genetic mechanism of Darwinian evolution. The fitness of each successive generation is evaluated and a number of selection strategies (elite selection, mutation, crossing) are used to generate the next generation until an optimal value of the fitness is reached.

The representative unit cell is taken to be a rectangular block with an aspect ratio of 10:1. The Young's modulus of mineral and protein are taken to be 100 GPa and 100 MPa, respectively. The effective failure strain is 1% for mineral and 150% for protein. The yield stress for protein is taken to be $\sigma_f^P = 20$ MPa, corresponding to $\gamma = 5$; E^0 and Γ^0 in the objective function are $E^0 = E^M \overline{V}/|Y|$ and $\Gamma^0 = \sigma_f^P \varepsilon_f^P \times lenx \times leny/4$. The volume fraction of mineral is taken to be 57%. The prescribed displacements are imposed on the lateral boundary of the unit cell in the horizontal direction to simulate the pulling-out deformation in front of a propagating crack.

The parameters used in GA are as follows: the initial population size is 150; the mutation rate is 0.09; the crossover fraction is 0.8; the elite count in each generation is 15. A maximum generation number (500) is used to terminate the optimization process.

Figure 10 shows selected snapshots of the topological evolution of a 7×3 discretized representative unit cell driven by the GA algorithm. The topological configurations in the initial population are randomly generated. The staggered structure is obtained and remains stable after about 80 generations of GA evolution. To investigate the effect of discretization on the optimization, we performed the same GA evolution also for 14×3 and 7×6 discretized representative cells. Figure 11 shows the selected snapshots of topological evolution from the 14×3 discretization. The staggered structure is again obtained, but the result shows non-uniform longitudinal spacing between neighbouring hard plates. Careful examination of calculations indicates that this is not a numerical error; the non-uniform arrangement of platelets can further increase the longitudinal stiffness in comparison with the usually assumed configuration of uniform spacing. Figure 12 shows the evolution of the 7×6

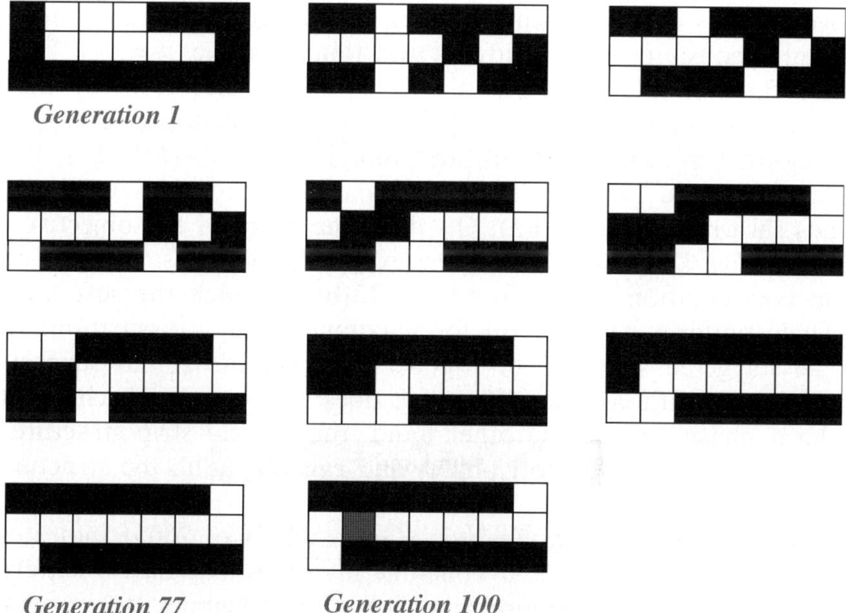

Figure 10. Selected snapshots of topological structural evolution in a 7×3 discretized unit cell. The staggered nanostructure is obtained and remains stable after about 80 generations of GA evolution. The red colored block indicates the position where the soft matrix first reaches its failure strain.

120   *H. Gao*

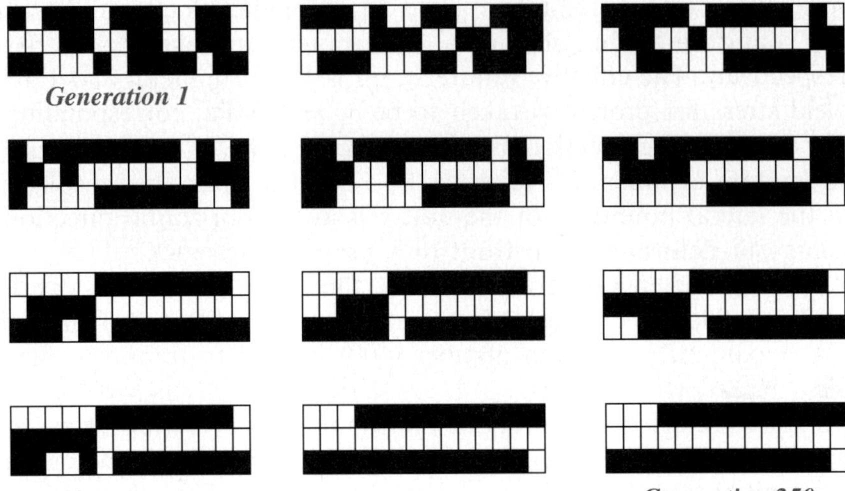

Figure 11. Selected snapshots of topological structural evolution in a 14×3 discretized unit cell. The staggered nanostructure is obtained and remains stable after about 200 generations of GA evolution. The results show that the longitudinal spacing between mineral platelets in the optimized structure is not uniform.

discretized unit cell. In all cases, the staggered structure is obtained and remains stable after a number of generations. We have also considered more finely discretized unit cells (e.g., 14×6 and 28×12) and found essentially the same results.

The above analysis indicates that the staggered structure emerges when the objective is to simultaneously optimize stiffness and toughness, i.e. $f = -(E^{\mathrm{eff}}/E^0)(\Gamma/\Gamma^0)$. The objective function plays an essential role in the final form of the structure. Different structures would be expected if the objective function is altered. For example, if the objective is to optimize stiffness alone, i.e. $f = -(E^{\mathrm{eff}}/E^0)$, the expected optimal structure would have the horizontal parallel strip configuration (the Voigt upper bound) shown in Figure 13(a). On the other hand, if the objective is to optimize toughness only, i.e. $f = -(\Gamma/\Gamma^0)$, the expected optimal structure would be the vertical strip configuration shown in Figure 13(b) in which the soft matrix undergoes completely uniform deformation for maximum energy dissipation. In comparison with the staggered structure of Figure 13(c), the horizontal strip structure in Figure 13(a) is too brittle as the soft matrix does not have much chance to deform before the hard phase fail; on the other hand, the vertical strip structure in Figure 13(b) is too soft (as soft as the soft phase) and can not fulfill the structural support function of bone.

In summary, our preliminary analysis (Guo and Gao, 2005) indicates that the staggered nanostructure of bone and bone-like materials may have been evolved with an objective to simultaneously optimize stiffness and toughness for mechanical support and flaw tolerance. More sophisticated optimization models should be developed for further study on the hierarchical structures of bone as well as other biological systems.

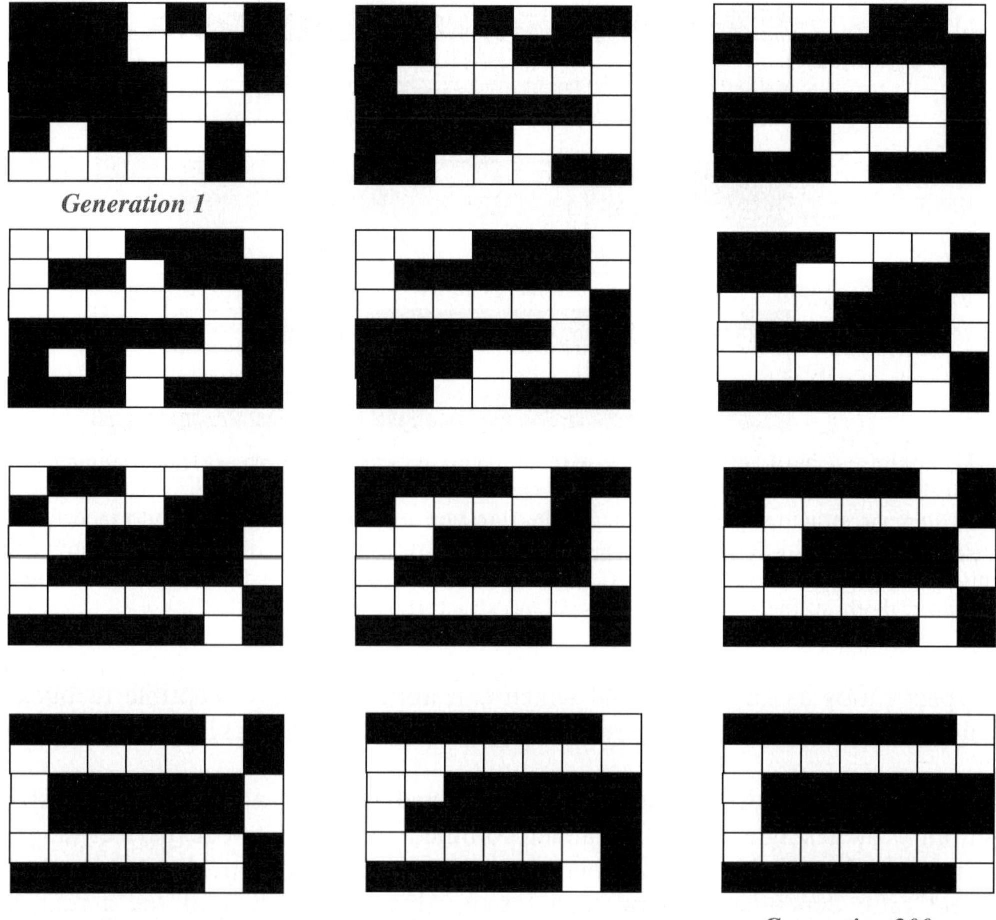

Figure 12. Selected snapshots of topological structural evolution in a 7×6 discretized unit cell. The staggered nanostructure is obtained and remains stable after about 200 generations of GA evolution.

7. The buckling question: how does nature prevent slender mineral crystals from buckling under compression?

During the 18th century, Swiss mathematician Léonard Euler found that a slender elastic rod under compression becomes unstable at a critical load (Figure 14a). For a rod with a square cross-section, the critical stress for buckling is

$$\sigma_{cr} = \frac{\pi^2 E}{12\rho^2}, \tag{33}$$

where E is the Young's modulus and ρ is the aspect ratio defined as the length divided by the thickness of the rod. Below this critical stress, the rod remains straight and is stable with respect to lateral perturbations. Above the buckling stress, the rod becomes unstable and tends to buckle into a curved profile which may subsequently lead to plastic deformation or fracture. The buckling stress decreases with increasing aspect ratio ρ. The buckling mode, as indicated in Figure 14(a), remains at the lowest wavenumber and is independent of the aspect ratio. The mineral crystals in bone

Figure 13. Optimized hard–soft structures with different objective functions. (a) The columnar structure parallel to the direction of loading is expected if the objective is to optimize the stiffness alone. (b) The columnar structure perpendicular to the direction of loading results in uniform deformation in the soft matrix and is expected to be the optimal structure if the objective is to optimize the amount of energy absorption. (c) The staggered structure is expected to be optimal if the objective is to optimize both stiffness and toughness of the structure.

have aspect ratios as large as 30–40 which can appear to be susceptible to buckling. How does nature prevent the slender mineral crystals from buckling under compression?

A first analysis of this question was conducted by Ji et al. (2004a) who studied the stability of a single mineral platelet confined in an otherwise perfect staggered nanostructure. It was found that there exists a transition of buckling strength from an aspect ratio dependent regime to a lower value independent of the aspect ratio. Interestingly, typical values of the aspect ratio of mineral platelets in bone and nacre fall in the aspect ratio independent region, which may be important from the point of view of structure robustness as the behavior of biocomposite should not depend sensitively on small variations (defects) in crystal size and shape.

We briefly summarize the main results from the analysis of Ji et al. (2004a). For a perturbed mineral platelet embedded in an otherwise perfect staggered structure, the buckling is found to occur at a critical wavenumber m which increases with the mineral aspect ratio. For large aspect ratios, the critical buckling stress can be expressed as

$$\sigma_{\mathrm{cr}} = E_{\mathrm{p}} \left[\frac{\pi^2 m^2}{12} \frac{E_{\mathrm{m}} \Phi}{E_{\mathrm{p}} \rho^2 (1-\nu^2)} + \frac{2}{m^2 \pi^2} \frac{\rho^2 \Phi^2}{1-\Phi} \right], \tag{34}$$

where E_{p} denotes the Young's modulus of protein. Minimizing Equation (34) with respect to m yields the critical buckling mode,

$$\hat{m} \approx \left(\frac{24\Phi E_{\mathrm{p}}(1-\nu^2)}{\pi^4 E_{\mathrm{m}}(1-\Phi)} \right)^{1/4} \rho. \tag{35}$$

Therefore, in the case of $m \gg 1$, the buckling mode increases proportionally with the aspect ratio of mineral (Figure 14b). This is in contrast to classical Euler buckling in

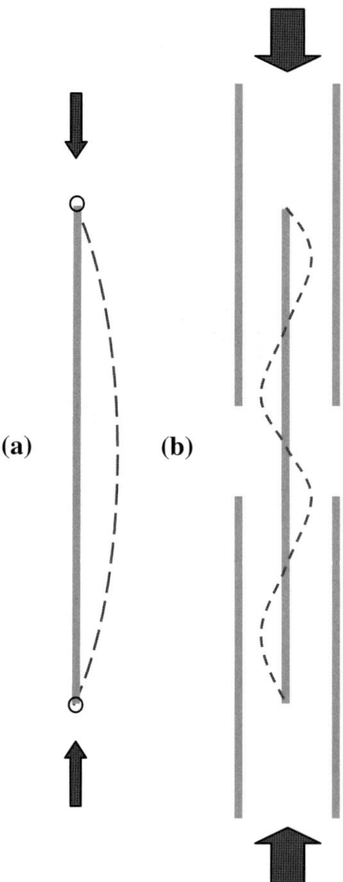

Figure 14. Buckling modes of a slender mineral plate when (a) free standing and (b) confined in a staggered hard-soft structure. The free-standing plate undergoes Euler buckling while the confined plate is coerced to buckle at higher mode due to the confinement of the protein matrix.

which the buckling mode is independent of the aspect ratio of the structure. The corresponding wavelength of the critical buckling mode is proportional to the thickness of the protein channel h_p according to

$$\lambda = \frac{L}{m} \approx \left(\frac{\pi^4 E_\mathrm{M} \Phi_\mathrm{M}^3}{24 E_\mathrm{P}(1-\nu^2)(1-\Phi_\mathrm{M})^3} \right)^{1/4} h_\mathrm{P}. \tag{36}$$

Equation (36) indicates that the structural confinement plays a critical role in determining the buckling mode of the embedded mineral platelets. For large aspect ratios, substitution of Equation (35) into Equation (34) gives the critical buckling stress of a confined mineral platelet as

$$\hat{\sigma}_\mathrm{cr} = \sqrt{\frac{2 E_\mathrm{m} E_\mathrm{p} \Phi^3}{3(1-\nu^2)(1-\Phi)}}. \tag{37}$$

In contrast to Euler buckling where the critical load decreases with increasing aspect ratio, the critical buckling stress in Equation (37) is independent of the aspect ratio

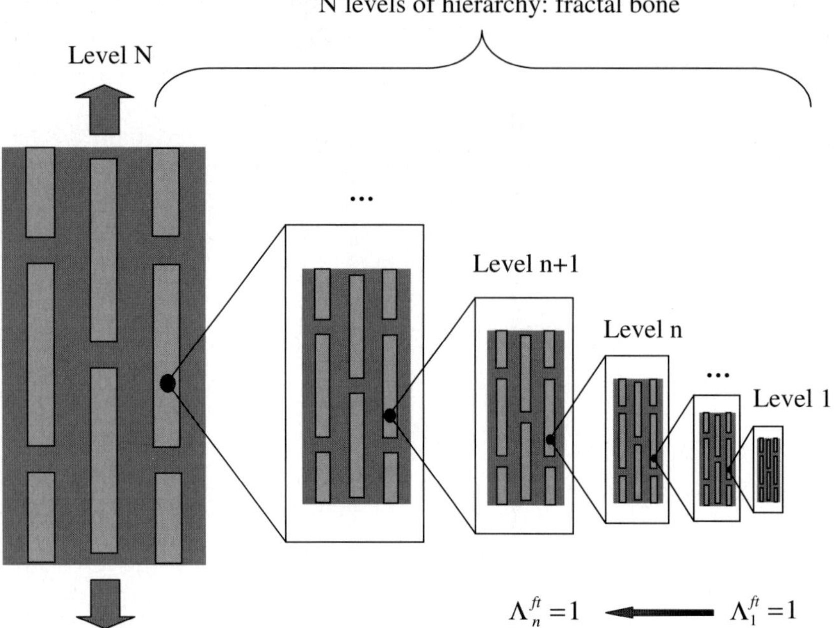

Figure 15. The N-level hierarchical structures of a fractal bone. Every level of structure is similar to the elementary structure of bone and nacre, with a slender hard phase aligned in a parallel staggered pattern in a soft matrix. The hard phase at the $(n+1)$th level is made of the hard–soft microstructure at the nth level. The principle of flaw tolerance is used to determine the characteristic size of all levels using a bottom up approach.

of mineral platelets, giving a lower threshold value below which the mineral platelets will never buckle no matter how large the aspect ratio is. Note that Equation (37) is not the compressive strength on the composite level. The latter requires further study on the cooperative buckling behaviors in the hierarchical structures of bone under compressive loads.

A short answer to the buckling question is that the protein matrix, despite of its very low elastic modulus in comparison with mineral, is nevertheless sufficient for stabilizing individual mineral crystals against localized buckling. The next challenge is to study the cooperative buckling behaviors in the hierarchical structures of bone. In particular, the concentric arrangement of mineral platelets in a mineralized collagen fibril may play an important role in stabilizing the coordinated buckling of biocomposites.

8.   The hierarchy question: what is the role of structural hierarchy?

In order to gain some preliminary understanding on the question of structural hierarchy, we consider a hypothetical hierarchical material with multiple levels of self-similar structures mimicking the staggered nanostructure of bone, as shown in Figure 15. The resulting "fractal bone" is still made of mineral and protein at some fixed volume fraction, but the material is now distributed in a highly non-homogeneous way to form a hierarchical material with different properties at different length scales. Instead of just one or two levels of structures as in most of the man-made composites, the fractal bone can have a large number of structural hierarchies and at each hierarchical level exhibits the same structure of slender hard plates arranged in a parallel staggered pattern in a soft matrix (Figure 15), similar to the nanostructure of bone.

The study described in Section 5 indicates that the staggered nanostructure is a result of optimization with respect to stiffness and toughness. It is hoped that the fractal bone would allow the optimized nanostructural properties to be extended toward macroscopic length scales. In the fractal bone model, the hard plates at each level are assumed to be made of a staggered hard-soft microstructure from one level below. The total number of hierarchical levels is N. At each level of hierarchy, the roles of the hard and the soft materials are similar to those that have been discussed for the nanostructure of bone, i.e. the slender hard plates provide structural rigidity while the soft matrix absorbs and dissipates fracture energy associated with crack-like flaws in the size range of the corresponding hierarchical level. The same flaw tolerance criterion is applied to all hierarchical levels using a bottom-up design approach. First, the characteristic size scale of the lowest level of structure, i.e. the nanostructure, is determined as in Section 2. Then the properties at the next level above are determined from the current level of structures, and the criterion of flaw tolerance is applied to determine the characteristic size at the next level. The same process of design is repeated until all N levels of structural hierarchy are determined.

Figure 16 illustrates the adopted bottom-up design approach. The properties of level n are used to determine properties at level $(n+1)$. The principle of flaw tolerance is used in each level to ensure robustness of the structure against crack-like flaws.

At hierarchical level n, the geometrical parameters are the thickness h_n and the length ℓ_n of the hard plates with aspect ratio $\rho_n = \ell_n/h_n \gg 1$. The volume fraction of the hard phase is denoted as φ_n. The total mineral volume fraction

$$\Phi = \varphi_1 \varphi_2 \cdots \varphi_N = \prod_1^N \varphi_n \tag{38}$$

is assumed to be a fixed system parameter. The properties[5] at level n are Young's modulus E_n, strength S_n and fracture energy Γ_n. In the bottom-up approach shown in Figure 16, the structures are designed in the following sequence

$$h_1 \to h_2 \to \cdots \to h_N = H. \tag{39}$$

8.1. MULTI-LEVEL STIFFNESSES OF THE FRACTAL BONE

The simple formula for the Young's modulus of the staggered nanostructure in Equation (16) can be generalized to the fractal bone as

$$\frac{1}{E_{n+1}} = \frac{4(1-\varphi_n)}{G_n^p \varphi_n^2 \rho_n^2} + \frac{1}{\varphi_n E_n}, \tag{40}$$

---

[5]Strictly speaking, the elastic properties of the hard plates at higher hierarchical levels are anisotropic. For simplicity, we do not consider this complication in this first simple analysis.

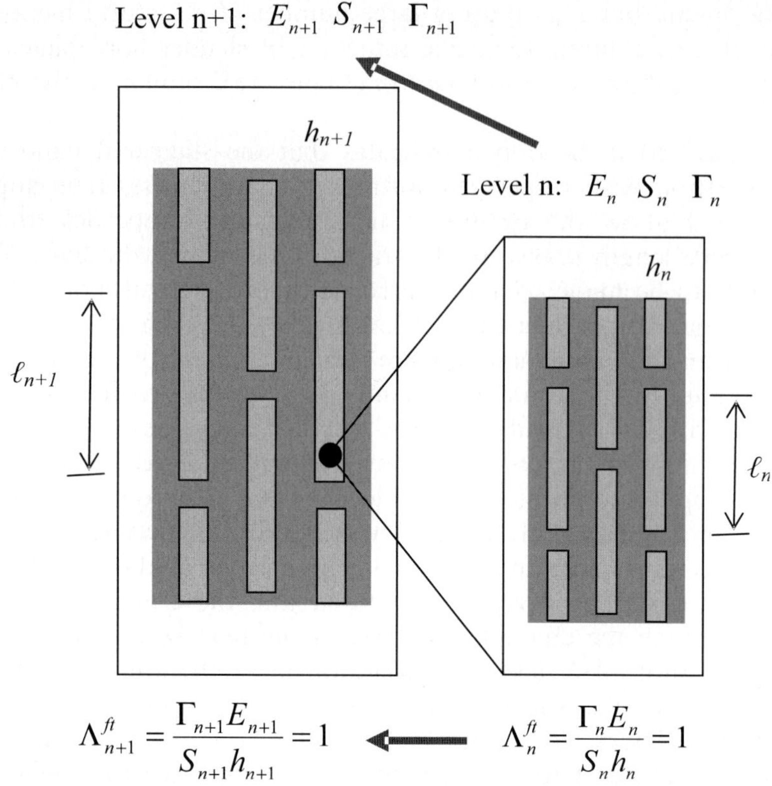

$$\Lambda_{n+1}^{ft} = \frac{\Gamma_{n+1}E_{n+1}}{S_{n+1}h_{n+1}} = 1 \quad \Longleftarrow \quad \Lambda_n^{ft} = \frac{\Gamma_n E_n}{S_n h_n} = 1$$

Figure 16. The bottom-up approach for designing the fractal bone. The properties of level n structure are used to determine those of level $(n+1)$. The principle of flaw tolerance is applied at each level, using the bottom-up design approach ($h_n \rightarrow h_{n+1}$), to determine the characteristic sizes of all levels.

where E_{n+1} is the Young's modulus at the $(n+1)$th level, G_n^{p} is the shear modulus of the soft phase at the nth level, and E_n, φ_n, ρ_n are the Young's modulus, volume fraction and aspect ratio of the hard phase at the nth level, respectively. In comparison, the Voigt upper bound of the composite stiffness at the $(n+1)$th level is

$$E_{n+1} = (1 - \varphi_n)\, E_n^{\mathrm{p}} + \varphi_n E_n \cong \varphi_n E_n, \tag{41}$$

where E_n^{p} denotes the Young's modulus of protein of the nth level. When the total volume fraction of mineral is fixed, increasing the number of hierarchy levels tends to increase φ_n, allowing E_{n+1} of Equation (40) to approach the Voigt bound of Equation (41). Increasing the number of hierarchical levels generally increases the overall stiffness of the composite.

8.2. MULTI-LEVEL STRENGTHS OF THE FRACTAL BONE

When the staggered nanostructure of bone is subjected to uniaxial tension, the mineral plates are primarily under tension with protein layers in-between transfer loads primarily via shear (Figure 2b). By means of the fractal bone design, this feature is assumed to carry over all hierarchical levels. The tensile limiting strength at the $(n+1)$th level depends on which phase (hard or soft) from the nth level fails first.

Generalizing Equation (25) to all hierarchical levels leads to

$$S_{n+1} = \min\left(\frac{\varphi_n \rho_n S_n^{\mathrm{p}}}{2}, \frac{\varphi_n S_n}{2}\right),$$
(42)

where S_n^p is the shear strength of the soft matrix (or the interface) and S_n is the tensile strength of the hard plates at the nth level. If the hard plates fail first, then $S_{n+1} = \varphi_n S_n/2$ corresponds to the tensile stress at the $(n+1)$th level when the hard plates at the nth level reach their limiting strength S_n. On the other hand, if the soft matrix fails first, then $S_{n+1} = \varphi_n \rho_n S_n^{\mathrm{p}}/2$ corresponds to the tensile stress at the $(n+1)$th level when the soft matrix at the nth level reaches its limiting strength S_n^{p}. From the energy dissipation point of view, it is important that the soft matrix undergoes large deformation and sliding before the hard plates fail in tension. An optimal design is that the soft matrix ultimately fails together with the hard plates, i.e.

$$\rho_n S_n^{\mathrm{p}} = S_n.$$
(43)

Under this condition, the hierarchical limiting strengths can be simply expressed as

$$S_{n+1} = \varphi_n S_n/2,$$
(44)

which indicates that each additional level of structural hierarchy would degrade the strength of the material by at least a factor of 2 since $0 < \varphi_n < 1$.

8.3. MULTI-LEVEL FRACTURE ENERGIES OF THE FRACTAL BONE

Fracture energies play a critical role in the optimal design of hierarchical structures. To keep discussions simple, we assume that the energy is primarily dissipated by the deformation of the soft matrix at any hierarchical level. The hard plates are assumed to remain intact during the fracture process and they would be pulled out of the soft matrix in order for a crack to propagate, as shown in Figure 17.

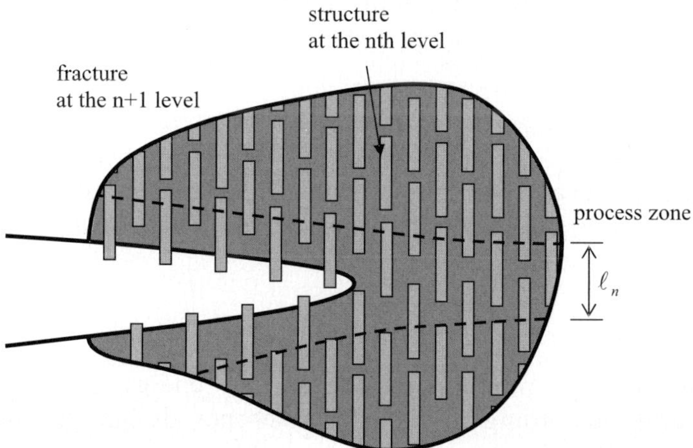

Figure 17. Pull-out of hard plates during fracture at the $(n+1)$th level of structural hierarchy. The hard plates are assumed to remain intact as the soft phase undergoes large deformation to failure. The width of the fracture process zone is assumed to be on the order of the length of hard plates.

The fracture energy associated with such a "fiber pull-out" process has already been considered in Section 4 for the staggered nanostructure of bone. Generalizing Equation (20) to all hierarchical levels, the fracture energy at the $(n+1)$th level can be estimated as

$$\Gamma_{n+1} = (1-\varphi_n)\,\ell_n \left(\int_{\text{softphase}} \tau\,d\varepsilon \right)_n = (1-\varphi_n)\,\rho_n h_n S_n^{\mathrm{p}}\Theta_n^{\mathrm{p}} = (1-\varphi_n)\,h_n S_n \Theta_n^{\mathrm{p}}, \tag{45}$$

where the integral is to be performed over the deformation history of the soft phase which undergoes large shear deformation (including sliding along the hard–soft interfaces); Θ_n^{p} denotes the effective strain which measures the range of deformation of the soft phase at the nth level.

8.4. MULTI-LEVEL FLAW TOLERANCE OF THE FRACTAL BONE – DETERMINATION OF STRUCTURE SIZES

The flaw tolerance criterion in Equation (9) will be used to determine the characteristic size of the hard phase at all hierarchical levels. We require

$$\Lambda_n^{\mathrm{ft}} = \frac{\Gamma_n E_n}{S_n^2 h_n} = 1 \tag{46}$$

for all structural levels $n = 1, 2, \ldots, N$. Inserting Equations. (44) and (45) into $\Lambda_{n+1}^{\mathrm{ft}} = 1$ yields an iterative equation

$$\frac{h_{n+1}}{h_n} = \frac{4\,(1-\varphi_n)\,\Theta_n E_{n+1}}{S_n \varphi_n^2}, \tag{47}$$

where E_{n+1} is to be determined from Equation (40). As the number of structural levels increases, the volume fraction of the hard plates increases at each individual level. Under this situation, the Young's modulus approaches the Voigt upper limit in Equations (41), and (47) can then be simplified as

$$\frac{h_{n+1}}{h_n} = \frac{4\,(1-\varphi_n)\,\Theta_n E_n}{S_n \varphi_n}. \tag{48}$$

At the nanostructural level, corresponding to $n=0$, the flaw tolerance condition can be expressed in terms of the materials properties of the mineral as

$$\Lambda_0^{\mathrm{ft}} = \frac{2\gamma E_{\mathrm{m}}}{\sigma_{\mathrm{th}}^2 h_0} = 1 \tag{49}$$

where E_{m} is the Young's modulus, γ the surface energy and σ_{th} the theoretical strength of mineral. According to the flaw tolerance design, the dimension of the mineral platelets is selected as

$$h_0 = \frac{2\gamma E_{\mathrm{m}}}{\sigma_{th}^2}. \tag{50}$$

For biominerals, we take $\gamma = 1\,\text{J/m}^2$, $E_\text{m} = 100\,\text{GPa}$ and $\sigma_\text{th} = E/30$, and find

$$h_0 = 18\,\text{nm}. \tag{51}$$

The nanostructure size h_0 becomes the basis for the bottom-up design of higher level structures in the sequence $h_1 \to h_2 \to \cdots \to h_N = H$, as in Equation (39).

8.5. NUMERICAL EXAMPLE

In order to demonstrate the potential of hierarchical material design, in the following we perform calculations based on some specific parameter choices. For simplicity, we will assume that the aspect ratio and volume fraction of the hard phase remain invariant for all hierarchical levels, i.e. $\rho_n = \rho$, $\varphi_n = \varphi$. Under these choices, the microstructure becomes completely self-similar and the volume fraction of the hard phase in each level is related to the total mineral content as

$$\Phi = \varphi^N \quad \text{or} \quad \varphi = \Phi^{1/N}. \tag{52}$$

In addition, we will assume that the soft phase in all hierarchical levels has the same elastic modulus ($G_n^\text{p} = G_p$, $E_n^\text{p} = E_\text{p}$) and the same range of deformation ($\Theta_n^\text{p} = \Theta_\text{p}$).

8.5.1. *Multi-level properties*

With these parameter selections, the multi-level stiffnesses of the fractal bone can be calculated from the iterative equation

$$\frac{1}{E_{n+1}} = \frac{4\left(1 - \Phi^{1/N}\right)}{G_p \Phi^{2/N} \rho^2} + \frac{1}{\Phi^{1/N} E_n}, \quad E_0 = E_m. \tag{53}$$

With increasing number of hierarchies, Equation (53) approaches the Voigt limit at each level

$$E_{n+1} \approx \Phi^{1/N} E_n, \tag{54}$$

which would yield an approximate solution

$$E_n \approx \Phi^{n/N} E_m. \tag{55}$$

This implies $E_N \approx \Phi E_\text{m}$.

The multi-level strengths are

$$S_{n+1} = \Phi^{1/N} S_n/2, \quad S_0 = \sigma_{th}, \tag{56}$$

which yields a simple solution

$$S_n = \Phi^{n/N} \sigma_{th}/2^n. \tag{57}$$

Note that the present choice of parameters implies

$$S_n^\text{p} = \frac{S_n}{\rho} = \frac{\Phi^{n/N} \sigma_\text{th}}{2^n \rho} \tag{58}$$

so that the strength of the soft matrix should decrease at higher structural levels, i.e. weaker proteins should be used at higher levels.

The multi-level fracture energies are

$$\Gamma_{n+1} = \left(1 - \Phi^{1/N}\right) h_n S_n \Theta_{\mathrm{p}}, \tag{59}$$

where h_n is to be determined from the flaw tolerance criterion

$$\frac{h_{n+1}}{h_n} = \frac{4\left(1 - \Phi^{1/N}\right)\Theta_{\mathrm{p}} E_{n+1}}{S_n \Phi^{2/N}}, \qquad h_0 = \frac{2\gamma E_{\mathrm{m}}}{\sigma_{\mathrm{th}}^2}. \tag{60}$$

If we approximate the Young's modulus by the Voigt solution in Equation (55), using Equation (57) leads to a simple solution

$$\frac{h_{n+1}}{h_n} = \frac{4\left(1 - \Phi^{1/N}\right)\Theta_{\mathrm{p}} E_{\mathrm{m}}}{\Phi^{1/N}\sigma_{th}} 2^n, \qquad h_0 = \frac{2\gamma E_{\mathrm{m}}}{\sigma_{\mathrm{th}}^2}. \tag{61}$$

This equation has a very simple solution

$$h_n = \left[\frac{4\left(1 - \Phi^{1/N}\right)\Theta_{\mathrm{p}} E_{\mathrm{m}}}{\sigma_{\mathrm{th}}}\right]^n \frac{2^{n(n-1)/2}}{\Phi} h_0. \tag{62}$$

By designing N levels of hierarchy, the overall dimension of the flaw tolerant material will reach

$$H = h_N = \left[\frac{4\left(1 - \Phi^{1/N}\right)\Theta_{\mathrm{p}} E_{\mathrm{m}}}{\sigma_{th}}\right]^N \frac{2^{N(N-1)/2}}{\Phi} h_0. \tag{63}$$

Note that $H \to \infty$ as $N \to \infty$ regardless of the values of the material properties E_{m}, Θ_{p}, σ_{th}, Φ. Therefore, with increasing hierarchical levels, the fractal bone can tolerate crack-like flaws without size limit.

Figures 18–20 and Table 1 show the calculated properties of the fractal bone as a function of the number of hierarchical levels. In the calculation, we assume typical materials properties of bone $\gamma = 1\,\mathrm{J/m^2}$, $\Phi = 0.45$, $E_{\mathrm{m}} = 100\,\mathrm{GPa}$, $\sigma_{\mathrm{th}} = E_{\mathrm{m}}/30$ and $E_{\mathrm{m}} = \mu_{\mathrm{p}}\rho^2 = 1000\mu_{\mathrm{p}}$. We consider two estimates $\Theta_{\mathrm{p}} = 25\%$ and $\Theta_{\mathrm{p}} = 100\%$ for the

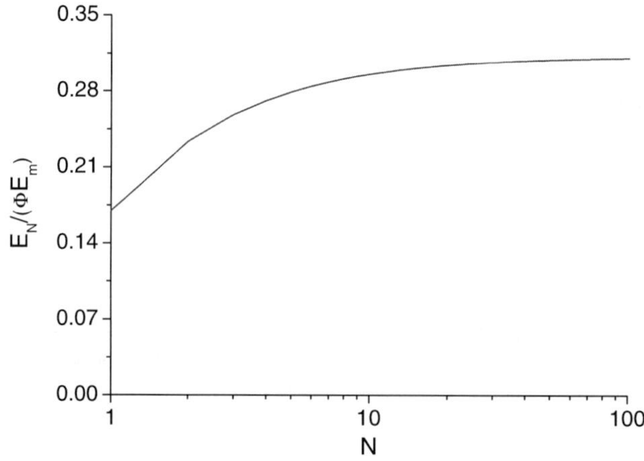

Figure 18. Variation of the overall Young's modulus of the fractal bone with the number of internal hierarchical levels.

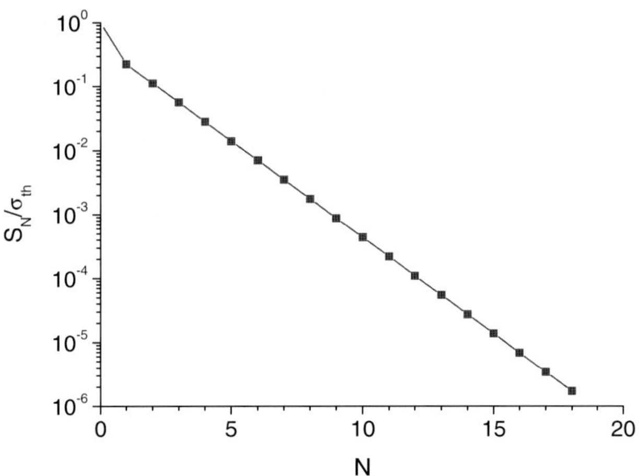

Figure 19. Variation of the overall strength of the fractal bone with the number of internal hierarchical levels.

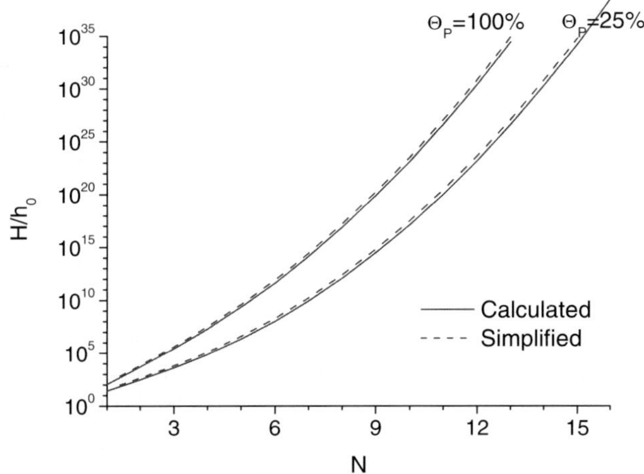

Figure 20. Variation of the overall size of the flaw tolerant fractal bone with the number of internal hierarchical levels.

Table 1. Variation of the normalized size H/h_0 of the flaw tolerant fractal bone with the number of internal hierarchical levels.

N	$\Theta_p = 25\%$	$\Theta_p = 100\%$
1	0.27×10^2	0.10×10^3
2	0.30×10^3	0.49×10^4
3	0.40×10^4	0.26×10^6
4	0.76×10^5	0.19×10^8
5	0.22×10^7	0.22×10^{10}
6	0.10×10^9	0.44×10^{12}
8	0.12×10^{13}	0.79×10^{17}
16	0.34×10^{39}	0.15×10^{49}

failure strain of protein. Figure 18 plots the overall stiffness of the fractal bone normalized by the Voigt upper bound of the composite. The result indicates that hierarchical design only results in a moderate increase in stiffness. After a few levels of hierarchy, the stiffness saturates at about 30% of the Voigt limit. Figure 19 shows that the strength of the fractal bone drops by roughly a factor of 2 with each added level of hierarchy, decreasing by about two orders of magnitude with 6 levels of hierarchy.

On the other hand, the hierarchical structures of the fractal bone exhibit very dramatic effects on the toughness of the composite. Figure 20 plots the normalized overall size H/h_0 of the fractal bone under flaw tolerance design. The solid lines are calculated from Equation (60) while the dashed lines correspond to the simplified solution in Equation (63). The results calculated from Equation (60) are also tabulated in Table 1. The results show that the flaw tolerance size of the material increases exponentially with the number of hierarchical levels. Under the selected material parameters, the flaw tolerance size of individual mineral platelets is estimated to be $h_0 = 18\,\text{nm}$. Depending upon the assumed failure strain Θ_p of protein, the flaw tolerance size of the fractal bone increases to about $1\,\mu\text{m}$ with only one level of hierarchy, 10–$100\,\mu\text{m}$ with two levels of hierarchy, 100–$10\,\text{mm}$ with 3 levels of hierarchy, $1\,\text{mm}$–$1\,\text{m}$ with 4 levels of hierarchy, $100\,\text{mm}$–$100\,\text{m}$ with 5 levels of hierarchy, $10\,\text{m}$–$10\,\text{km}$ with 6 levels of hierarchy, and 10^2–$10^6\,\text{km}$ with 8 levels of hierarchy. With 16 levels of hierarchy, the dimension of the fractal bone reaches astronomical sizes towards the edge of universe! These calculations demonstrate the potential of a bottom-up design methodology on improving the capability of materials against crack-like flaws.

There are several serious flaws or limitations to the theoretical arguments made above. In reality, one will never be able to design flaw tolerant materials of unlimited size scales. This is because the soft phase will eventually fail by localized deformation, rather than sustaining uniform shear deformation until rupture as we have implicitly assumed in estimating the hierarchical fracture energies for all levels. Also, the soft phase needs to have nearly perfect energy absorbing capability to prevent delamination cracks to propagate along the interfaces between the hard and soft phases. These assumptions inevitably break down at sufficiently large scales, at which point localized failure in the soft phase or along the soft–hard interfaces eventually replaces hard phase fracture as the dominant failure mode. In other words, localized failure modes in the soft phase will set another upper limit to hierarchical structural design. This point is worth pursuing in the future.

There is yet another limitation in the present consideration. The hard plates at the higher structural levels should be anisotropic with different elastic constants in the longitudinal and transverse directions. It is not clear how to improve this aspect of the calculation. One option is simply to treat the hard plates at each structural level as a homogenized anisotropic elastic medium. Another option is to treat them more like fiber bundles in a soft medium. Further study is needed to clarify these issues.

Despite various limitations mentioned above, our analysis based on the fractal bone model does suggest that structural hierarchy has the potential of optimizing topological distributions of organic and inorganic components to achieve flaw tolerance even at macroscopic length scales. The bottom-up designed fractal bone seems to be capable of scaling the superior properties of the staggered nanostructure of bone up to macroscopic lengths.

9. Outlook

The studies discussed in this paper have been aimed at illustrating some of the basic mechanical design principles of bone-like materials using simple analytical and numerical models. In this respect, the research described in this paper falls in the realm of bio-inspired mechanics of hierarchical materials. At this point of time, the study of hierarchical materials is still at a very primitive and premature stage. Much further research will be needed to understand the hierarchical mechanics of biological as well as bio-inspired materials. Increasingly sophisticated models may be necessary to take into account increasing biological complexities. Several important aspects are missing in the modeling effort described in this paper. First of all, bone-like materials and most biological systems exhibit strong anisotropy at various levels of structural hierarchy. In human bone, the degree of anisotropy seems to be the largest at the nanostructural level (Ji and Gao, 2006; Liu et al., 2006) and then decreases at higher levels of structures (Weiner and Wagner, 1998). Also, the mechanical properties of bone, like all protein-rich materials, are usually time-dependent. Viscoelastic or poroelastic models may be needed to understand the time-dependent mechanical behaviors of bone-like materials. This may be important for biomedical applications in developing materials to match the time-dependent behaviors of bone for viscoelastic biocompatibility (Mano, 2005). The unique properties of protein-rich soft matrix play an essential role in the superior properties of biocomposites. Therefore, better constitutive laws of protein may be needed in more sophisticated models. Mechanical properties of biocomposites will also be significantly influenced by the properties of organic–inorganic interfaces.

It will be especially interesting and challenging to study the relationship between the hierarchical structures in biology and their mechanical functions and properties. In this respect, we are still at a very primitive stage of research. The discussions in this paper have mostly focused on the staggered nanostructure of bone subjected to uniaxial loading. It is not even clear what kind of hierarchical structures would be needed for uniaxial compression, bending, torsion, or more complex three dimensional loads. We have shown that the staggered nanostructure of bone can be understood as a result of simultaneous optimization with respect to stiffness and toughness under uniaxial tension. It will be interesting to extend this line of research to including multiple levels of structural hierarchy with multiple objective functions including not only stiffness and toughness as has been done here, but also other functions such as strength, weight, stability, conductivity, transport, etc. Genetic algorithms and other optimization methods may be combined to address the hierarchical and multifunctional optimization problems in biological as well as bio-inspired systems. This area of research may have immediate applications in materials science and nanotechnology.

The study of how the hierarchical structures of bone are related to its mechanical properties at different length scales can have significant biomedical applications. Such study may help understand how aging, diseases and drugs influence the hierarchical structures and properties of bone, and may eventually help reproduce nature's material design through tissue engineering in the laboratory. It remains an outstanding challenge to identify, describe and understand the mechanical and biological features and characteristics that determine the ability of bone to resist fracture, and to use this information to identify new therapeutic targets and develop better biomarkers and noninvasive imaging techniques (Bouxsein, 2003). To achieve this goal, it may be important to study biological mechanisms that influence the hierarchical structures of bone. An example is the study by Roschger et al. (2004) on how the transcription factor Fra-1 influences the bone mass regeneration. These authors examined the mineralization density distribution and nanostructure of bone matrix generated in Fra-1 transgenic mice. Their analysis revealed an up to 5-fold increase in bone volume with no abnormalities in nanostructure for Fra-1 transgenic compared to wild type. The results provided a rationale for the development of therapeutic applications involving Fra-1-induced bone formation.

Hierarchical biomechanics can also impact the development of hierarchical materials science. In conventional materials theory, each material (metal, ceramic, polymer, etc.) is characterized with definite properties such as stiffness and toughness. Although it is known that material properties can be significantly altered by composite and microstructure design, there has been so far no systematic way to design hierarchical composite materials. Bio-inspired principles such as flaw tolerance may provide a systematic way of designing hierarchical materials with properties optimized at all relevant length scales. Nature does not distinguish between material and structure. In biological systems, mechanics (structure) and chemistry (material) are used with equal importance to achieve optimized properties at all scales. Understanding the engineering principles of nature's design of multi-functional and hierarchical materials may provide guidance on the development of novel materials with unique properties. In this paper, we have discussed properties such as stiffness, strength and toughness. Similar issues may exist with other properties such as adhesion, friction, hydrophobicity, corrosion, fatigue, and more generally, also electrical, optical and chemical properties of materials. It can be expected that bottom-up designed materials with properties optimized at multiple length scales with respect to multiple objective functions will have strong impact on materials technology in the future.

Acknowledgments

The author gratefully acknowledges stimulating discussions with Peter Fraztl and Eduard Arzt and their research groups on mechanics of biological materials. A number of past and present members of my own research group (Markus Buehler, Shaohua Chen, Xu Guo, Baohua Ji, Bin Liu, Xiang Wang, Haimin Yao, Lixian Zhang) have made significant contributions to research reviewed in this paper.

References

Bao, G. and Suo, Z. (1992). Remarks on crack-bridging concepts. *Applied Mechanics Review* **45**, 355–366.

Barenblatt, G.I. (1985). The formation of equilibrium cracks during brittle fracture: Rectilinear cracks in plane plates. *Journal of Applied Mathematics and Mechanics* **23**, 622–636.

Bazant, Z.P. (1976). Instability, ductility and size effect in strain-softening concrete. *Journal of the Engineering Mechanics Division-ASCE* **102**, 331–344.

Bazant, Z.P. and Cedolin, L. (1983). Finite element modeling of crack band propagation. *Journal of Structural Engineering-ASCE* **109**, 69–92.

Bazant, Z.P. and Planas, J. (1998). *Fracture and Size Effect in Concrete and Other Quasibrittle Materials.* CRC Press, Boca Raton, FL.

Bilby, B.A., Cottrell, A.H. and Swinden, K.H. (1963). The spread of plastic yield from a notch. *Proceedings of the Royal Society of London A* **272**, 304–314.

Bouxsein, M.L. (2003) Bone quality: where do we go from here? *Osteoporosis International* **14**, S118–S127.

Brett, C. and Waldron, K. (1981). *Physiology and Biochemistry of Plant Cell Walls*. Chapman & Hall, London.

Camacho, G.T. and Ortiz, M. (1996) Computational modeling of impact damage in brittle materials. *International Journal of Solids and Structures* **33**, 2899–2938.

Carpinteri, A. (1982). Notch sensitivity in fracture testing of aggregative materials. *Engineering Fracture Mechanics* **16**, 467–481.

Carpinteri, A. (1997). *Structural Mechanics: A Unified Approach*. Chapman & Hall, London.

Cox, B.N. and Marshall, D.B. (1994). Concepts for bridged cracks in fracture and fatigue. *Acta Metallurgica et Materialia* **42**, 341–363.

Currey, J.D. (1977). Mechanical properties of mother of pearl in tension. *Proceedings of the Royal Society of London B* **196**, 443–463.

Currey, J.D. (1984). *The Mechanical Adaptations of Bones*. Princeton University Press, Princeton, NJ, pp. 24–37.

Drugan, W.J. (2001). Dynamic fragmentation of brittle materials: analytical mechanics-based models. *Journal of the Mechanics and Physics of Solids* **49**, 1181–1208.

Dugdale, D.S. (1960). Yielding of steel sheets containing slits. *Journal of the Mechanics and Physics of Solids* **8**, 100–104.

Evans, A.G. (1990) perspective on the development of high-toughness ceramics. *Journal of the American Ceramic Society* **73**, 187–206.

Fantner, G.E., Birkedal, H., Kindt, J.H., Hassenkam, T., Weaver, J.C., Cutroni, J.A., Bosma, B.L., Bawazer, L., Finch, M.M., Cidade, G.A.G., Morse, D.E., Stucky, G.D. and Hansma, P.K. (2004). Influence of the degradation of the organic matrix on the microscopic fracture behavior of trabecular bone. *Bone* **35**, 1013–1022

Fengel, D. and Wegener, G. (1984). *Wood Chemistry, Ultrastructure, Reaction*. Walter de Gruter, Berlin.

Fratzl, P., Gupta, H.S., Paschalis, E.P. and Roschger, P. (2004a) Structure and mechanical quality of the collagen–mineral nano-composite in bone. *Journal of Materials Chemistry* **14**, 2115–2123.

Fratzl, P., Burgert, I. and Gupta, H.S. (2004b). On the role of interface polymers for the mechanics of natural polymeric composites. *Physical Chemistry Chemical Physics* **6**, 5575–5579.

Gao, H. and Chen, S. (2005). Flaw tolerance in a thin strip under tension. *Journal of Applied Mechanics* **72**, 732–737.

Gao, H. and Ji, B. (2003). Modeling fracture in nanomaterials via a virtual internal bond method. *Engineering Fracture Mechanics* **70**, 1777–1791.

Gao, H. and Yao, H. (2004). Shape insensitive optimal adhesion of nanoscale fibrillar structures. *Proceedings of the National Academy of Sciences of the United States of America* **101**, 7851–7856.

Gao, H., Ji, B., Jäger, I.L., Arzt, E. and Fratzl, P. (2003). Materials become insensitive to flaws at nanoscale: lessons from nature. *Proceedings of the National Academy of Sciences of the United States of America* **100**, 5597–5600.

Gao, H., Ji, B., Buehler, M.J. and Yao, H. (2004). Flaw tolerant bulk and surface nanostructures of biological systems. *Mechanics and Chemistry of Biosystems* **1**, 37–52.

Gao, H., Wang, X., Yao, H., Gorb, S. and Arzt, E. (2005). Mechanics of hierarchical adhesion structure of gecko, *Mechanics of Materials* **37**, 275–285.

Goldberg, D. (1989), *Genetic Algorithm in Search, Optimization, and Machine Learning*. Addison Wesley.

Guo, X. and Gao, H. (2005). Bio-inspired material design and optimization. IUTAM Symposium on topological design optimization of structures, machines and materials – status and perspectives, October 26–29, 2005, Rungstedgaard, Copenhagen, Denmark.

Hassenkam, T., Fantner, G.E., Cutroni, J.A., Weaver, J.C., Morse, D.E. and Hansma, P.K. (2004). High-resolution AFM imaging of intact and fractured trabecular bone. Bone **35**, 4–10.

Hillerborg, A., Modeer, M. and Petersson, P.E. (1976). Analysis of crack formation and crack growth in concrete by means of fracture mechanics and finite elements. *Cement and Concrete Research* **6**, 773–782.

Jackson, A.P., Vincent, J.F.V. and Turner, R.M. (1988). The mechanical design of nacre. *Proceedings of the Royal Society of London B* **234**, 415–440.

Jäger, I. and Fratzl, P. (2000). Mineralized collagen Mbrils: a mechanical model with a staggered arrangement of mineral particles. *Biophysical Journal* **79**, 1737–1746.

Ji, B. and Gao, H. (2004a). Mechanical properties of nanostructure of biological materials. *Journal of the Mechanics and Physics of Solids* **52**, 1963–1990.

Ji, B. and Gao, H. (2004b). A study of fracture mechanisms in biological nano-composites via the virtual internal bond model. *Materials Science and Engineering A* **366**, 96–103.

Ji, B. and Gao, H. (2006) Elastic properties of nanocomposite structure of bone. *Composite Science and Technology*, in press.

Ji B., Gao H. and Hsia, K.J. (2004a). How do slender mineral crystals resist buckling in biological materials? *Philosophical Magazine Letters* **84**, 631–641.

Ji, B., Gao, H. and Wang, T.C. (2004b). Flow stress of biomorphous metal–matrix composites. *Materials Science and Engineering A* **386**, 435–441.

Jiang, H.D., Liu, X.Y., Lim, C.T., and Hsu, C.Y. (2005). Ordering of self-assembled nanobiominerals in correlation to mechanical properties of hard tissues. *Applied Physics Letters* **86**, 163901.

Kamat, S., Su, X., Ballarini, R. and Heuer, A.H. (2000). Structural basis for the fracture toughness of the shell of the conch Strombus gigas. *Nature* **405**, 1036–1040.

Karihaloo, B.L. (1979). A note on complexities of compression failure. *Proceedings of the Royal Society of London A* **368**, 483–493.

Kauffmann, F., Ji, B., Dehm, G., Gao, H. and Arzt, E. (2005). A quantitative study of the hardness in a superhard nanocrystalline titanium nitride/silicon nitride coating. *Scripta Materialia* **52**, 1269–1274.

Kendall, K. (1978). Complexities of compression failure. *Proceedings of the Royal Society of London A* **361**, 245–263.

Kessler, H., Ballarini, R., Mullen, R.L., Kuhn, L.T. and Heuer, A.H. (1996). A biomimetic example of brittle toughening: (I) steady state multiple cracking. *Computational Materials Science* **5**, 157–166.

Kotha, S.P., Kotha, S. and Guzelsu, N. (2000). A shear-lag model to account for interaction effects between inclusions in composites reinforced with rectangular platelets. *Composites Science and Technology* **60**, 2147–2158.

Landis, W.J. (1995). The strength of a calcified tissue depends in part on the molecular structure and organization of its constituent mineral crystals in their organic matrix. *Bone* **16**, 533–544.

Landis, W.J., Hodgens, K.J., Song, M.J., Arena, J., Kiyonaga, S., Marko, M., Owen, C., and McEwen, B.F. (1996). Mineralization of collagen may occur on fibril surfaces: evidence from conventional and high voltage electron microscopy and three dimensional imaging. *Journal of Structural Biology* **117**, 24–35.

Liu, B., Zhang, L. and Gao, H. (2006). Poisson ratio can play a crucial role in mechanical properties of biocomposites. *Mechanics of Materials*, in press.

Mano, J.F. (2005). Viscoelastic properties of bone: mechanical spectroscopy studies on a chicken model. *Materials Science and Engineering C* **25**, 145–152.

Massabo, R. and Cox, B.N. (1999). Concepts for bridged mode II delamination cracks. *Journal of the Mechanics and Physics of Solids* **47**, 1265–1300.

Menig, R., Meyers, M.H., Meyers, M.A. and Vecchio, K.S. (2000). Quasi-static and dynamic mechanical response of *Haliotis rufescens* (abalone) shells. *Acta Materialia* **48**, 2383–2398.

Menig, R., Meyers, M.H., Meyers, M.A. and Vecchio, K.S. (2001). Quasi-static and dynamic mechanical response of *Strombus gigas* (conch) shells. *Materials Science and Engineering A* **297**, 203–211.

Mori, T. and Tanaka, K. (1973). Average stress in matrix and average elastic energy of materials with misfitting inclusion. *Acta Metalurgica* **21**, 571–574.

Mulmule, S.V. and Dempsey, J.P. (2000). LEFM size requirement for the fracture testing of sea ice. *International Journal of Fracture* **102**, 85–98.

Needleman, A. (1987) A continuum model for void nucleation by inclusion debonding. *Journal of Applied Mechanics* **54**, 525–531.

Neves, N.M. and Mano, J.F. (2005). Structure/mechanical behavior relationships in crossed-lamellar sea shells. *Materials Science and Engineering C* **25**, 113–118.

Okumura, K. and de Gennes, P.-G. (2001). Why is nacre strong? Elastic theory and fracture mechanics for biocomposites with stratified structures. *European Physical Journal E* **4**, 121–127.

Pugno, N.M. and Ruoff, R.S. (2004). Quantized fracture mechanics. *Philosophical Magazine* **84**, 2829–2845.

Rho, J.Y., Kuhn-Spearing, L. and Zioupos, P. (1998). Mechanical properties and the hierarchical structure of bone. *Medical Engineering & Physics* **20**, 92–102.

Rice, J.R. (1980). *The Mechanics of Earthquake Rupture*. International School of Physics "E. Fermi", Course 78, 1979: Italian Physical Society/North Holland Publ. Co.

Roschger, P., Grabner, B.M., Rinnerthaler, S., Tesch, W., Kneissel, M., Berzlanovich, A., Klaushofer, K. and Fratzl, P. (2001). Structural development of the mineralized tissue in the human L4 vertebral body. *Journal of Structural Biology* **136**, 126–136.

Roschger, P., Matsuo, K., Misof, B.M., Tesch, W., Jochum, W., Wagner, E.F., Fratzl, P. and Klaushofer, K. (2004) Normal mineralization and nanostructure of sclerotic bone in mice overexpressing Fra-1. *Bone* **34**, 776–782.

Smith, B.L., Schaeffer, T.E., Viani, M., Thompson, J.B., Frederick, N.A., Kindt, J., Belcher, A., Stucky, G.D., Morse, D.E. and Hansma, P.K. (1999). Molecular mechanistic origin of the toughness of natural adhesive, fibres and composites. *Nature* **399**, 761–763.

Song, F., Soh, A.K. and Bai, Y.L. (2003). Structural and mechanical properties of the organic matrix layers of nacre. *Biomaterials* **24**, 3623–3631.

Suo, Z., Ho, S. and Gong, X. (1993). Notch ductile-to-brittle transition due to localized inelastic band. *Journal of Engineering Materials and Technology* **115**, 319–326.

Tada, J., Paris, P.C. and Irwin, G.R. (1973). *The Stress Analysis of Cracks Handbook*. Del Research Corporation, St. Louis (2nd edition, 1985).

Tang, R.K., Wang, L.J., Orme, C.A., Bonstein, T., Bush, P.J. and Nancollas, G.H. (2004). Dissolution at the nanoscale: Self-preservation of biominerals. *Angewandte Chemie-International Edition* **43**, 2697–2701.

Tang, T., Hui, C.-Y. and Glassmaker, N.J. (2005) Can a fibrillar interface be stronger and tougher than a non-fibrillar one? *Journal of the Royal Society Interface* **2**, 505–516.

Tesch, W., Eidelman, N., Roschger, P., Goldenberg, F., Klaushofer, K. and Fratzl, P. (2001). Graded microstructure and mechanical properties of human crown dentin. *Calcified Tissue International* **69**, 147–157.

Thompson, J.B., Kindt, J.H., Drake, B., Hansma, H.G., Morse, D.E. and Hansma, P.K. (2001) Bone indentation recovery time correlates with bond reforming time. *Nature* 414, 773–776.

Tvergaard, V. and Hutchinson, J.W. (1992) The relation between crack growth resistance and fracture process parameters in elastic–plastic solids. *Journal of the Mechanics and Physics of Solids* **40**, 1377–1397.

Wang, R.Z., Suo, Z., Evans, A.G., Yao, N. and Aksay, I.A. (2001). Deformation mechanisms in nacre. *Journal of Materials Research* **16**, 2485–2493.

Wang, L.J., Tang, R.K., Bonstein, T., Orme, C.A., Bush, P.J. and Nancollas, G.H. (2005). A new model for nanoscale enamel dissolution. *Journal of Physical Chemistry B* **109**, 999–1005.

Warshawsky, H. (1989). Organization of crystals in enamel. *Anatomical Record* **224**, 242–262.

Weiner, S. and Wagner, H.D. (1998). The material bone: structure–mechanical function relations. *Annual Review of Materials Science* **28**, 271–298.

Xu, X.P. and Needleman, A. (1994). Numerical simulations of fast crack-growth in brittle solids. *Journal of the Mechanics and Physics of Solids* **42**, 1397–1434.

Yao, H. and Gao, H. (2006). Mechanics of robust and releasable adhesion in biology: bottom-up designed hierarchical structures of gecko. *Journal of the Mechanics and Physics of Solids*, in press.

International Journal of Fracture (2006) 138:139–166
DOI 10.1007/s10704-006-0035-1

© Springer 2006

Development of the local approach to fracture over the past 25 years: theory and applications

A. PINEAU

Centre des Matériaux, Ecole des Mines de Paris, UMR CNRS 7633, BP 87, 91003 Evry Cedex, France (E-mail: andrepineau.@ensmp.fr)

Received 1 March 2005; accepted 1 December 2005

Abstract. This review paper is devoted to the local approach to fracture (LAF) for the prediction of the fracture toughness of structural steels. The LAF has been considerably developed over the past two decades, not only to provide a better understanding of the fracture behaviour of materials, in particular the failure micromechanisms, but also to deal with loading conditions which cannot easily be handled with the conventional linear elastic fracture mechanics and elastic–plastic fracture mechanics global approaches. The bases of this relatively newly developed methodology are first presented. Both ductile rupture and brittle cleavage fracture micromechanisms are considered. The ductile-to-brittle transition observed in ferritic steels is also briefly reviewed. Two types of LAF methods are presented: (i) those assuming that the material behaviour is not affected by damage (e.g. cleavage fracture), (ii) those using a coupling effect between damage and constitutive equations (e.g. ductile fracture). The micromechanisms of brittle and ductile fracture investigated in elementary volume elements are briefly presented. The emphasis is laid on cleavage fracture in ferritic steels. The role of second phase particles (carbides or inclusions) and grain boundaries is more thoroughly discussed. The distinction between nucleation and growth controlled fracture is made. Recent developments in the theory of cleavage fracture incorporating both the effect of stress state and that of plastic strain are presented. These theoretical results are applied to the crack tip situation to predict the fracture toughness. It is shown that the ductile-to-brittle transition curve can reasonably be well predicted using the LAF approach. Additional applications of the LAF approach methods are also shown, including: (i) the effect of loading rate and prestressing; (ii) the influence of residual stresses in welds; (iii) the mismatch effects in welds; (iv) the warm-prestressing effect. An attempt is also made to delineate research areas where large improvements should be made for a better understanding of the failure behaviour of structural materials.

Key words: Cleavage, ductile-brittle transition, ductile fracture, fracture toughness, micromechanisms, scatter, statistics.

1. Introduction

The assessment of the mechanical integrity of any flawed mechanical structure requires the development of approaches which can deal not only with simple situations, such as small-scale yielding (SSY) under pure mode I isothermal loading, but also with much more complex situations, including large-scale plasticity, mixed-mode cracking, and non-isothermal loading. Two types of approaches have been developed for that purpose. The first one, referred to as the "global" approach, is essentially based on linear elastic fracture mechanics (LEFM) and elastic–plastic fracture mechanics (EPFM). In this methodology it is assumed that the fracture resistance can be measured in terms of a single parameter, such as K_{IC}, J_{IC} or CTOD. More recently global approaches incorporating a second parameter (T and Q stresses) have been introduced (see e.g. O'Dowd and Shih, 1991, 1992). This methodology is extremely useful and absolutely necessary, but it has also a number of limitations such as the absence of any prediction of size effects observed in brittle fracture or the application to non-isothermal loading conditions. This is the reason why another approach has been developed since the 80s. This is the so-called "local approach" to fracture (LAF) in which the modelling of fracture toughness is based on local fracture criteria usually established from tests on volume elements, in particular notched specimens (see, e.g. Pineau, 1982; Beremin, 1983; Mudry, 1987; Pineau and Joly, 1991). These criteria are applied to the crack tip, as schematically shown in Figure 1, which describes the methodology applied in the LAF approach. This methodology requires that two conditions are fulfilled: (i) micromechanistically based models must be established; (ii) a perfect knowledge of the crack tip stress-strain field must be known.

A considerable research effort has been devoted to the development of the local approaches over the past 25 years. A book devoted to this topic has been published recently by Besson (2004). The first approaches assumed that the material obeyed a conventional behaviour supplemented by models of local fracture processes. This is the situation for brittle fracture (Beremin, 1983; Mudry, 1987). Further approaches similar to those used in continuum damage mechanics (CDM), involving constitutive equations with a softening effect due to "damage" have been introduced later. This is the present situation for ductile fracture.

The LAF introduces new parameters which have to be determined experimentally. This raises the problem of the strategy which has to be used to determine these parameters (see Pineau, 2003). For instance, the seminal work by Beremin (1983) introduces the Weibull stress, σ_w, as a probabilistic fracture parameter. The Beremin model predicts the evolution of the Weibull stress with macroscopic applied stress or the stress intensity factor, K, to define the conditions leading to local material failure. The transferability models of elastic–plastic fracture toughness values rely on the notion of a Weibull stress as a crack driving force (see, e.g. Ruggieri and Dodds,

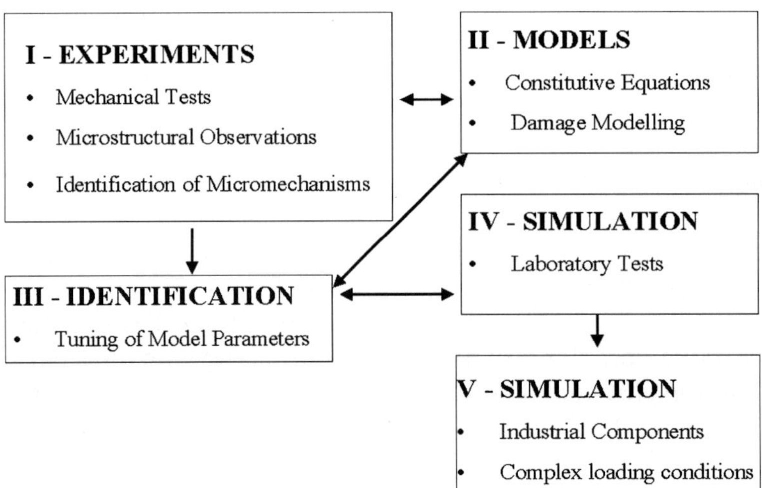

Figure 1. Sketch showing the methodology followed in the local approach to fracture.

1996; Ruggieri et al., 1998; Petti and Dodds, 2004). In this approach to brittle fracture the simple axiom is that unstable cleavage crack propagation occurs at a critical value of the Weibull stress (for a given probability to failure). It is worth noting that the Weibull stress, σ_w, can be calculated for any type of loading conditions, including non-isothermal large scale yielding (LSY).

In this review paper, the emphasis is laid on brittle cleavage fracture which is the most deleterious failure mode observed in ferritic steels at low temperature. However ductile fracture involving cavity formation is also briefly considered. After reviewing some recent developments in the knowledge of the fracture micromechanisms, the paper deals with modelling fracture toughness. A number of applications of the local approach to fracture are also shown.

2. Micromechanisms of fracture

Cleavage fracture and ductile rupture are considered successively.

2.1. CLEAVAGE FRACTURE

2.1.1. *Introduction*
In many ferritic steels, it has been found that the cleavage stress, σ_c, is independent of temperature. This strongly suggests that in these materials the mechanism of cleavage fracture is growth controlled (see, e.g. Curry and Knott, 1979; Pineau, 1982, 1992): cleavage microcracks are progressively nucleated under the influence of plastic strain. These microcracks are arrested at microstructural barriers and fracture occurs when the longest crack reaches the Griffith stress given by:

$$\sigma_c = \sqrt{\frac{2E\gamma_S}{\alpha a}} \tag{1}$$

where E is the Young's modulus, γ_S the "effective" surface energy, α a numerical constant depending of the crack shape, and a the size of the longest microcrack. In this equation all terms are almost independent of temperature, except the term γ_S which is much higher than the true surface energy because of the plasticity accompanying crack propagation.

However, this theory is too simple since it does not recognize the different steps encountered during microcrack initiation and microcrack propagation (see, e.g. Martin-Meizoso et al., 1994). Moreover it does not include the statistical aspects which are well known to play a key role. This is the reason why the situation must be analysed in more detail.

Cleavage fracture of ferritic steels most frequently occurs by the dynamic propagation of microcracks initiated by slip-induced cracking of brittle second phase particles (i.e. carbides in steels) or inclusions. Fracture results from the successive occurrence of three elementary events (Figure 2):
– slip-induced cracking of a brittle particle,
– propagation of the microcrack on a cleavage plane of the neighbouring matrix grain across the particle/matrix interface under the local stress state,
– propagation of the grain-sized crack to neighbouring grains across the grain boundary.

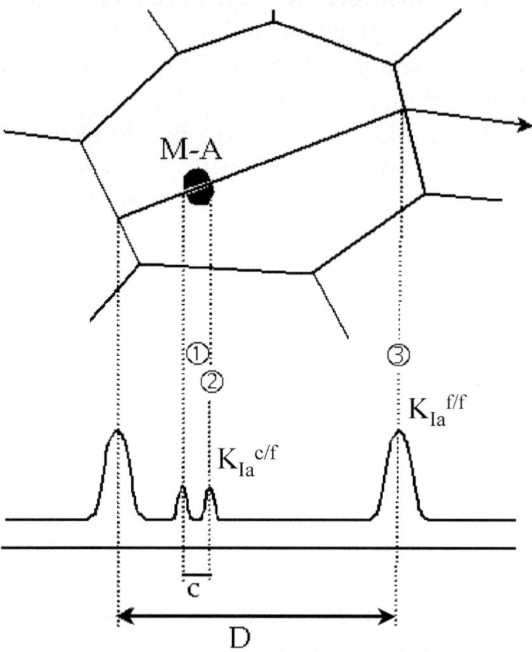

Figure 2. Initiation of a cleavage microcrack from a particle (M-A constituent). The crack is arrested at the interface c/f, then propagates through the matrix and is arrested at the grain boundaries.

The first event corresponding to brittle fracture of particles is governed by a critical stress, σ_d when the particle size is larger than ~ 0.1–$1\,\mu m$ (see, e.g. Pineau, 1992). Below this size a dislocation based theory must be used. It has been shown that the stress σ_d is related to the maximum principal stress, σ_1, the equivalent von Mises stress, σ_{eq} and the yield stress, σ_{YS}, by:

$$\sigma_1 + k\left(\sigma_{eq} - \sigma_{YS}\right) = \sigma_d \tag{2}$$

where k is a function of particle shape (Beremin, 1981a; François and Pineau, 2001). This simple expression which is similar to that used by Margolin in his model for brittle fracture initiated from particles (Margolin et al., 1998) shows that for a given stress state the strain necessary to nucleate microcracks strongly increases with temperature. It shows also that crack nucleation is favoured by stress state conditions, such as those of a crack tip, leading to high stress triaxiality ratio. The value of σ_d is *a priori* statistically distributed. The values of the local fracture toughness, $k_I^{c/f}$ and $k_I^{f/f}$ are also statistically distributed. Recent studies have shown that, for instance, in bainitic steels the effective packets which lead to local crack arrest at grain boundaries are those for which the misorientation between the packets is large (Bouyne et al., 1998; Gourgues et al., 2000; Lambert-Perlade et al., 2004). The role of particle and grain size distributions have therefore to be also taken into account, as schematically shown in Figure 3. In this figure (Martin-Meizoso et al., 1994) the critical values of the particle and grain size C^*, D^* corresponding to the different steps in brittle fracture are simply related to the local value of the maximum principal stress, σ_1, by:

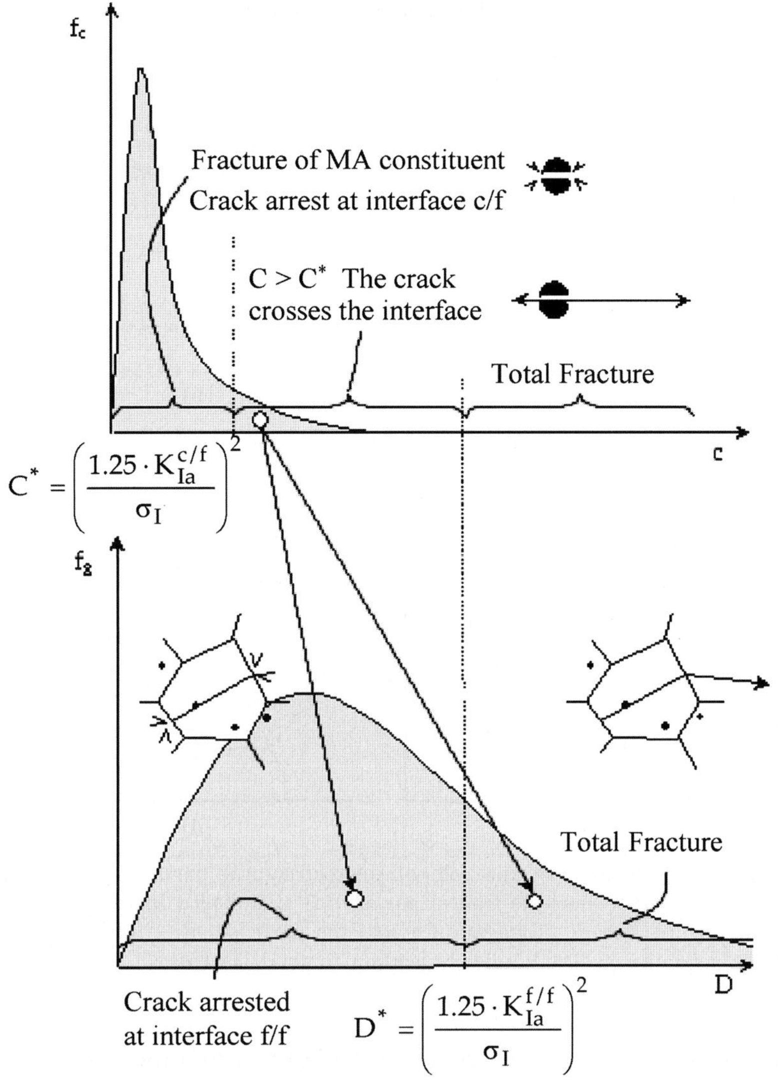

$$C^* = \left(\frac{1.25 \cdot K_{Ia}^{c/f}}{\sigma_I} \right)^2$$

$$D^* = \left(\frac{1.25 \cdot K_{Ia}^{f/f}}{\sigma_I} \right)^2$$

Figure 3. Multiple barrier model. Three events are schematically shown (Martin-Meizoso et al., 1994).

$$C^* = \left(\frac{\delta . K_I^{c/f}}{\sigma_1} \right)^2 \quad \text{and} \quad D^* = \left(\frac{\delta . K_I^{f/f}}{\sigma_1} \right)^2 \tag{3}$$

where δ is a numerical factor close to 1.

Recent experiments on bainitic microstructures simulating the heat-affected zone (HAZ) of welds indicated that cleavage fracture was initiated from tiny ($\cong 1\,\mu m$) brittle particles formed by martensite-austenite (M-A) constituents (Lambert-Perlade et al., 2004). It was shown that at low temperature the critical step corresponds to the nucleation of microcracks from these M-A particles. When the temperature was increased the critical step is the propagation of packet size microcracks through grain boundaries, as schematically shown in Figure 4 and as illustrated in Figure 5 on a bainitic steel. Acoustic emission technique has been useful to reach this conclusion.

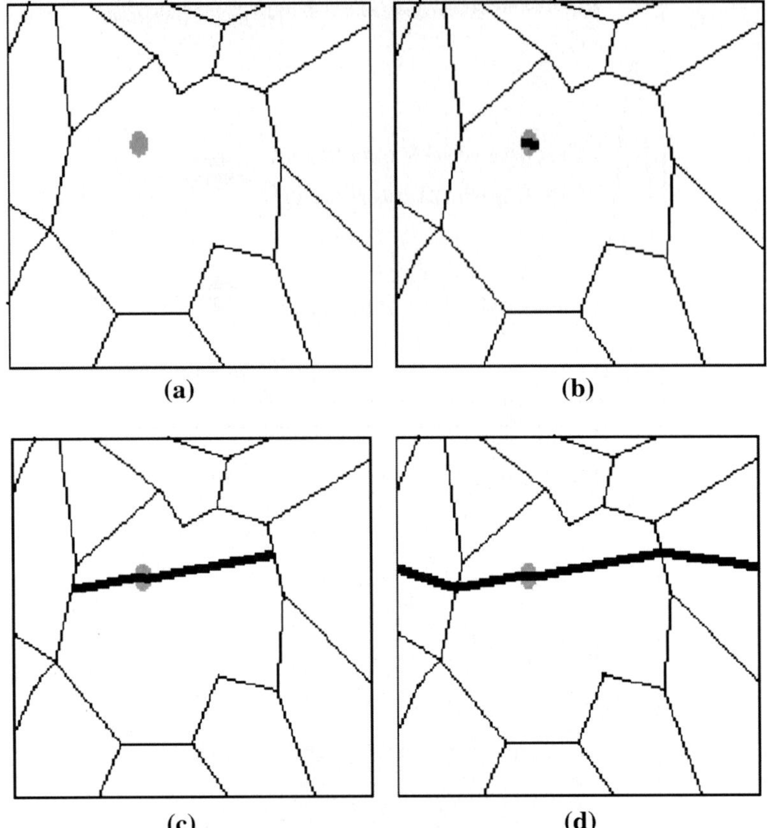

Figure 4. Schematic representation of the role of microstructural barriers on cleavage micromechanisms. The crack is assumed to nucleate from a particle (a) undamaged material; (b) microcrack initiation in the particle, (c) microcrack propagation across the particle/matrix interface and (d) microcrack propagation across a bainite packet boundary leading to final fracture.

This situation observed in one specific steel is likely to be more general. These observations strongly suggest that the micromechanisms operating during fracture toughness measurements at increasing temperature are not necessarily the same. In such conditions it would seem preferable to involve a multiple-barrier (MB) model to account for the temperature dependence of fracture toughness, as shown below.

2.1.2. *Models of cleavage fracture*

In this part the statistical aspects of cleavage fracture are modelled. Then some comments are made on the existence of a threshold stress for cleavage fracture and the effect of plastic strain. It is also shown how multiple barrier models can be developed.

2.1.2.1. *Statistics.* Rather surprisingly although the scatter in cleavage stress measurements is well known, it was only in the 1980s that models have been proposed to account for this scatter (for a review, see, e.g. Wallin, 1991a, b). Nowadays the most largely used models are those derived from the work by Beremin, 1983). Assuming that the material contains a population of microdefects of a size, a, (particles or grain

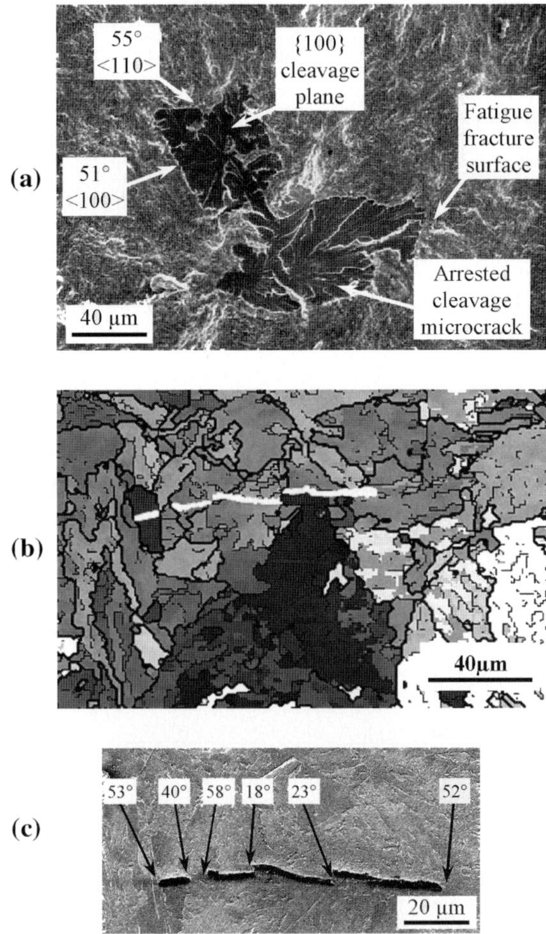

Figure 5. Bainitic steel. Arrested cleavage microcracks obtained with interrupted tests on a notched specimen. (a) Fracture surface after subsequent fatigue crack propagation. Indexations of crack arresting boundaries are given. (b) EBSD map. Thin lines and thick lines denote low-angle and high-angle boundaries, respectively. (c) SEM observations and disorientation analysis showing microcrack arrest at high-angle boundaries (same area as *b*) (Lambert-Perlade et al., 2004).

sized microcracks) distributed according to a simple (power or exponential) law, $p(a)$, the weakest link theory tells us that the probability to failure $P(\sigma)$ of a representative volume element, V_u, is given by:

$$P(\sigma) = \int_{a_c(\sigma)}^{\infty} p(a)\,\mathrm{d}a \tag{4}$$

where a_c is simply given by Equation (1), i.e. :

$$a_c = \frac{2E\gamma_s}{\alpha\sigma^2} \tag{5}$$

Knowing the distribution $p(a)$ it is therefore possible to calculate $P(\sigma)$.

146 *A. Pineau*

In a volume V which is uniformly loaded and which contains V/V_u statistically independent elements the probability to failure can be expressed as:

$$P_R = 1 - \exp\left[-\frac{V}{V_u} P(\sigma)\right] \tag{6}$$

As a general rule the function $p(a)$ is not known. However when the critical step for cleavage fracture is the propagation of microcracks initiated from particles, the distribution $p(a)$ may be determined experimentally when assuming that all the particles participate to fracture. Two types of laws are usually proposed:
- a power law given by

$$p(a) = \gamma a^{-\beta}, \tag{7}$$

- an exponential law including eventually a cut-off parameter (see, e.g. Carassou et al., 1998; Lee et al., 2002) such as the cumulative probability is given by:

$$p(\text{size} > a) = \exp\left[-\left(\frac{a - a_u}{a_o}\right)^n\right]. \tag{8}$$

The simple power law leads to the well-known Weibull expression:

$$P_R = 1 - \exp\left[-\frac{V}{V_o}\left(\frac{\sigma}{\sigma_u}\right)^m\right] \tag{9}$$

with the Weibull shape factor $m = 2\beta - 2$ and

$$\sigma_u = (m/2\gamma)^{1/m} (2E\gamma_S/\alpha)^{1/2}. \tag{10}$$

It should be noted that within a first approximation, m and σ_u are temperature independent. Similarly the exponential law Equation (8) leads to Tanguy et al. (2003):

$$P_R = 1 - \exp\left\{-\frac{V}{V_o}\left[-\left(\frac{\frac{1}{\sigma^2} - \frac{1}{\sigma_u^2}}{\frac{1}{\sigma_o^2}}\right)^n\right]\right\}, \tag{11}$$

where

$$\sigma_o = (2E\gamma_S/\alpha a_o)^{1/2} \quad \text{and} \quad \sigma_u = (2E\gamma_S/\alpha a_u)^{1/2}. \tag{12}$$

Equation (9) is a simplified expression since no threshold is introduced. In three dimensions (3D) and in the presence of smooth stress gradients, this equation can be written as:

$$P_R = 1 - \exp\left[-\frac{\int_{PZ} \sigma_1^m dV}{\sigma_u^m V_u}\right], \tag{13}$$

where the volume integral is extended over the plastic zone (PZ). This equation can be rewritten as:

$$\sigma_w = \sqrt[m]{\int_{PZ} \frac{\sigma_1^m dV}{V_u}} = \sigma_u \sqrt[m]{\text{Ln}\left[\frac{1}{1 - P_R}\right]}, \tag{14}$$

where σ_w is referred to as the "Weibull stress".

2.1.2.2. *Threshold.*   Some investigators have introduced a threshold stress, σ_{th}, directly into Equation (14) (Bakker and Koers, 1991; Xia and Cheng, 1997). One proposal for the integrand of Equation (13) has the form $[(\sigma_1 - \sigma_{\text{th}}) / (\sigma_u - \sigma_{\text{th}})]^m$. However rational calibration procedures for σ_{th} remain an open issue. To avoid these difficulties, Gao et al. (1999) proposed a modified form of Equation (14) given by:

$$P_R = 1 - \exp\left[-\left(\frac{\sigma_w - \sigma_{w\,\text{min}}}{\sigma_u - \sigma_{w\,\text{min}}}\right)^m\right],\tag{15}$$

where $\sigma_{w\,\text{min}}$ represents the minimum value of σ_w at which cleavage fracture becomes possible. (For a full discussion, see also Gao et al., 1998; Gao and Dodds, 2000).

In the original Beremin model (Beremin, 1983), an implicit threshold Weibull stress was also included. In the literature this is something which is somewhat overlooked. In this model, it is assumed that cleavage fracture cannot occur in the absence of plastic deformation, i.e. below the yield strength, σ_{YS}. This means that the PZ size must be larger than a critical size, X_c, or otherwise stated that there exists a threshold, $K_{\text{I min}}$, below which cleavage fracture cannot occur.

This threshold is given by:

$$K_{\text{I min}} \approx \sigma_{\text{YS}}\sqrt{3\pi X_c}.\tag{16}$$

It should also be noted that the exponential law with a cut-off parameter used to describe the particle distribution Equation (8) leads also to a threshold stress given by Equation (11).

2.1.2.3. *Multiple-barrier models.*   As stated previously the Beremin model is essentially based on the description of the propagation of an existing critical defect belonging to a single population. This is a simplification which might explain why in the application of this model over a wide range of temperatures, a number of investigators have reported that it was necessary to assume the normalizing stress, σ_u, to be an increasing function of temperature (see, e.g., Tanguy et al., 2005a, b). This might simply reflect the existence of critical different steps depending on temperature, as indicated earlier.

Multiple-barrier models would therefore appear more satisfactory to account for the variation of cleavage fracture toughness over a wide temperature range. In particular MB models reflecting the situation schematically shown in Figure 4 have been proposed (Martin-Meizoso et al., 1994; Lambert-Perlade et al., 2004). These models are also based on the weakest link theory. The nature of these barriers depends on temperature. The application of these models requires the knowledge of a certain number of metallurgical factors including the nucleating particle size distribution and the grain (packet) size distribution. These factors were measured in one specific steel in which the brittle particles were formed by M-A constituents (Lambert-Perlade et al., 2004). The application of these models also necessitates the knowledge of the local fracture toughness, $K_I^{c/f}$ and $K_I^{f/f}$, and that of the cleavage fracture stress of particles, σ_d. There are very few results on these values in the literature. However a number of results are reported in Table 1. In the study devoted to a bainitic steel containing M-A particles it was assumed that the local values of fracture toughness, $K_I^{c/f}$ and $K_I^{f/f}$ were not temperature dependent, which is a crude approximation. In spite of this approximation a

Table 1. Parameters of multiple barrier models.

Parameter	Present study value	Literature data		
		Value	Microstructural unit	Reference
σ_{d} (MPa)	2112	–	–	–
$K_{\mathrm{I}}^{\mathrm{c/f}}$ (MPa m$^{1/2}$)	7.8	2.5–5.0	Carbides	Martin-Meizoso et al. (1994)
		2.5	Globular carbides	Hahn (1984)
		1.8	TiN particles	Rodrigues-Ibabe (1998)
$K_{\mathrm{I}}^{\mathrm{f/f}}$ (MPa m$^{1/2}$)	CGHAZ-25	5.0–7.0	Bainite packets	Martin-Meizoso et al. (1994)
	28	7.0	Bainite packets	Martin-Meizoso et al. (1994)
		7.5	Ferrite grains	Hahn (1984)
	ICCGHAZ-25	4.8	Bainite packets	Rodrigues-Ibabe (1998)
	18	15.2	Bainite packets	Rodrigues-Ibabe (1998)

good agreement was found between the experimental values of the fracture toughness and those inferred from this MB barrier model, as shown later.

2.1.2.4. *Strain correction.*   The usual Weibull expression in Equations (13) and (14) of the product of a stress function and a volume (PZ) is based on the assumption of a microcrack population nucleated at the onset of plastic deformation, which then remains active over the entire loading history. Detailed examinations on a number of materials, in particular those to which the MB model has been applied have shown that the situation is far more complex. A number of studies (see, e.g. Kaechele and Tetelman, 1969) have shown that the number of nucleated microcracks was an increasing function of plastic strain and increased when the temperature was lowered. From this observation one should be tempted to include in the expression of the Weibull stress a function increasing with plastic strain and decreasing with temperature. The increasing strain dependence of the Weibull stress was the basis of the modification in the Beremin model introduced by Kroon and Faleskog (2002) and recently applied by Faleskog et al. (2004). The effect of plastic strain on cleavage fracture has also been discussed by Chen et al. (2003). Still more recently this effect was also introduced in a model proposed by Bordet et al. (2005a, b). Moreover these authors have also attempted to account for the temperature dependence on the nucleation of microcracks from particles. Here it is worth mentioning that Equation (2) also contains both effects implicitly.

   A number of other observations have also clearly shown that microcracks arrested at grain boundaries are blunted with further straining, making them very unlikely to propagate again. In other words, these observations strongly suggest that only freshly nucleated carbide microcracks are expected to act as effective nucleation sites. This time, plastic strain appears to be beneficial or otherwise stated leads to an increase of the normalizing stress factor, σ_{u}, introduced in Equations (13) and (14). This was the basis of the "strain correction" originally introduced in the Beremin model where the probability to failure was expressed as:

$$P_{\mathrm{R}} = 1 - \exp\left[-\frac{\int_{\mathrm{PZ}} \sigma_1^m \exp\left(-m\varepsilon_1/\gamma\right) \cdot \mathrm{d}V}{\sigma_{\mathrm{u}}^m V_{\mathrm{u}}} \right], \tag{17}$$

where ε_1 is the plastic strain in the direction of the highest principal stress in a given element of volume, dV, and γ is a constant assumed to be close to 2. Recently a new modification to the Beremin model was introduced (Bernauer et al., 1999). These authors assumed that the nucleation rate of cavities from particles could be described by the Chu and Needleman law (see Needleman and Tvergaard, 1987). Once a cavity is nucleated from a particle, it is assumed that this particle can no longer act as a nucleation site for a cleavage microcrack. Stöckl et al. (2000) have applied this model to interpret results on warm-prestress effect.

Two opposite effects of plastic strain have therefore been invoked : (i) a deleterious effect through the continuous nucleation of microcracks, and (ii) a beneficial effect associated with the blunting of already initiated microcracks which are no longer active. Very recently, Bordet et al. (2005a, b) have attempted to incorporate those two opposite effects writing that the probability to cleavage, P_{cl}, could generally be expressed as:

$$P_{\mathrm{cl}} = P_{\mathrm{nuc}} \times P_{\mathrm{prop}} \tag{18}$$

where P_{nuc} is the probability of nucleating a microcrack while P_{prop} is the probability of propagating this nucleated microcrack. These authors have proposed the following modified expression for the Weibull stress , σ_{w}^{*}:

$$\sigma_{\mathrm{w}}^{*} = \left\{ \int_{PZ} \left[\int_{0}^{\varepsilon_{\mathrm{p,u}}} \frac{\sigma_{\mathrm{YS}}}{\sigma_{\mathrm{YS,o}}} \left(\sigma_1^m - \sigma_{\mathrm{th}}^m \right) \exp\left(-\frac{\sigma_{\mathrm{YS}}}{\sigma_{\mathrm{YS,o}}} \cdot \frac{\varepsilon_{\mathrm{p}}}{\varepsilon_{\mathrm{p,o}}} \right) d\varepsilon_{\mathrm{p}} \right] \frac{dV}{V_o} \right\}^{1/m}, \tag{19}$$

where σ_{YS} is the yield strength at a given temperature and for a given strain rate, and $\sigma_{\mathrm{YS,o}}$ is a reference yield strength. Similarly $\varepsilon_{\mathrm{p,o}}$ is a reference plastic strain. If the ratio $\sigma_{\mathrm{YS,o}} \times \varepsilon_{\mathrm{p,o}}/\sigma_{\mathrm{YS,o}}$ is large, that is when the diminution of potential nucleation sites can be neglected, σ_{w}^{*} can be simplified as :

$$\sigma_{\mathrm{w}}^{*} = \left\{ \int_{PZ} \left[\int_{0}^{\varepsilon_{\mathrm{p,u}}} \frac{\sigma_{\mathrm{YS}}}{\sigma_{\mathrm{YS,o}}} \left(\sigma_1^m - \sigma_{\mathrm{th}}^m \right) d\varepsilon_{\mathrm{p}} \right] \frac{dV}{V_o} \right\}^{1/m}. \tag{20}$$

This expression is very similar to that of Beremin, as expected.

All these modifications to the original Beremin model have been introduced recently and they have not yet been largely tested. They lead to more sophisticated expressions requiring the identification of supplementary parameters.

3. Ductile rupture and ductile-to-brittle transition

3.1. INTRODUCTION

Ductile fracture of metallic materials involves cavity nucleation and void growth to coalescence. Many studies have been devoted to the micromechanisms accompanying void nucleation from second-phase particles (see, e.g. Garrison and Moody, 1987). However large improvements have to be made to include all the details related to these particles in particular the effect of inhomogeneous spatial distribution. Recently a number of attempts have been made in this field. Similarly recent advances have also been made for a better modelling of void growth and coalescence.

For a long time ductile rupture has been approached using uncoupled models (see, e.g. Mc Clintock, 1971; Beremin, 1981a). In these models it was assumed that fracture occurred when the calculated volume fraction of cavities reached a critical value, f_c. More recently coupled models in which the effect of growing cavities on the constitutive equations of porous materials have been introduced (Rousselier, 1987; Gurson, 1977). The so-called Gurson–Tvergaard–Needleman (GTN) (1984, 1990) model involves two parameters, f_c and an acceleration factor, δ introduced to "simulate" void coalescence. It has been shown that the f_c and δ parameters have no unique values to fit the experimental results (Zhang and Niemi, 1994, 1995; Benzerga, 2002). This motivated further improvements which are briefly summarized.

3.2. Heterogeneous void nucleation

The effect of the inhomogeneity in spatial void distribution has been thoroughly investigated in two materials: (i) a cast duplex (ferrite + austenite) stainless steel (Devillers-Guerville et al., 1997) in which cavities in the austenite phase are nucleated from cleavage microcracks initiated in the thermally embrittled ferrite, and (ii) a plain carbon steel (A48) in which cavities are initiated from MnS inclusions (Bauvineau, 1996; Decamp et al., 1997). In both materials it was shown that the strain to failure was largely scattered and decreased when the specimen size was increased. It has been possible to partly account for this scatter and this size effect without using the δ accelerating factor provided that the inhomogeneity in the spatial distribution of cavity nucleation sites was taken into account. This suggests that the use of this parameter to simulate coalescence is not essential when the spatial distribution of initiation sites for ductile rupture is properly taken into account. Here again, this requires detailed metallographical observations.

3.3. Cavity growth

Much progress has been made since the pioneering studies by Berg (1962), Mc Clintock (1968), Rice and Tracey (1969) and Gurson (1977). In particular recent models have been proposed to account for plastic anisotropy and cavity shape anisotropy.

Following the method used by Gurson (1977) a yield surface for a plastically anisotropic material described by Hill quadratic criterion has been derived (Benzerga, 2000; Benzerga and Besson, 2001), assuming that the cavities remain spherical:

$$\phi_{GA} = \left(\frac{\sigma_H}{\sigma_o}\right)^2 + 2f \cosh\left(\frac{1}{h}\frac{\sigma_{kk}}{\sigma_o}\right) - 1 - f^2 = 0, \tag{21}$$

where σ_H is the Hill equivalent stress defined with six coefficients h_{ij}. In this equation, the parameter h has been expressed as a function of the h_{ij} coefficients. For an isotropic material $h = 2$ and Equation (21) corresponds to the Gurson potential.

An extension of the Gurson model has also been proposed by Gologanu, Leblond and Devaux (GLD) (1993, 1994) to account for cavity shape. The GLD model applies to axisymmetric ellipsoïdal cavities characterized by their aspect ratio, S. The model which is expressed in terms of a Gurson-like plastic potential is therefore limited to transversally isotropic porous plastic materials. This model has been extensively used by a number of authors (see, e.g. Pardoen and Hutchinson, 2000).

3.4. CAVITY COALESCENCE

Significant progress has also been made in modelling the onset of void coalescence by internal necking in ductile materials (Pardoen and Hutchinson, 2000; Benzerga, 2002; Benzerga et al., 2002). This last stage of ductile rupture is modelled using an extension of the Thomason model (Thomason, 1985) in which it is assumed that fracture occurs when the plastic limit load criterion originally proposed by this author is reached. In addition to the GLD potential these models give a set of constitutive equations including a closed form of the yield surface after void coalescence with appropriate laws for void shape and the size of the ligament between cavities. Details about these models can be found elsewhere (see, e.g. Pardoen and Besson, 2004). In Figure 6 the transition from elastic behaviour (a) to plastic void growth (b, c) to void coalescence in terms of the variation of the yield surfaces and current loading conditions are shown. In this figure σ_{eq} and σ_m represent the equivalent von Mises stress and the hydrostatic stress, respectively, while σ_f represents the flow stress of the material. In these models the load bearing capacity of the elementary volume decreases as a natural outcome of the void spacing reduction without introducing an *a priori* value for the δ coefficient in the GTN potential. These models are very promising since, theoretically, they are able to predict the strain to failure as a function of stress triaxiality when the initial microstructural parameters of the material (volume fraction of cavities, shape of the cavity initiation sites and void spacing) are known. The comparison between experimental results and theoretical results inferred from these sophisticated models are still limited (see Benzerga et al., 1999, 2004a, b), since these new models require quantitative information about the detailed microstructure of the materials.

3.5. DUCTILE-TO-BRITTLE TRANSITION

Charpy *V*-notch impact tests are still widely used to study the fracture properties of steels and, in particular, to define a ductile-to-brittle transition temperature (DBTT). Instrumented Charpy devices are increasingly used which allow to measure the whole force-displacement curve from which much information can be gained.

The first numerical simulation of the Charpy test was proposed by Norris (1979) using plane strain conditions. Effects of material rate sensitivity (Tvergaard and Needleman, 1986), of temperature dependence (Tvergaard and Needleman, 1988), of the 3D geometry (Mathur et al., 1993, 1994) and of specimen size (Benzerga et al., 2002) were studied in a series of theoretical papers lacking of comparisons with experiments. Such comparisons were carried out by other authors for pressure vessel steels (Böhme et al., 1992; Schmitt et al., 1997; Floch and Burdekin, 1999; Rossoll et al., 2002).

More recently a finite element simulation of the Charpy test has been developed in order to model the DBTT curve of A508 steel which is the material used for the fabrication of French nuclear pressure vessel reactors (Tanguy et al., 2005a, b). The

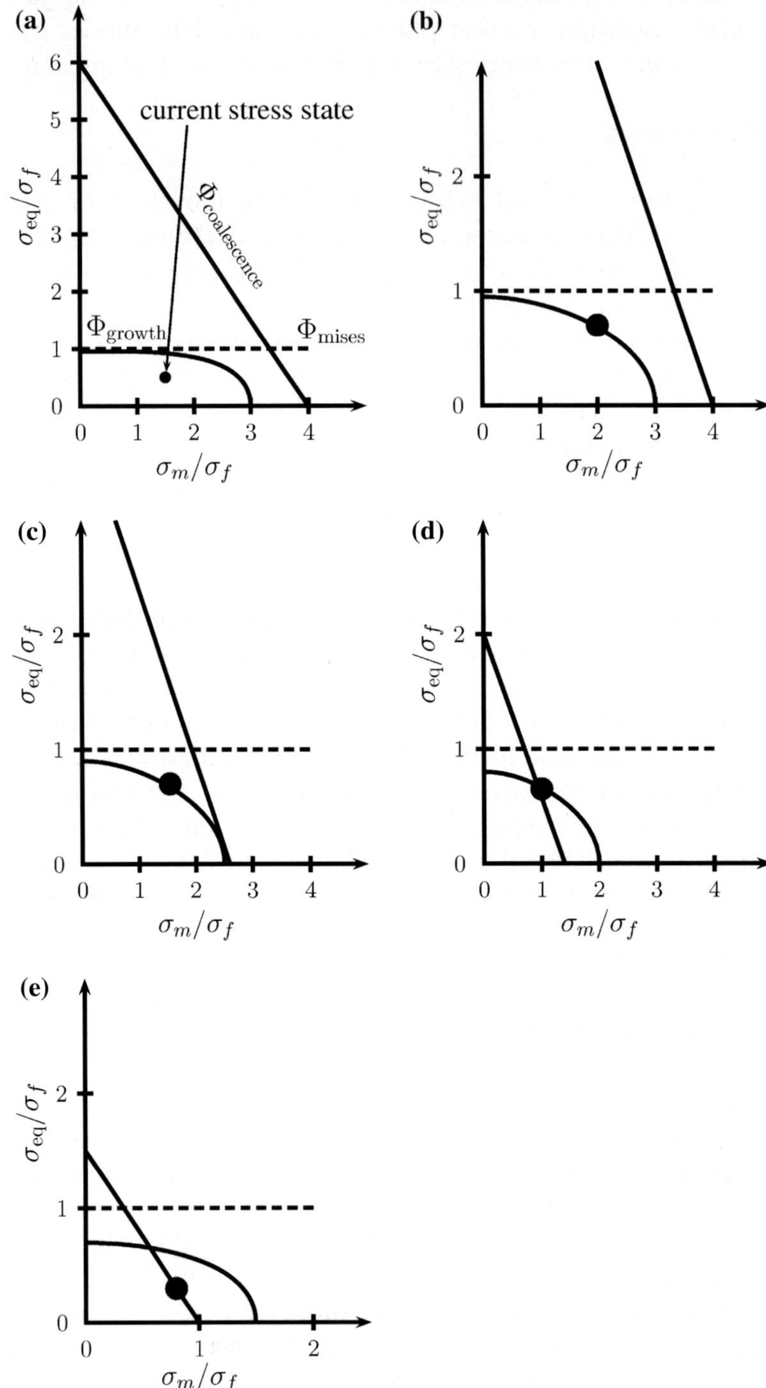

Figure 6. Ductile Fracture. Transition from (a) elastic behaviour, to (b) and (c) plasticity/void growth, to (d) and (e) plasticity/void coalescence. Yield surfaces and loading point as a function of increasing deformation. Calculations performed for a constant stress triaxiality equal to 3 (Pardoen and Besson, 2004).

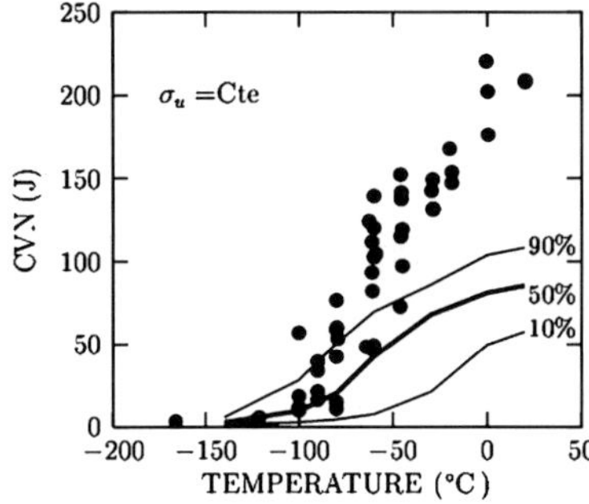
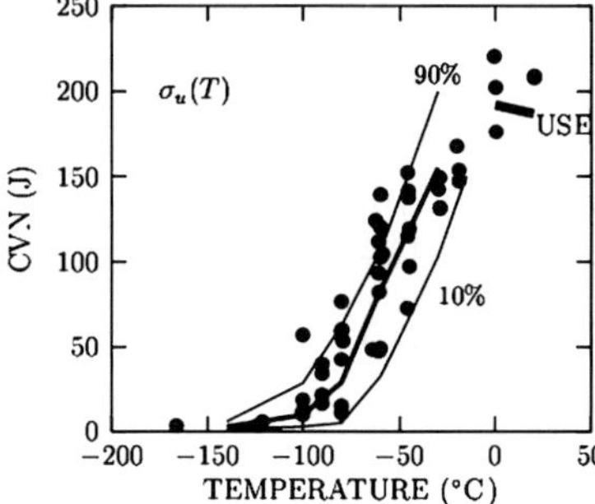

Figure 7. A 508 Steel. Charpy *V* notch energy as a function of temperature. Experimental results and numerical simulations using modified Rousselier model for ductile fracture and Beremin model for cleavage fracture. Three calculated curves corresponding to $P_R = 0.10$, 0.50 and 0.90 are shown. Results obtained with (a) $\sigma_u = $ cte, (b) σ_u increasing with temperature (Tanguy et al., 2005a, b).

simulation included a detailed description of the material viscoplastic behaviour over a wide temperature range. Ductile fracture was modelled using modified Rousselier model, while the Beremin model was used to simulate cleavage fracture. The Charpy test was modelled using a full 3D mesh and accounting for adiabatic heating and contact between the specimen, the striker and the anvil. Figure 7 shows that it has been possible to correctly model the DBTT curve assuming that the Beremin stress parameter σ_u (see Equation (9)) was slightly temperature dependent. As indicated previously, this temperature dependence might be related to a change in the nature of the microstructural barriers controlling cleavage with temperature.

4. Fracture toughness

4.1. INTRODUCTION. MAIN CHARACTERISTICS OF BRITTLE FRACTURE TOUGHNESS

The first characteristics of brittle fracture in ferritic steels is the scatter in test results, as illustrated in Figure 8. The absolute value of this scatter increases with test temperature but not necessarily the relative value, although at low temperature the scatter tends to be lower.

Another important aspect of brittle cleavage fracture is the size effect. For instance, Iwadate et al. (1985) have determined the fracture toughness, K_{IC}, of a pressure vessel steel using CT specimens with various thicknesses and in-plane dimensions. These authors showed that the mean value of fracture toughness was a decreasing function of specimen size. These results may reflect the effect of changing constraint conditions but, as shown later, the theory for brittle fracture predicts that the fracture toughness, K_{IC}, varies with specimen thickness, B, with a universal slope, such as the product $K_{IC}^4 B$ is constant for a given temperature.

A third factor important in brittle fracture of ferritic steels is the so-called "short crack" effect or, more generally, the geometrical dependence of K_{IC} or K_{JC}. This dependence has been thoroughly investigated and modelled (see, e.g. Dodds et al., 1991; Petti and Dodds, 2004).

A fourth factor associated with cleavage fracture is the effect of metallurgical factors, such as the grain size, the nature of packet boundaries in bainitic steels, the amount and the size of second-phase particles. In particular, as shown later, the fracture toughness of mutipass steel welds is largely dependent on the amount of local brittle zones (LBZs) which are found in the coarse grain heat affected zones (CGHAZs) and in the intercritically reheated CGHAZ (ICCGHAZs) (see, e.g. Kenney et al., 1997; Zhou and Lin, 1998; Lambert-Perlade et al., 2004).

4.2. BEREMIN MODEL FOR CLEAVAGE FRACTURE TOUGHNESS AND ITS EXTENSIONS

4.2.1. *Small-scale yielding conditions*

The stress–strain field ahead of the crack tip under SSY conditions is simply scaled by the ratio $x/(J/\sigma_{YS})$ where x is the distance from the crack front. Under these conditions the application of the Beremin model (Equation (13)) is straightforward (Pineau, 1982, 1992, 2003; Beremin, 1983, Mudry, 1987). The probability to fracture of a specimen containing a 2D crack expressed in terms of K_{IC} can simply be written as:

$$P_R = 1 - \exp\left[-\frac{K_{IC}^4 B \sigma_{YS}^{m-4} C_m}{V_u \sigma_u^m} \right],\tag{22}$$

where C_m is a numerical factor which is function of the work-hardening exponent, n, of the material $\left(\sigma = K' \varepsilon^n\right)$. It is well to remember that the term $K_{IC}^4\, B$ simply arises from the volume of material which is plastically deformed, i.e. the product of the square plastic zone size $(K_{IC}/\sigma_o)^2$ and the specimen thickness, B. The same size dependence is expected whatever the statistical law is used provided that the weakest link theory does apply.

Equation (22) can be extended to a 3D crack of length ℓ for which the stress intensity factor, K, is a function of the curvilinear abscissa, s, along the crack front. It can easily be shown that, for a given probability to fracture, the following expression is satisfied:

$$\int K_I^4(s)\,\mathrm{d}s = K_{IC}^4 \ell\tag{23}$$

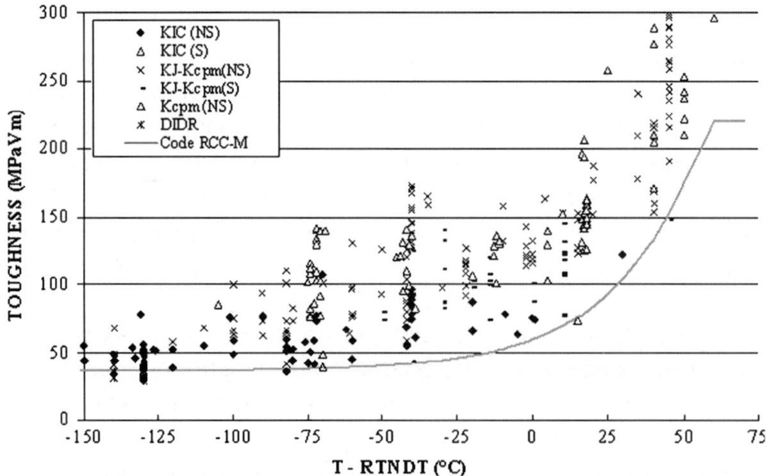

Figure 8. A 508 steel. Scatter in fracture toughness tests (Naudin et al., 2001).

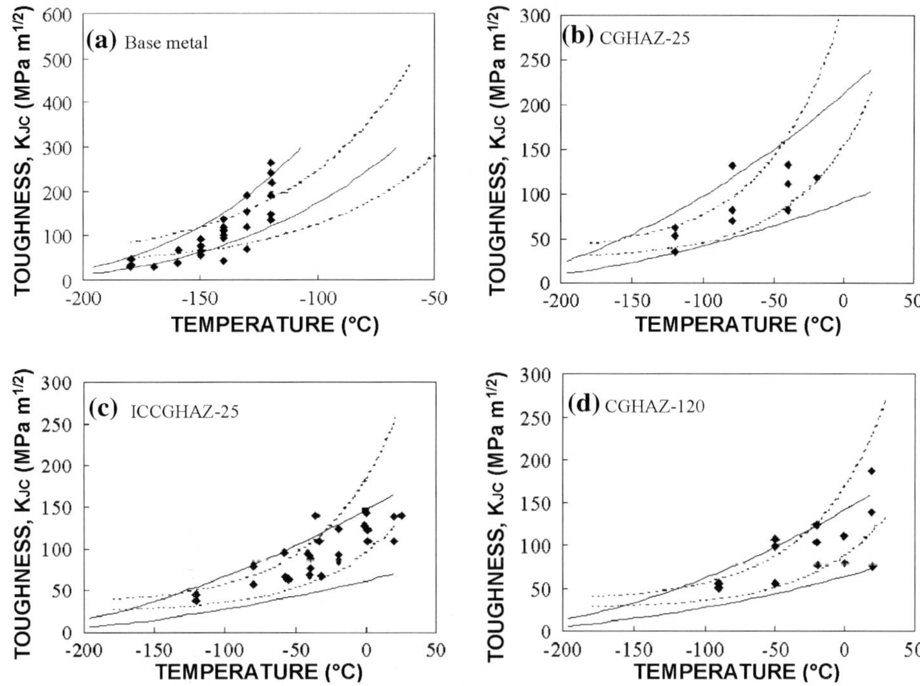

Figure 9. Bainitic steel. Ductile-to-brittle toughness transition (a) Base metal, (b) CGHAZ-25, (c) ICCGHAZ-25 and (d) CGHAZ-120 microstructures. Solid lines (respectively, dotted lines) show fracture probabilities of 10 and 90 pct given by the Beremin model (respectively, by the "Master curve" approach (Lambert-Perlade, et al., 2004).

Equations (14) and (22) contain only two independent parameters, as shown previously. However in this simple theory there is a hidden supplementary condition, which is the initiation of plastic deformation as a prerequisite to initiate cleavage. This condition given in Equation (16) does not appear explicitly in Equation (22). However this does not mean as often claimed in the literature that Equation (22) implies that fracture can occur even for infinitely small values of K (Pineau, 1982).

The Beremin model has now been applied to a large amount of steels. Recent results on a bainitic steel heat-treated in order to simulate different welding conditions are shown in Figure 9 (Lambert-Perlade et al., 2004).

In this figure, we have included the results obtained in the as-received conditions (ferrite + perlite) and those corresponding to three welding conditions which produced various amounts of LBZs formed by M-A constituents in a bainitic matrix. As indicated previously, cleavage micro-cracks are initiated from these M-A particles. Figure 9 shows that the Beremin model is able to describe the evolution of K_{JC} with temperature and the scatter in test results. The values of the parameters, σ_u and m for Beremin model are given in Table 2. These values are typical of those found in other steels (Beremin, 1983; Mudry, 1987). In Figure 9, we have also included the theoretical curves derived from the Master Curve approach which will be briefly described hereafter.

A number of authors (see, e.g. Bakker and Koehrs, 1991; Xia and Shih, 1996; Petti and Dodds, 2004 and others) have introduced a threshold stress, σ_{th} into the calculation of the Weibull stress Equation (14). As stated previously the rational calibration

Table 2. High-strength low alloy steel. Parameters of the Beremin model [Unit volume $V_u = (100\mu m)^3$] and of the "Master Curve" approach for four investigated microstructures; T_0 is the temperature for which $K_{JC} = 100\,\mathrm{MPa}\sqrt{m} \cdot T_{100}^*$ is the value of T_0 determined experimentally.

Microstructure	Beremin model		Master curve approach	Experimental values
	σ_u (MPa)	m	fitted value of T_o (°C)	of T_{100}^* (°C)
Base metal	2158	27	−130	−140
CGHAZ-25	2670	20	−45	−55
ICCGHAZ-25	2351	20	−12	−20
CGHAZ-120	2085	20	−6	−10

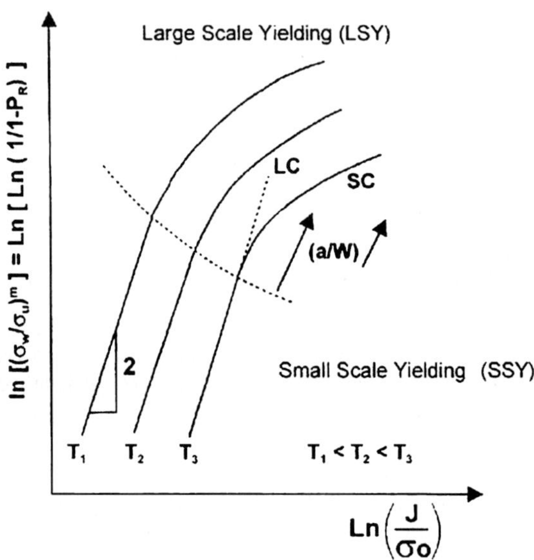

Figure 10. Sketch showing the variation of the probability to fracture with loading (J/σ_o) at various temperatures. The transition from SSY to LSY conditions is indicated by a dotted line. The difference between long-cracks (LC) and short-cracks (SC) related to constraint effect is shown.

procedures to determine σ_{th} remain an open issue. To introduce an explicit threshold toughness into Weibull stress model, Gao et al. (1998) have proposed a modified form for Equation (14), by introducing another parameter, $\sigma_{w\ min}$, as also stated previously.

4.2.2. *Large-scale yielding conditions*

Equation (22) cannot be directly applied when SSY conditions are no longer prevailing. The influence of the loss of constraint effect on the Weibull stress for LSY conditions is schematically shown in Figure 10.

In this figure, the increase in the fracture toughness of specimens tested at increasing temperatures is schematically shown. A quantitative analysis of these effects associated with the loss of constraint effect requires the use of finite element method (FEM) calculations. A recent study has presented a practical approach to compare directly the two most commonly tested specimens, the single-edge notched bend, SE(B) and the compact tension, $C(T)$ specimens (Petti and Dodds, 2004). Some of

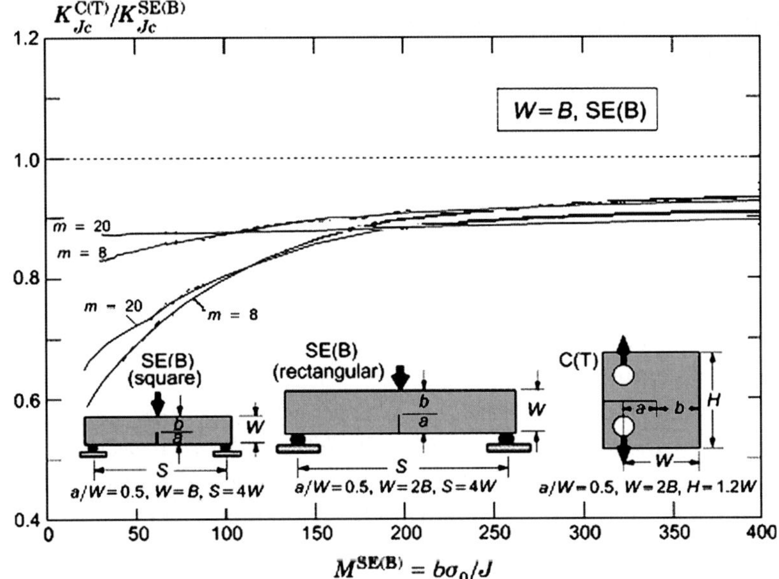

Figure 11. Comparison of K_{JC} for $C(T)$ specimens to K_{JC} values for square and rectangular SE(B) specimens. Results for material flow properties $\varepsilon = (\sigma_0/E)\,(\sigma/\sigma_0)^n$ (Petti and Dodds, 2004).

the results obtained by these authors and established for specific values of m (8 and 20) and the ratio $1/n$ (10) are shown in Figure 11.

This figure shows the toughness scaling with $K_{JC}^{C(T)}/K_{JC}^{SE(B)}$ plotted versus the non-dimensional loading parameter for the square ($W = B$) and the rectangular ($W = 2B$) SE (B) specimen with $M = b\sigma_0/J$. These results show that the ratio of $K_{JC}^{C(T)}/K_{JC}^{SE(B)}$ remains fairly constant when $M > 200$. On the other hand this ratio decreases to the range 0.60–0.65 for the $1/n = 10$ material when $M \cong 30$. Similar types of analysis can therefore provide the basis of calibration procedures when testing materials with different specimen geometries.

As indicated earlier a number of researchers have shown that the initiation of cleavage fracture in a deformed material was made more difficult by strain. This is why in the Beremin model a strain correction was introduced (see Equation (17)). This "correction" or those proposed by others produces an increase in the curvature of the K_{IC} versus temperature curve. This is often necessary to account for the strong increase of fracture toughness in the ductile to brittle transition (DBT) regime.

4.2.2.1. *Master Curve concept.* The ASTM E1921 (2002) testing standard employs a weakest link based model to characterize cleavage fracture in ferritic steels over the low-to-mid portion of the DBT curve. This standard relies on the studies made by Wallin (1991a, b) and Wallin et al. (1984). This author has expressed the fracture toughness scatter as:

$$P_R = 1 - \exp\left[-\frac{B}{B_o} \left(\frac{K_{IC} - K_{I\,min}}{K_o - K_{I\,min}} \right)^4 \right], \tag{24}$$

where B_o is an arbitrary (normalized) thickness, K_o is a parameter depending on temperature while $K_{I\,min}$ is a limiting value below which fracture is impossible

($K_{\mathrm{I\,min}} \simeq 20\,\mathrm{MPa\,m}^{1/2}$). The scale parameter K_o corresponds to a 63% cumulative probability level for specimen failure by cleavage. In the MC concept the shape of the median K_{JC} toughness, K_{JC} (medium) for $1T$ specimens is assumed to be described by an universal law:

$$K_{\mathrm{JC}}(\mathrm{med}) = 30 + 70\exp[0.019(T - T_\mathrm{o})], \tag{25}$$

where K is in $\mathrm{MPa\,m}^{1/2}$ and T in °C. In this expression, T_o corresponds to the temperature at which the mean (median) fracture toughness for a $25\,\mathrm{mm}$ thick specimen has the value of $100\,\mathrm{MPa\,m}^{1/2}$. Details about the determination of T_o can be found in ASTM E 1921 standards. The MC concept which relies also on the weakest link-based scaling model requires theoretically that strict plane-strain SSY conditions exist along the entire crack front at fracture. The validity of this assumption has been discussed recently by Petti and Dodds (2004). As most often written, Equation (25) gives the fracture probability for $1T$ size specimens. When the test specimen has other than $1T$ thickness, the as tested xT size toughness scales to $1T$ thick specimen as follows:

$$K_{\mathrm{JC}}^{1T} = K_{\mathrm{I\,min}} + \left(K_{\mathrm{JC}}^{xT} - K_{\mathrm{I\,min}}\right)\left(B_{xT}/B_{1T}\right)^{1/4} \tag{26}$$

The comparison between the theoretical expressions derived from Beremin model (Equation (22)) and the MC concept (Equation (24)) shows that the fracture toughness dependences with specimen thickness are similar. Moreover when $K_{\mathrm{I\,min}}$ can be neglected in Equation (25), i.e. when the temperature is in the mid range of the DBT curve, the implicit threshold implied in the application of the Beremin model is also much lower than the fracture toughness. Equations (22) and (25) can then be directly compared. Both models predict the same toughness temperature dependence provided that the ratio of $K_\mathrm{o}(T)/\sigma_\mathrm{o}(T)^{1-m/4}$ is constant. This ratio was calculated in the steel containing M.A. constituents described earlier (Lambert-Perlade et al. 2004). It was found that for K_I and K_o values much higher than $K_{\mathrm{I\,min}}$ (i.e. $T > -150°\mathrm{C}$ for the as-received metal and $T > -100°\mathrm{C}$ for the simulated HAZ microstructures) this ratio varies by less than 20%. This means that in this particular case both the Beremin model and the MC approach lead to similar results, as shown in Figure 9. The values of parameter T_o used in Equation (25) are given in Table 2 with those of σ_u and m of the Beremin model. In Figure 9 it is difficult to conclude that a model is better than another one. The MC approach leads to a less-satisfactory agreement with the base metal due to the $K_{\mathrm{I\,min}}$ threshold of $20\,\mathrm{MPa\,m}^{1/2}$. On the other hand a close examination to the theoretical curves shows that the MC curves have a higher slope than those inferred from the Beremin model. This might be partly due to the fact that the 3PB fracture toughness specimens have been calculated using 2D FEM modelling and assuming strict plane strain conditions.

4.3. MULTIPLE BARRIER MODEL

This model has been applied to the steel in which various HAZs were simulated. According to MB model (see Section 2.1.2) cleavage fracture occurs by following three steps successively, as shown in Figure 4. Using the weakest link-based theory the fracture probability, P_R is then given by the combined conditional probabilities of the three steps involved during fracture. A number of authors have shown that the critical local stress necessary to cross the grain boundaries is lowest at low temperature but increases with temperature and eventually becomes larger than the stress necessary to propagate a crack nucleated from a particle (see, e.g. Narström and Isacsson, 1999). The effect of temperature on fracture toughness can thus be described by considering four temperature ranges.

(i) At very low temperatures the critical step is the nucleation of a microcrack from M-A particles.

(ii) At somewhat higher temperatures microcracks initiate at particles and stop at the particle/matrix interface. The critical step is then the propagation of these particle sized microcracks.

(iii) At higher temperatures microcracks are stopped at grain boundaries and the critical step is the propagation of these arrested grain sized microcracks.

(iv) At still higher temperatures ductile fracture eventually nucleated from particles occurs before cleavage fracture. This temperature range is not considered.

The input parameters of the MB model are therefore:

(i) The fracture probability $p(c)$ of a M-A particle of size (C). It was simply assumed that this initiation process occurred for a single valued of the critical stress (see Equation (2)). It was assumed that $\sigma_d = 2112\,\mathrm{MPa}$ (see Table 1).

(ii) The parameters $f_c(C)$ and $f_g(D)$ giving the distribution functions of particles and bainitic packets. These functions were experimentally determined. It was shown that they could be approximated by log-normal functions.

(iii) The critical size for cracked MA particles and cracked bainitic packets given by Equation (3). For the sake of simplicity, it was assumed that $K_I^{c/f}$ and $K_I^{f/f}$ were independent of temperature as indicated previously. Therefore, in this MB model, the temperature dependence of fracture toughness arises mainly from the variation of yield strength with temperature, as in the Beremin model.

The probability to failure can then simply be expressed using these input parameters, as shown elsewhere (Lambert-Perlade et al., 2004). The numerical values of the input parameters have already been given in Table 1. The results showing the application of the MB model to one specific condition (ICCGHAZ) are reported in Figure 12. A good quantitative agreement is observed when the comparison between the theory (solid lines) and the experimental values is made. In particular the model is able to account for the dispersion which is not trivial since the calculated scatter derives directly from the experimental size distribution of second phase particles and bainitic packets. In Figure 12 the lowest value of fracture toughness (open symbols) corresponding to the first detection of microcrack events detected by acoustic emission occurs for stress intensity factors close to $30-40\,\mathrm{MPa\,m}^{1/2}$. These values agree with the calculated probability for a cleavage microcrack to propagate across the particle/matrix boundary which is shown by dotted lines. The model predicts not only the evolution of fracture toughness with temperature but also the critical stress intensity factor for the development of temporarily stable grain-size microcracks. Clearly this is a superiority of this physically based model.

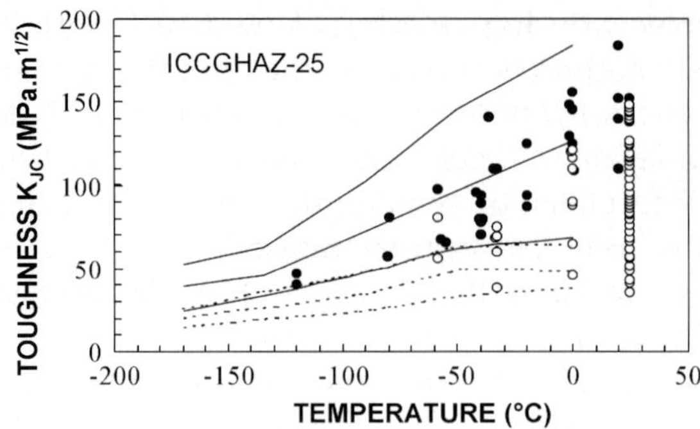

Figure 12. Bainitic steel. Prediction of the variation of fracture toughness with temperature obtained from multiple barrier model. Open circles denote microcrack events detected by acoustic emission; solid circles correspond to final fracture. Solid (respectively, dotted) lines represent 10, 50 and 10 pct probabilities for the specimen to fracture (respectively, for a cleavage microcrack to propagate across a particle/matrix boundary) as given by the MB model. Numerical values used in this model are given in Table 1. (Lambert-Perlade et al., 2004).

4.4. APPLICATIONS

The LAF methodology has received an increasing number of applications over the last past 25 years (for a review, see Pineau, 2003). Three main domains have been explored. They are briefly considered below.

4.4.1. *Metallurgical improvements of materials*

It is becoming clear that in the upper part of the DBT curve, the fracture toughness is related to the propagation of microcracks which are arrested at grain boundaries in ferritic steels or at packet boundaries in bainitic steels.

Recent studies have shown that in bainitic steels the effective packet boundaries are those with a high angle misorientation (see, e.g. Bouyne et al., 1998; Gourgues et al., 2000; Lambert-Perlade et al., 2004). The improvement of these materials requires the development of these favourable boundaries. This rises a new research field in which the metallurgical variables controlling the formation of these specific boundaries are investigated. Fundamental studies on the crystallography and the microstructure of bainitic transformations must therefore be encouraged, using either chemical and thermomechanical manipulations or other tools like phase transformations under intense magnetic fields which have not yet been explored in much detail (see, e.g. Jaramillo et al., 2005).

4.4.2. *Mechanical and numerical testing*

A large number of researches over the past 25 years have yielded a sufficient quantitative understanding of the DBT behaviour to develop engineering approaches that characterize the scatter and temperature dependence of macroscopic fracture toughness measured in terms of J_C or equivalently K_{JC}. These efforts have led to the development of the ASTM Standard Test Method E 1921 which, to some extent, can be considered as the engineering application of the local approach theory. The ASTM test method has largely contributed to the development of the procedures for quantitative evaluation of cleavage fracture toughness in through-crack, high-constraint laboratory test specimens, e.g. deep-notch SE(B)s and $C(T)$s. The transferability of this method to real applications requires additional models that account for large constraint differences and variations in local J values along a crack front. Micromechanical models for brittle cleavage fracture as those presented previously offer the most promising approach at present to understand toughness transferability issues and to develop quantitative frameworks. Calibration of the Weibull stress scale parameter, σ_u, using the MC approach has been presented recently by Petti and Dodds (2005). These authors have used both experimental results and numerical calculations performed on cracked specimens to determine the σ_u parameter in Equation (22) assuming that the shape factor m was independent of temperature.

4.4.3. *Complex thermomechanical loading conditions*

One example of complex thermomechanical loading conditions is that of earth quakes. Major factors promoting brittle fracture during earth quakes are prestraining and dynamic loading. It is well known that an increase in loading rate leads to a shift of the DBT curve towards higher temperatures (see, e.g. Henry et al., 1985). The effect of loading rate and prestraining on the fracture toughness of a structural steel has been investigated by Minami and Arimochi (2001). These authors showed that the Beremin model was able to account for the significant decrease of the fracture toughness measured in terms of CTOD when prestraining and dynamic loading were applied.

Similarly a number of researchers have used the local approach to cleavage fracture to predict the failure behaviour of heat affected zones (HAZ) in welds (see, e.g. Cardinal et al., 1996; Matos and Dodds, 2001; Lambert-Perlade et al., 2004).

Many studies have been devoted to the so-called mismatch effect on fracture toughness of welded joints (see, e.g. Kim and Schwalbe, 2001a, b). This effect is related to the difference in strength between the base material and the weld metal. This effect has been investigated by many authors (see, e.g. Ohata et al., 1996). These authors showed also that the Beremin model was able to account for their experimental results.

The Beremin model has also been applied to warm prestressing (WPS) effect (Beremin, 1981b; Roos et al., 1998; Hadidi-moud et al., 2004). This effect plays a key role in the structural integrity assessment of nuclear pressure vessel (see, e.g. Lefèvre et al., 2002). A specimen or a component prestressed in the upper-shelf domain of the DBT curve exhibits an apparent increase in fracture toughness when subsequently tested at lower temperature. The application of this model to load-unload-cool-fracture (LUCF) condition is extremely sensitive to the constitutive equations which are used to calculate the residual stresses after the load-unload part of the cycle (see Lefevre et al., 2002). This might explain why the Beremin model was shown to underestimate the fracture toughness results obtained with LUCF loading cycle (Stöckl et al., 2000).

4.4.4. *Component testing*

The LAF methodology has also been applied to assess the brittle fracture of large structural components. Two examples including: testing of a large mock-up and thermal shock on large cylinders are given elsewhere (Pineau, 2003).

5. Conclusions

This review paper has attempted to show that the local approach to fracture has largely contributed to a better understanding of the fracture behaviour of metallic materials. Micromechanical models for brittle cleavage fracture, such as the Beremin model and its extensions, offer the most promising approach to tackle with toughness transferability issues. The development of standards should largely contribute to the application of the local approach to fracture.

Many research areas remain to be explored. They are related to both fracture micromechanisms and fracture toughness modelling. In particular, it is expected that MB multiple barrier type models can largely help in a better understanding of the DBTT curve. These models must rely on a firm basis of experimental results. Acoustic emission technique and other experimental techniques such as X-ray tomography should largely contribute to the development of this basis. However for most practical applications, due to the complexity of the problem, it is necessary to simplify the analysis and in particular, to develop and test simple criteria.

The local approach methodology could also be applied to other engineering materials, such as metal matrix composites and concrete or to other failure modes in steels, such as intergranular fracture due to the segregation of impurities along grain boundaries.

Acknowledgements

The author would like to acknowledge all the former Ph.D students of his research group. Thanks also to Drs. J. Besson, S. Bordet, A.-F. Gourgues and B. Tanguy.

References

ASTM E1921 (2002). *Test Method for the Determination of Reference Temperature To for Ferritic Steels in the Transition Range.* American Society for Testing and Materials, Philadelphia.

Bakker, A. and Koers, R.W.I. (1991). Prediction of cleavage fracture events in the brittle-ductile transition region of a ferritic steel. In: *Defect Assessment in Components-Fundamentals and Applications, ESIS/EG9* (Edited by Blauel, J.G. and Schwalbe, K.-H.). Mechanical Engineering Publications, London, 613–632.

Bauvineau, L. (1996). Approche locale de la rupture ductile: applications à un acier Carbone-Manganèse. PhD thesis, Ecole des Mines de Paris.

Benzerga, A.A. (2000). Rupture ductile des tôles anisotropes. PhD thesis, Ecole des Mines de Paris.

Benzerga, A.A. (2002). Micromechanics of coalescence in ductile fracture. *Journal of the Mechanics and Physics of Solids* **50**, 1331–1362.

Benzerga, A., Besson, J. and Pineau, A. (1999). Coalescence-controlled anisotropic ductile fracture. *Journal of Engineering Materials and Technology* **121**, 221–229.

Benzerga, A., Tvergaard, V. and Needleman, A. (2002). Size effects in the Charpy V-notch test. *International Journal of Fracture* **116**, 275–296.

Benzerga, A.A. and Besson, J. (2001). Plastic potentials for anisotropic porous solids. *European Journal of Mechanics A/Solids* **20**, 397–434.

Benzerga, A.A., Besson, J. and Pineau, A. (2004a). Anisotropic ductile fracture: Part I: experiments. *Acta Materialia* **52**, 4623–4638.

Benzerga, A.A., Besson, J. and Pineau, A. (2004b). Anisotropic ductile fracture. Part II: theory. *Acta Materialia* **52**, 4639–4650.

Beremin, F.M. (1981a). Cavity formation from inclusions in ductile fracture of A508 steel. *Metallurgical Transactions A* **12A**, 723–731.

Beremin, F.M. (1981b). Numerical modeling of warm prestress effect using a damage function for cleavage fracture. In: *ICF5, Advances in Fracture Research*, (Edited by François, D.). Pergamon, New York, Vol II 825–832.

Beremin, F.M. (1983). A local criterion for cleavage fracture of a nuclear pressure vessel steel. *Metallurgical Transactions A* **14**, 2277–2287.

Berg, A. (1962). Proceedings of the motion of cracks in plane viscous deformation. In: 4th US National Congress of Applied Mechanics, (Edited by Rosenberg, R.M.) University of California, CA, 885–892.

Bernauer, G., Brocks, W. and Schmitt, W. (1999). Modification of the Beremin model for cleavage fracture in the transition region of a ferritic steel. *Engineering Fracture Mechanics* **64**, 305–325.

Besson, J. (2004). Local Approach to Fracture. *Les Presses, Ecole des Mines, Paris, France.*

Böhme, W., Sun, D., Schmitt, W. and Höning, A. (1992). Application of micromechanical material models to the evaluation of Charpy tests. In: *Advances in Fracture/Damage Models for the Analysis of Engineering Problems.* (Edited by Giovanola, J.). *ASME,* 203–216.

Bordet, S.R., Karstensen, A.D., Knowles, D.M. and Wiesner, C.S. (2005a). A new statistical local criterion for cleavage fracture in steel. Part I - Model presentation. *Engineering Fracture Mechanics* **72**, 435–452.

Bordet, S.R., Karstensen, A.D., Knowles, D.M. and Wiesner, C.S. (2005b). A new statistical local criterion for cleavage fracture in steel. Part II – Application to an offshore structural steel. *Engineering Fracture Mechanics* **72**, 453–474.

Bouyne, E., Flower, H.M., Lindley, T.C. and Pineau, A. (1998). Use of EBSD technique to examine microstructure and cracking in a bainitic steel. *Scripta Materialia* **39**, 295–300.

Carassou, S., Renevey, S., Marini, B. and Pineau, A. (1998). Modelling of the ductile to brittle transition of a low alloy steel. In: *Fracture from Defects, ESIS/ECF12*, (Edited by Brown, M.W., de los Rios, E.R. and Miller, K.J.) EMAS, Chameleon Press, London **2**, 691–696.

Cardinal, J., Wiesner, C.S., Goldthorpe, M.C. and Bannister, A.C. (1996). Application of the local approach to cleavage fracture to failure prediction of heat affected zones. Euromech-Mecamat. *Journal de Physique IV* **6**, C6–C6–194.

Chen, J.H., Wang, Q., Wang, G.Z. and Li, Z. (2003). Fracture behavior at crack Tipp – a new framework for cleavage mechanism of steel. *Acta Materialia* **51**, 1841–1855.

Curry, D.A. and Knott, J.F. (1979). Effect of microstructure on cleavage fracture toughness of quenched and tempered steels. *Metal Science* **13**, 341–345.

Decamp, K., Bauvineau, L., Besson, J. and Pineau, A. (1997). Size and geometry effect on ductile rupture of notched bars in a C-Mn steel : experiments and modelling. *International Journal of Fracture* **88**, 1–18.

Devillers-Guerville, L., Besson, J. and Pineau, A. (1997). Notch fracture toughness of a cast duplex stainless steel: modelling of experimental scatter and size effects. *Nuclear Engineering and Design* **168**, 211–225.

Dodds, R.H., Anderson, T.L. and Kirk, M.T. (1991). A framework to correlate a/w ratio effects on elastic-plastic fracture toughness (Jc). *International Journal of Fracture* **48**, 1–22.

Faleskog, J., Kroon, M. and Oberg, H. (2004). A probabilistic model for cleavage fracture with a length scale-parameter estimation and predictions of stationary crack experiments. *Engineering Fracture Mechanics* **71**, 57–79.

Floch, L. and Burdekin, F. (1999). Application of the coupled brittle ductile model to study correlation between Charpy energy and fracture toughness values. *Engineering Fracture Mechanics* **63**, 57–80.

François, D. and Pineau, A. (2001). Fracture of metals. Part II: ductile fracture. In: *Physical Aspects of Fracture*, (Edited by Bouchaud, E., Jeulin, D., Prioul, C. and Roux, S.) Kluwer, Dordrecht, 125–146.

Gao, X. and Dodds, R.H. (2000). Constraint effects on the ductile-to-brittle transition temperature of ferritic steels : a Weibull stress model. *International Journal of Fracture* **102**, 43–69.

Gao, X., Dodds, R.H., Tregoning, R.L., Joyce, J.A. and Link, R.E. (1999). A Weibull stress model to predict cleavage fracture in plates containing surface cracks. *Fatigue and Fracture of Engineering Materials and Structures* **22**, 481–493.

Gao, X., Ruggieri, C. and Dodds, R.H. (1998). Calibration of Weibull stress parameters using fracture toughness data. *International Journal of Fracture* **92**, 175–200.

Garrison, W.M. and Moody, N.R. (1987). Ductile rupture. *Journal of Physics and Chemistry of Solids* **48**, 1035–1074.

Gologanu, M., Leblond, J.-B. and Devaux, J. (1993). Approximate models for ductile metals containing non-spherical voids – Case of axisymmetric prolate ellipsoidal cavities. *Journal of the Mechanics and Physics of Solids* **41**, 1723–1754.

Gologanu, M., Leblond, J.-B. and Devaux, J. (1994). Approximate models for ductile metals containing non-spherical voids - Case of axisymmetric oblate ellipsoidal cavities. Trans. ASME. *Journal of Engineering Materials and Technology* **116**, 290–294.

Gourgues, A.F., Flower, H.M. and Lindley, T.C. (2000). Electron backscattering diffraction study of acicular ferrite, bainite, and martensite steel microstructures. *Materials Science and Technology* **16**, 26–40.

Gurson, A. (1977) Continuum theory of ductile rupture by void nucleation and growth: Part I – Yield criteria and flow rules for porous ductile media. *Journal of Engineering Materials and Technology* **99**, 2–15.

Hadidi-moud, S., Mirzaee-Sisan, A., Truman, C.E. and Smith, D.J. (2004). A local approach to cleavage fracture in ferritic steels following warm pre-stressing. *Fatigue and Fracture of Engineering Materials and Structures* **27**, 931–942.

Hahn, G.T. (1984) The influence of microstructure on brittle fracture toughness. *Metallurgical Transactions* **15A**, 947–959.

Henry, M., Marandet, B., Mudry, F. and Pineau, A. (1985). Effets de la température et de la vitesse de chargement sur la ténacité à rupture d'un acier faiblement allié – Interprétation par des critères locaux. *Journal de Mécanique Théorique et Appliquée* **4**, 741–768.

Iwadate, T., Tanaka, Y., Takemata, H. and Kabutomori, T. (1985). Elastic–plastic fracture toughness behavior of heavy section steels for nuclear pressure vessels. *Nuclear Engineering and Design* **87**, 89–99.

Jaramillo, R.A., Babu, S.S., Ludtka, G.M., Kisner, R.A., Wilgren, J.B., Mackiewicz-Ludtka, G., Nicholson, D.M., Kelly, S.M., Murugananth, M. and Bhadeshia, H.K.D.H. (2005). Effect of 30T magnetic field on transformations in a novel bainitic steel. *Scripta Materialia* **52**, 461–466.

Kaechele, L.E. and Tetelman, A.S. (1969). A statistical investigation of microcrack formation. *Acta Metallurgica* **17**, 463–475.

Kenney, K.L., Reuter, W.G., Reemsnyder, H.S. and Matlock, D.K. (1997). Fatigue and Fracture Mechanics, ASTM STP 1321, (Edited by Underwood, J.H., Macdonald, B.D. and Mitchell, M.R.). *ASTM*, Philadelphia, PA, **28**, 427–449.

Kim, Y.J. and Schwalbe, K.-H. (2001a). Mismatch effect on plastic yield loads in idealised weldments: I. Weld center cracks. *Engineering Fracture Mechanics* **68**, 163–182.

Kim, Y.J. and Schwalbe, K.-H. (2001b). Mismatch effect on plastic yield loads in idealised weldments: II. Heat affected zone cracks. *Engineering Fracture Mechanics* **68**, 183–199.

Kroon, M. and Faleskog, J. (2002). A probabilistic model for cleavage fracture with a length scale-effect of materials parameters and constraint. *International Journal of Fracture* **118**, 99–118.

Lambert-Perlade, A., Gourgues, A.F., Besson, J., Sturel, T. and Pineau, A. (2004). Mechanisms and modeling of cleavage fracture in simulated heat-affected zone microstructures of a high-strengh low alloy steel. *Metallurgical and Materials Transactions A* **35A**, 1039–1053.

Lee, S., Kim, S., Hwang, B., Lee, B. and Lee, C. (2002). Effect of carbide distribution on the fracture toughness in the transition temperature region of an SA 508 steel. *Acta Materialia* **50**, 4755–4762.

Lefèvre, W., Barbier, G., Masson, R. and Rousselier, G. (2002). A modified Beremin model to simulate the warm pre-stress effect. *Nuclear Engineering and Design* **216**, 27–42.

Margolin, B.Z., Gulenko, A.G. and Shvetsova, V.A. (1998). Improved probabilistic model for fracture toughness prediction for nuclear pressure vessel steels. *International Journal of Pressure Vessel and Piping* **75**, 843–855.

Martin-Meizoso, A., Ocana-Arizcorreta, I., Gil-Sevillano, J. and Fuentes-Pérez, M. (1994). Modelling cleavage fracture of bainitic steels. *Acta Metallurgica et Materialia* **42**, 2057–2068.

Mathur, K., Needleman, A. and Tvergaard, V. (1993). Dynamic 3D analysis of the Charpy V-notch test. Modelling simulation. *Materials Science and Engineering* **1**, 467–484.

Mathur, K., Needleman, A. and Tvergaard, V. (1994). 3D analysis of failure modes in the Charpy impact test. Modelling Simulation. *Materials Science and Engineering* **2**, 617–635.

Matos, C.G. and Dodds, R.H. (2001). Modelling the effects of residual stresses on cleavage fracture in welded steel frames. In: *ICF 10 Conference*, Hawaï.

Mc Clintock, F.A. (1968). A criterion for ductile fracture by the growth of holes. *Journal of Applied Mechanics* **35**, 363–371.

Mc Clintock, F.M. (1971). Plasticity aspects of fracture. In: *Fracture*, (Edited by Liebowitz, H.) Academic Press, New-York and London, 3, 47–225.

Minami, F. and Arimochi, K. (2001). Evaluation of prestraining and dynamic loading effects on the fracture toughness of structural steels by the local approach. *Journal of Pressure Vessel Technology* **123**, 362–372.

Mudry, F. (1987). A local approach to cleavage fracture. *Nuclear Engineering and Design* **105**, 65–76.

Narström, T. and Isacsson, M. (1999). Microscopic investigation of cleavage initiation in modified A 508 B pressure vessel steel. *Materials Science and Engineering* A **271**, 224–231.

Naudin, C., Pineau, A. and Frund, J.M. (2001). Toughness modeling of PWR vessel steel containing segregated zones. In: *10th Conference on Environmental Degradation of Materials in Nuclear Power Systems-water Reactors*, Lake Tahoe, Nevada, USA.

Needleman, A. and Tvergaard, V. (1987). An analysis of ductile rupture modes at a crack tip. *Journal of the Mechanics and Physics of Solids* **35**, 151–183.

Norris, D. (1979). Computer simulation of the Charpy V-notch toughness test. *Engineering Fracture Mechanics* **11**, 261–274.

O'Dowd, N.P. and Shih, C.F. (1991). Family of Crack-Tip Fields characterized by a Triaxiality Parameter : Part I – Structure of Fields. *Journal of the Mechanics and Physics of Solids* **39**, 989–1015.

O'Dowd, N.P. and Shih, C.F. (1992). Family of Crack-Tip Fields characterized by a Triaxiality Parameter : Part II – Fracture Applications. *Journal of the Mechanics and Physics of Solids* **40**, 939–963.

Ohata, M., Minami, F. and Toyoda, M. (1996). Local approach to strength mismatch effect on cleavage fracture of notched material. Euromech-Mecamat. *Journal de Physique IV* **6**, C6–C6–278.

Pardoen, T. and Hutchinson, J.W. (2000). An extended model for void growth and coalescence. *Journal of the Mechanics and Physics of Solids* **48**, 2467–2512.

Pardoen, T. and Besson, J. (2004). Micromechanics - based constitutive models of ductile fracture. *Local Approach to Fracture*. (Edited by Besson, J.). Les Presses, Ecole des Mines de Paris, 221–264.

Petti, J.P. and Dodds, R.H. (2004). Constraint comparisons for common fracture specimens : C(T)s and SE(B)s. *Engineering Fracture Mechanics* **71**, 2677–2683.

Petti, J.P. and Dodds, R.H. (2005). Calibration of the Weibull stress scale parameter, σ_u, using the Master Curve. *Engineering Fracture Mechanics* **72**, 91–120.

Pineau, A. (1982). Review of fracture micromechanisms and a local approach to predicting crack resistance in low strength steels. In: *Advances in Fracture Research, ICF5* (Edited by François, D.). Pergamon, Oxford, UK, 553–577.

Pineau, A. (1992). Global and local approaches of fracture - Transferability of laboratory test results to components. In: *Topics in Fracture and Fatigue*, (Edited by Argon, A.S.). Springer, New-York, 197–234.

Pineau, A. (2003). Practical Application of Local Approach Methods. In: *Comprehensive Structural Integrity* (Edited by Ainsworth, R.A. and Schwalbe, K.-H.). Elsevier, Amsterdam, 7, 177–225.

Pineau, A. and Joly, P. (1991). Local versus global approaches to elastic-plastic fracture mechanics: Application to ferritic steels and a cast duplex stainless steel. In *Defect Assessment in Components - Fundamentals and Applications ESIS/EGF9*, (Edited by Blauel, J.G. and Schwalbe, K.-H.). Mechanical Engineering Publications, London, 381–414.

Rice, J.R. and Tracey, D.M. (1969). On the ductile enlargement of voids in triaxial stress fields. *Journal of the Mechanics and Physics of Solids* **17**, 201–217.

Rodriguez-Ibabe, J.M. (1998). The role of microstructure in toughness behaviour of microalloyed steels. *Materials Science Forum, Microalloying in Steels* **284–286**, 51–62.

Roos, E., Alsmann, U., Elsässer, K., Eisele, W. and Seidenfuss, M. (1998). Experiments on warm prestress effect and their numerical simulation based on local approach. In: *Fracture from Defects*. ECF 12 (Edited by Brown, M.W., de los Rios, E.R. and Miller, K.J.). EMAS (Engineering Materials Advisory Services LTD), UK. Vol. II, 939–944.

Rossoll, A., Berdin, C. and Prioul, C. (2002). Determination of the fracture toughness of a low alloy steel by the instrumented Charpy impact test. *International Journal of Fracture* **115**, 205–226.

Rousselier, G. (1987). Ductile fracture models and their potential in local approach of fracture. *Nuclear Engineering and Design* **105**, 97–111.

Ruggieri, C. and Dodds, R.H. (1996). A transferability model for brittle fracture including constraint and ductile tearing effects : A probabilistic approach. *International Journal of Fracture* **79**, 309–340.

Ruggieri, C., Dodds, R.H. and Wallin, K. (1998). Constraint effects on reference temperature To for ferritic steels in the transition region. *Engineering Fracture Mechanics* **60**, 19–36.

Schmitt, W., Sun, D. and Blauel, J. (1997). Recent advances in the application of the Gurson model to the evaluation of ductile fracture toughness. In: *Recent Advances in Fracture. TMS*, 77–87.

Stöckl, H., Böschen, R., Schmitt, W., Varfolomeyev, I. and Chen, J.H. (2000). Quantification of the warm prestressing effect in a shape welded 10 Mn Mo Ni 5-5 material. *Engineering Fracture Mechanics* **67**, 119–137.

Tanguy, B., Besson, J. and Pineau, A. (2003). Comment on "Effect of carbide distribution on the fracture toughness in the transition temperature region of an SA 508 steel". *Scripta Materialia* **49**, 191–197.

Tanguy, B., Besson, J., Piques, R. and Pineau, A. (2005a). Ductile-to-brittle transition of an A 508 steel characterized by Charpy impact test : Part I – experimental results. *Engineering Fracture Mechanics* **72**, 49–72.

Tanguy, B., Besson, J., Piques, R. and Pineau, A. (2005b). Ductile-to-brittle transition of an A 508 steel characterized by Charpy impact test. Part II: modeling of the Charpy transition curve. To appear in *Engineering Fracture Mechanics* **72**, 413–434.

Thomason, P.F. (1985). Three-dimensional models for the plastic limit – loads at incipient failure of the intervoid matrix in ductile porous solids. *Acta Metallurgica* **33**, 1079–1085.

Tvergaard, V. (1990). Material failure by void growth to coalescence. *Advances in Applied Mechanics* **27**, 83–151.

Tvergaard, V. and Needleman, A. (1984). Analysis of cup-cone fracture in a round tensile bar. *Acta Metallurgica* **32**, 157–169.

Tvergaard, V. and Needleman, A. (1986). An analysis of the temperature and rate dependence of Charpy V-notch test. *Journal of the Mechanics and Physics of Solids* **34**, 213–241.

Tvergaard, V. and Needleman, A. (1988) An analysis of the temperature and rate dependence of Charpy V-notch energies for a high nitrogen steel. *International Journal of Fracture* **37**, 197–215.

Wallin, K. (1991a). Fracture toughness transition curve shape for ferritic structural steels. In *Proceedings of the Joint FEFG/ICF International Conference on Fracture of Engineering Materials and Structures*, (Edited by Teoh, S.H. and Lee, K.H.). Elsevier, London, 83–88.

Wallin, K. (1991b). Statistical modelling of fracture in the ductile-to-brittle transition region. In *Defect Assessment in Components – Fundamentals and Applications. ESIS/EGF9*, (Edited by Blauel, J.G. and Schwalbe, K.-H.). Mechanical Engineering Publications, London, 415–445.

Wallin, K., Saario, T. and Torronen, K. (1984). Statistical model for carbide induced brittle fracture in steel. *Metal Science* **18**, 13–16.

Xia, L. and Cheng, L. (1997). Transition from ductile tearing to cleavage fracture. A cell model approach. *International Journal of Fracture* **87**, 289–305.

Xia, L. and Shih, F.C. (1996). Ductile crack growth:III. Transition to cleavage fracture incorporating statistics. *Journal of the Mechanics and Physics of Solids* **44**, 603–639.

Zhang, Z.L. and Niemi, E. (1994). Analyzing ductile fracture using dual dilational constitutive equations. *Fatigue and Fracture of Engineering Materials and Structures* **17**, 695–707.

Zhang, Z.L. and Niemi, E. (1995). A new failure criterion for the Gurson-Tvergaard dilational constitutive model. *International Journal of Fracture* **70**, 321–334.

Zhou, Z.L. and Lin, S.H. (1998). Influence of local brittle zones on the fracture toughness of high strength low-alloyed multipass weld metals. *Acta Metallurgica Sinica* **11**, 87–92.

International Journal of Fracture (2006) 138:167–195
DOI 10.1007/s10704-006-7158-2

© Springer 2006

The effect of hydrogen on fatigue properties of metals used for fuel cell system

Y. MURAKAMI

Department of Mechanical Engineering Science, Kyushu University, 6-10-1 Hakozaki, Higashi-ku, Fukuoka, 812-8581 Japan (E-mail: ymura@mech.kyushu-u.ac.jp)

Received 1 March 2005; accepted 1 December 2005

Abstract. The effect of hydrogen on the fatigue properties of alloys which are used in fuel cell (FC) systems has been investigated. In a typical FC system, various alloys are used in hydrogen environments and are subjected to cyclic loading due to pressurization, mechanical vibrations, etc. The materials investigated were three austenitic stainless steels (SUS304, SUS316 and SUS316L), one ferritic stainless steel (SUS405), one martensitic stainless steel (0.7C-13Cr), a Cr-Mo martensitic steel (SCM435) and two annealed medium-carbon steels (0.47 and 0.45%C). In order to simulate the pick-up of hydrogen in service, the specimens were charged with hydrogen. The fatigue crack growth behaviour of charged specimens of SUS304, SUS316, SUS316L and SUS405 was compared with that of specimens which had not been hydrogen-charged. The comparison showed that there was a degradation in fatigue crack growth resistance due to hydrogen in the case of SUS304 and SUS316 austenitic stainless steels. However, SUS316L and SUS405 showed little degradation due to hydrogen. A marked increase in the amount of martensitic transformation occurred in the hydrogen-charged SUS304 specimens compared to specimens without hydrogen charge. In case of SUS316L, little martensitic transformation occurred in either specimens with and without hydrogen charge. The results of *S-N* testing showed that in the case of the 0.7C–13Cr stainless steel and the Cr–Mo steel a marked decrease in fatigue resistance due to hydrogen occurred. In the case of the medium carbon steels hydrogen did not cause a reduction in fatigue behaviour. Examination of the slip band characteristics of a number of the alloys showed that slip was more localized in the case of hydrogen-charged specimens. Thus, it is presumed that a synergetic effect of hydrogen and martensitic structure enhances degradation of fatigue crack resistance.

Key words: Crack growth, Cr-Mo steel, ductility loss, fatigue, hydrogen, medium carbon steel, stainless steel, slip behaviour.

1. Introduction

In order to enable the "hydrogen society" (Figure 1) (Shinohara, 2004, Private Communication) in the near future, a number of pressing technical problems (Figure 2) (Murakami, 2004) must be solved. One important task for mechanical engineers and material scientists is the development of materials and systems which are capable of withstanding the effects of cyclic loading in hydrogen environments. In the past much research has been concentrated on the phenomenon known as hydrogen embrittlement (Farrell and Quarrell, 1964; Brass and Chene, 1998; Shih et al., 1988). Hydrogen effects on slip localization (Birnbaum and Sofronis, 1994; Brass and Chene, 1998; Shih et al., 1998), softening and hardening (Heller, 1961; Au and Birnbaum, 1973; Dufresne and Seeger, 1976; Kimura and Matsui, 1979; Matsui and

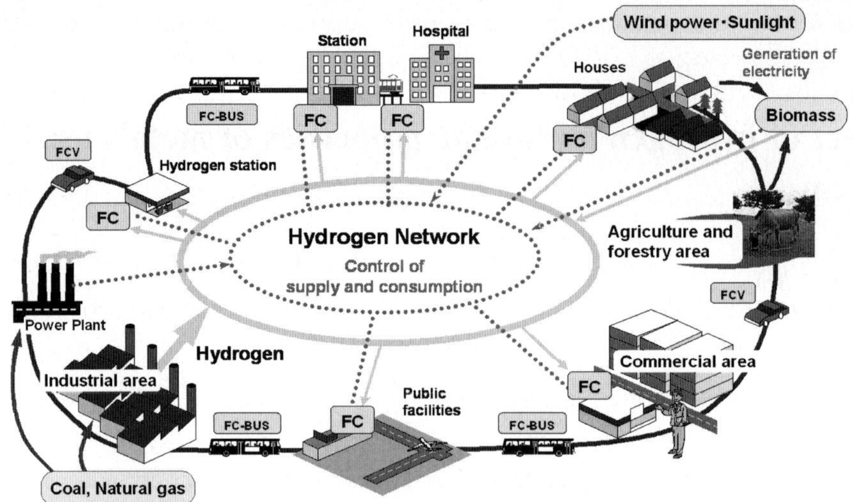

Figure 1. Image of the hydrogen society (Shinohara, 2004, Private Communication) (FC: Fuel Cell; FVC: Fuel Cell Vehicle).

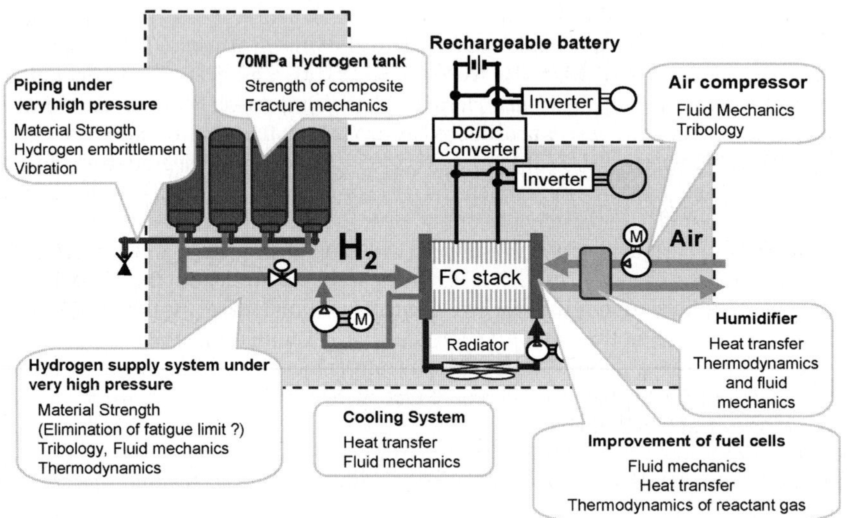

Figure 2. Layout of the automotive fuel cell system and problems of mechanical engineering and material science (Murakami, 2004).

Kimura, 1979; Hirth, 1980; Birnbaum and Sofronis, 1994; Senkov and Jonas, 1996; Magnin et al., 2001), hydrogen-dislocation interactions (Clum 1975; Birnbaum et al., 2000; Magnin et al., 2001) and creep (Mignot et al., 2004) have been also reported. However, the influence of hydrogen on the fatigue of metals has not been studied in sufficient detail. In order to produce components which must perform satisfactorily in service for up to 15 years, there is an urgent need for basic, reliable data on the fatigue behaviour of candidate materials in hydrogen environments.

Two typical fuel cell (FC) systems are the stationary FC system and the automotive FC system. Figure 2 shows the layout of the automotive FC system and problems of mechanical engineering and material science. In the automotive FC system, many components such as the liner of high-pressure hydrogen storage tank, valves, pressure sensors, hydrogen accumulators, pipes, etc, are exposed to hydrogen environment for a long period up to 15 years. Sufficient data have not been obtained on the content of hydrogen which diffuses into metals during a long period of exposure to hydrogen. "How much hydrogen is contained in components in the fuel cell related system?" is a very important question. But this question is difficult to answer. Table 1 shows the hydrogen content in three steels which were exposed to 45 MPa hydrogen for 40 days at 85°C. We need to know not only the hydrogen content but also the effect of hydrogen content on fatigue properties. The effect of hydrogen on fatigue properties of metals is different depending on materials, i.e. microstructures.

The effect of hydrogen on high cycle fatigue was revealed by chance in course of the research to study the elimination phenomenon of the conventional fatigue limit in high strength steels. In high strength steels cycled into the very high cycle range fatigue cracks are often initiated at sub-surface inclusions. Murakami et al. (1998) pointed out that in these steels, hydrogen trapped by nonmetallic inclusions can play a crucially important role in the early stages of the fatigue crack growth process. Observations made with the aid of an optical microscope of the fracture origin revealed that an Optically Dark Area (ODA) was present around inclusions, as shown in Figure 3. The hydrogen trapped by the inclusion at fracture origin was clearly revealed by the Secondary Ion Mass Spectrometry (SIMS) and the tritium autoradiography, as shown in Figure 4 (Murakami et al., 2000) and Figure 5 (Kawazoe et al., 2005). Table 2 presents the experimental evidence for the influence of hydrogen on ODA formation. It is noted that the ODA is formed only around nonmetallic inclusions. It is not formed around artificial defects and regions of microstructural cleavage as these are not able to trap hydrogen. It is significant that a decrease in hydrogen content of a high strength steel leads to a reduction in the size of the ODA as well as to an improvement in high cycle fatigue properties (Murakami et al., 1999, 2000, 2001; Nagata and Murakami, 2003; Shiina et al., 2004). In addition, it has been also reported that ODA is hardly formed at high R ratio regardless of heat treatment (Murakami et al., 2000; Shiina et al., 2004).

In the present paper, the fatigue properties obtained in hydrogen environments of several steels which are candidate materials for use in FC systems are presented. In addition, a number of interesting aspects of the effect of hydrogen on fatigue properties and fatigue mechanisms will be discussed.

2. Materials and experimental methods

To investigate the influence of hydrogen environments and the resultant diffusion of hydrogen into alloys, eight alloys were tested. The eight alloys included three solution

Table 1. Hydrogen content in three steels exposed to 45 MPa hydrogen for 40 days at 85°C.

	0.45%C steel	SUS304L	SUS316L
Before exposure	0.01 ppm	3.88 ppm	1.51 ppm
After exposure	1.09 ppm	11.48 ppm	9.56 ppm

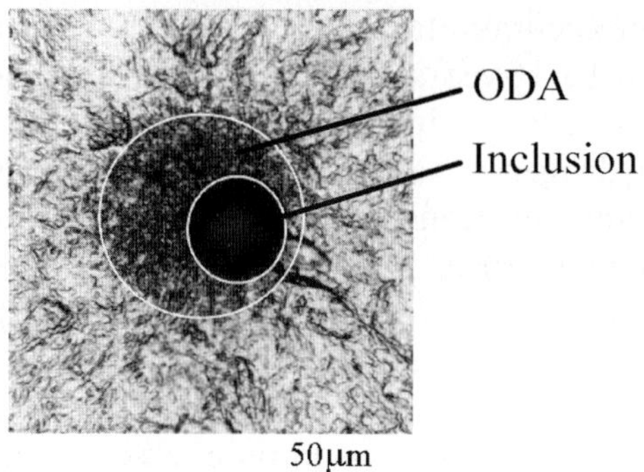

Figure 3. Optical micrograph of the fracture origin (SCM435, $\sigma_a = 561\,$MPa, $N_f = 1.11 \times 10^8$). ODA: *Optically Dark Area.*

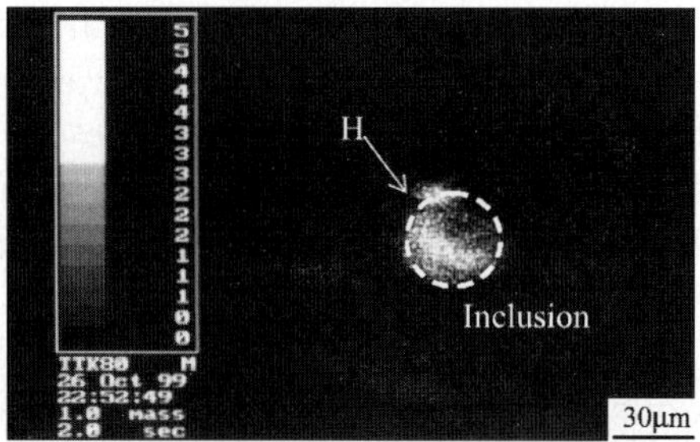

Figure 4. Hydrogen trapped by a nonmetallic inclusion ($Al_2O_3(CaO)_x$) at fracture origin observed by SIMS (Murakami et al., 2000). (SCM435, $\sigma_a = 561\,$MPa, $N_f = 5.17 \times 10^7$).

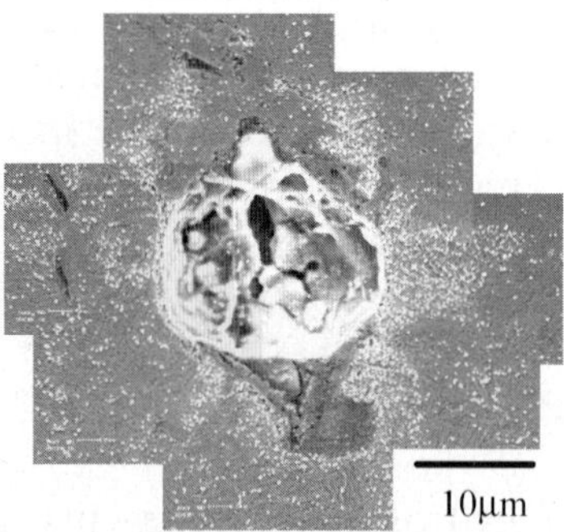

Figure 5. Hydrogen trapped by a nonmetallic inclusion ($Al_2O_3(CaO)_x$) at fracture origin observed by tritium autoradiography (SCM435, $\sigma_a = 480\,$MPa, $N_f = 9.61 \times 10^8$, Courtesy of Kawazoe et al., 2005).

treated austenitic stainless steels (SUS304, SUS316 and SUS316L), one annealed ferritic stainless steel (SUS405), one martensitic stainless steel (0.7C–13Cr), a Cr–Mo martensitic steel (SCM435) and annealed medium-carbon steels (0.47 and 0.45%C). Table 3(a)–(c) list the chemical compositions of the alloys together with their Vickers hardnesses. In the case of the SCM435 steel, since the specimens were prepared

Table 2. Experimental evidence of influence of hydrogen on ODA formation (Murakami et al., 1999, 2000, 2001; Nagata and Murakami, 2003; Shiina et al., 2004; Murakami and Nagata, 2005).

Material	Fracture origin	Heat treatment	ODA
SAE52100(SUJ2) & SCM435	Inclusion	RX gas (Hydrogen)	Dense
SAE52100(SUJ2) & SCM435	Inclusion	Annealed in a vacuum at 300°C	Light
SAE52100(SUJ2) & SCM435	Inclusion	Heat treated in a vacuum	Light
SAE52100(SUJ2)	Bainite	RX gas or conventional	No
0.45C-13Cr-0.14Ni (Martensitic stainless steel)	Small artificial hole	Conventional	No
Materials tested at high R ratio	Inclusion	Regardless of heat treatment	Light

ODA: *Optically Dark Area* observed in the neighborhood of the nonmetallic inclusion by optical microscope. The ODA looks dark by the observation of optical microscope due to the rough morphology where fatigue crack is presumed to grow very slowly [not cycle by cycle (Murakami et al., 1999, 2000, 2001; Nagata and Murakami, 2003; Shiina et al., 2004; Murakami and Nagata, 2005)].

from two lots, A and B, their slightly different chemical compositions are presented separately. The SCM435 specimens were quenched and tempered after carbonitriding.

Figure 6a–d show the shapes and dimensions of the fatigue test specimens together with the types of fatigue test (tension-compression or rotating bending) and methods of surface finishing. The fatigue tests were carried out by using specimens into which hydrogen was artificially charged by either a cathodic charging method or a soaking method. Table 4 shows the conditions of hydrogen charging for each material, together with the hydrogen content of uncharged and hydrogen-charged specimens. The solution used for the cathodic charging was a dilute sulfuric acid (PH=3.5) and the current density was $i = 27\,\mathrm{A/m^2}$. The solution used in the soaking method was a 15–20 mass% ammonium thiocyanate solution (NH_4SCN). Hydrogen contents and hydrogen desorption behaviour were measured by Thermal Desorption Spectrometry (TDS). In the following, hydrogen content throughout the entire cross section of specimen is denoted by C_H.

The influence of hydrogen charging on crack growth and slip bands formation was investigated on SUS304, SUS316, SUS316L and SUS405 charged and uncharged specimens which contained a small hole of diameter $d = 100\,\mu$m and depth $h = 100\mu$m, as shown in Figure 7. Fatigue crack growth behaviour was determined by means of the plastic replica method. The tension-compression *S-N* characteristics of the 0.7C–13Cr martensitic stainless steel, the SCM435 Cr-Mo steel and the 0.47%C steel were determined in hydrogen-charged and uncharged conditions. Crack initiation behaviour of 0.45%C steel was compared in hydrogen charged and uncharged specimens. All the fatigue test were carried out at stress ratio $R = -1$.

Fatigue tests were performed with four servo-hydraulic tension-compression fatigue testing machines with different frequencies up to 1000 Hz. Figure 8 shows an example of temperature rise during fatigue test for SUS304. An appropriate test frequency ranging from 1 to 1000 Hz was chosen to minimize any temperature rise that might occur during cyclic loading.

Table 3. Chemical compositions.

	C	Si	Mn	P	S	Ni	Cr	Cu	Mo	Al
(a) Stainless steels (wt.%)										
SUS304 (Austenitic, HV=176)	0.06	0.36	1.09	0.03	0.02	8.19	18.7	–	–	–
SUS316 (Austenitic, HV=161)	0.05	0.27	1.31	0.03	0.03	10.2	17.0	–	2.08	–
SUS316L (Austenitic, HV=157)	0.02	0.78	1.40	0.04	0.01	12.1	17.0	–	2.04	–
SUS405 (Ferritic, HV=159)	0.04	0.37	0.38	0.02	0.01	0.21	13.2	–	–	0.26
0.7C–13Cr (Martensitic, HV=675)	0.66	0.30	0.64	0.02	0.01	0.22	12.5	0.11	0.13	–

	C	Si	Mn	P	S	Ni	Cr	Mo	Cu	O
(b) SCM435 (wt.%)										
Lot A (HV=561)	0.36	0.19	0.77	0.014	0.006	0.08	1.00	0.15	0.13	8 (ppm)
Lot B (HV=563)	0.36	0.28	0.76	0.024	0.014	0.02	1.00	0.17	0.01	11 (ppm)

	C	Si	Mn	Cr	Al	S	P	Cu	Ni
(c) Medium carbon steels (wt.%)									
0.47%C (HV=170)	0.47	0.19	0.7	0.12	0.024	0.024	0.012	–	–
0.45%C (HV=185)	0.45	0.25	0.79	0.18	–	0.01	0.01	0.09	0.03

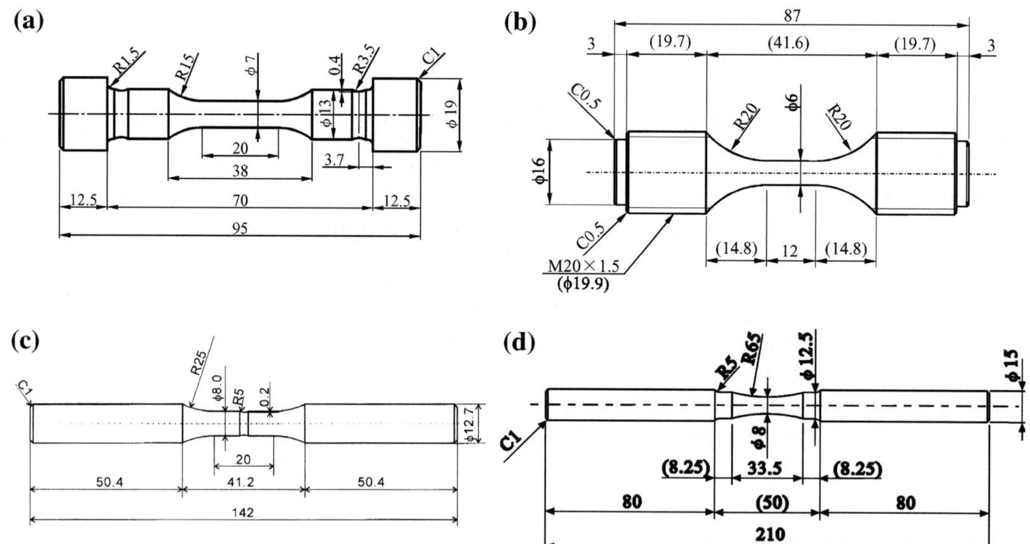

Figure 6. Shapes and dimensions of fatigue test specimens, in mm. (a) SUS304, SUS316, SUS316L, SUS405 and SCM435 (Tension-compression, finished with #2000 emery paper) (b) 0.7C–13Cr steel and SCM435 (Tension-compression, finished with #2000 emery paper) (c) 0.47%C steel (Tension-compression, finished with buff) (d) 0.45%C steel (Rotating bending, finished with electropolishing).

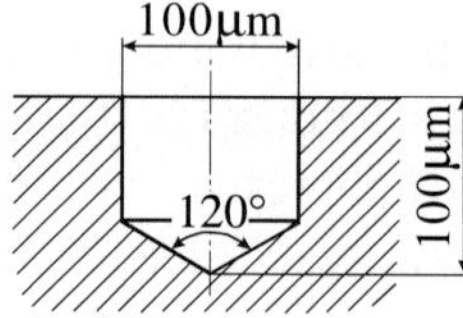

Figure 7. Small hole introduced onto specimen surface.

Table 4. Hydrogen charging conditions for fatigue test specimens.

	Charging method	Charging time (h)	Temperature of solution (°C)	Hydrogen content of uncharged specimen (ppm)	Hydrogen content for entire cross section of hydrogen-charged specimen C_H	
SUS304	Cathodic	672	50	2.2	3.7 ppm	(14 h after HC)
SUS316	Cathodic	672	50	3.4	5.5 ppm	(102 h after HC)
SUS316L	Cathodic	672	50	2.7	4.3 ppm	(81 h after HC)
	Cathodic	504	85		11.4 ppm	(100 h after HC)
SUS405	Cathodic	336	50	0.1	2.1 ppm	(27 h after HC)
0.7C-13Cr	Soaking	24	50	0.2	2.4 ppm	(2 h after HC)
SCM435(B)	Soaking	24	50	0.26	10 ppm	(1.5 h after HC)
0.47%C	Soaking	24	35–40	0.05	0.52 ppm	(2 h after HC)
0.45%C	Soaking	24	40	0.1–0.2	0.3 ppm	(2 h after HC)

HC: Hydrogen charging Hydrogen content of stainless steels charged with hydrogen was measured immediately after the fatigue test.

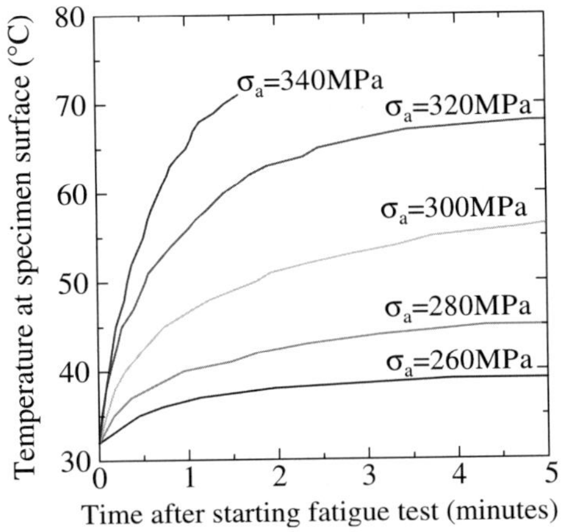

Figure 8. Example of temperature rise during tension-compression fatigue test (SUS304, Test frequency: 1 Hz).

3. Results and discussion

3.1. AUSTENITIC AND FERRITIC STAINLESS STEELS

Figure 9a–c show the hydrogen content distribution in specimens of austenitic stainless steels (SUS304, SUS316 and SUS316L) after 504–672 h of charging (Kanezaki et al., under preparation). The hydrogen content throughout the entire cross section of a specimen was measured by progressive removal of thin layers of material by polishing. The local hydrogen content was obtained from the determinations of hydrogen content before and after polishing. In the austenitic stainless steels, hydrogen diffuses only into a very thin surface layer (100–200 μm). Figure 10 shows the decrease in hydrogen content in hydrogen-charged SUS304 and SUS316 specimens

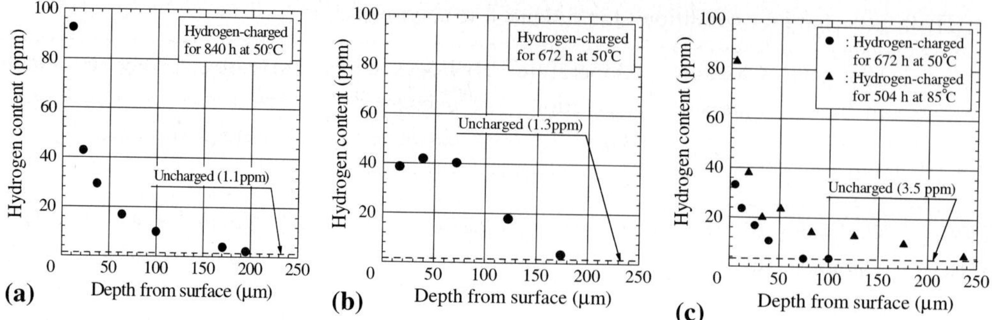

Figure 9. Hydrogen content distribution in hydrogen-charged austenitic stainless steels (Kanezaki et al., under preparation). C_H: Hydrogen content throughout the entire cross section of specimen measured immediately after hydrogen charging. (a) SUS304 (C_H: 3.5 ppm) (b) SUS316 (C_H: 4.2 ppm) (c) SUS316L (C_H: 4.0–4.8 ppm).

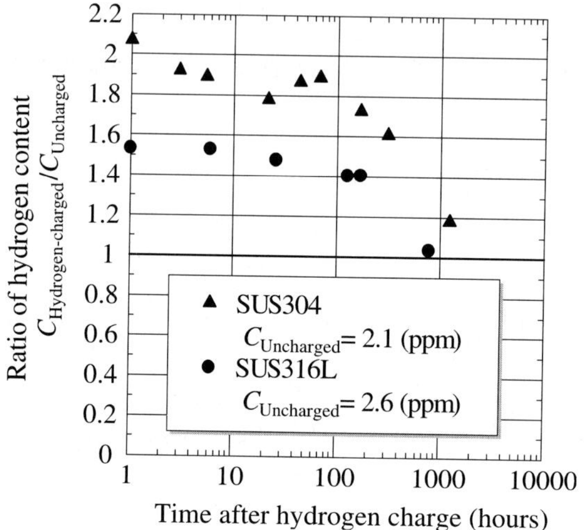

Figure 10. Decrease in hydrogen content in hydrogen-charged SUS 304 and SUS316 specimens at room temperature (Kanezaki et al., under preparation).

at room temperature as a function of time after hydrogen charging (Kanezaki et al., under preparation). The hydrogen content of charged specimens recovered to the initial level before charging after about 1000 h. Figure 11 shows the hydrogen content distribution in hydrogen-charged specimens of the ferritic stainless steel (SUS405) (Kanezaki et al., under preparation). In contrast to the austenitic stainless steels, hydrogen had diffused to a depth of more than 5 mm after only 336 h of hydrogen-charging. Clearly the depth of hydrogen diffusion is dependent on the crystal structure, and in fact it has been reported that the hydrogen diffusion coefficient in austenitic stainless steels is about four orders of magnitude smaller than that of ferritic and martensitic steels (Sakamoto and Katayama, 1982).

Figure 12a–d show the fatigue crack growth curves starting from a hole in uncharged and hydrogen-charged specimens of SUS304, SUS316, SUS316L and SUS405 (Kanezaki et al., under preparation). The particular test conditions and hydrogen contents measured immediately after the fatigue test are also included. For both SUS304 and SUS316, hydrogen charging led to a marked increase in fatigue

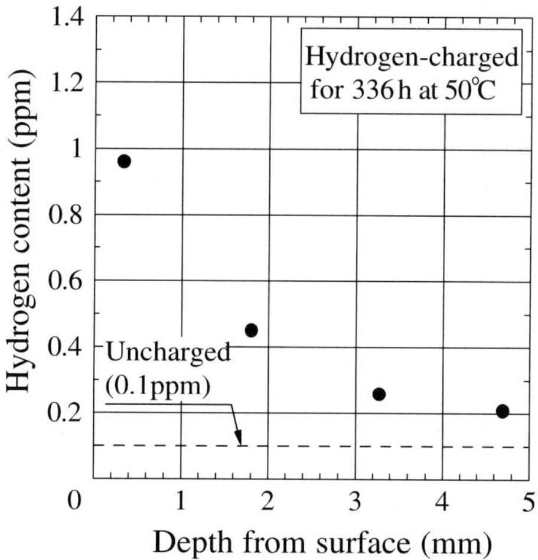

Figure 11. Hydrogen content distribution in hydrogen-charged SUS405 (Kanezaki et al., under preparation).

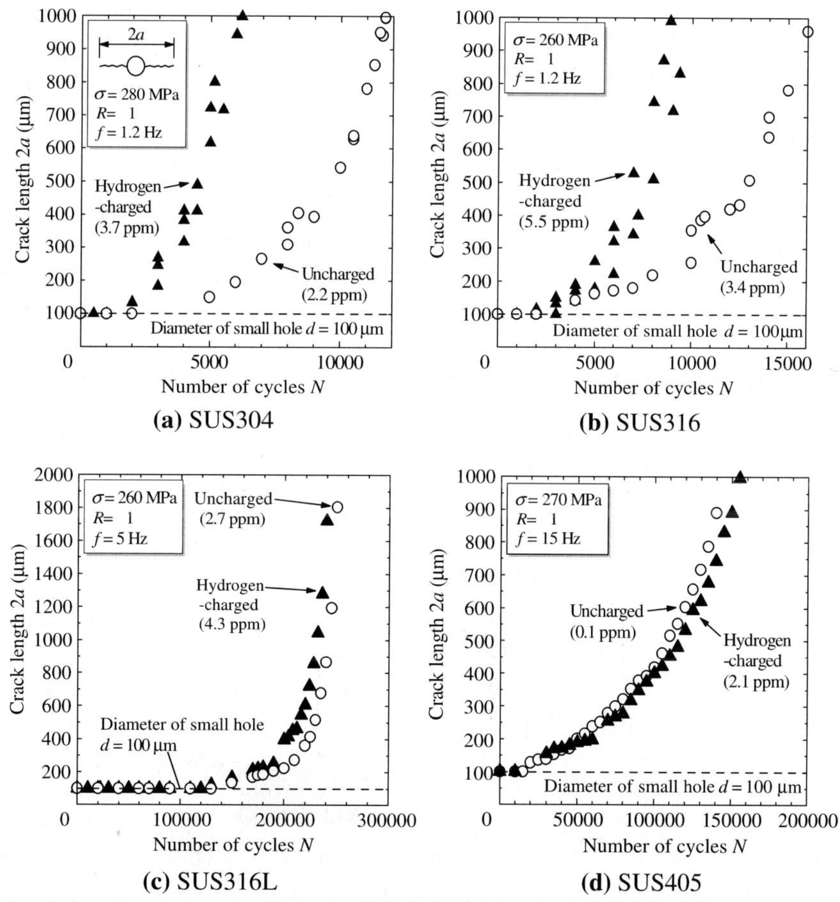

Figure 12. Influence of hydrogen charging on crack growth from 100μm hole (Kanezaki et al., under preparation). (Hydrogen content of specimen was measured immediately after fatigue test.)

crack growth rate. However for both SUS316L and SUS405 there was little if any effect of hydrogen charging on fatigue crack growth.

Figure 13 shows the relationship between the crack length $2a$ and the aspect ratio b/a of fatigue crack measured in uncharged and hydrogen-charged specimens of SUS304 and SUS316 (Kanezaki et al., 2004). In the hydrogen-charged specimens, hydrogen diffusion was detected to a depth of about 200μm. The hydrogen-charged specimens showed smaller aspect ratio than the uncharged specimens. This is due to the higher crack growth rate near surface layer with high hydrogen content.

Metastable austenitic stainless steels are subject to a phase transformation to martensite when stressed. Figure 14 give a comparison of the amount of martensitic transformation on the fracture surface of three austenitic stainless steels, which was measured by X-ray diffraction technique (Kanezaki et al., under preparation). The fracture surface of SUS316L showed less martensitic transformation than that of SUS304 or SUS316. These results suggest that the extent of transformation from austenite to martensite at the fatigue crack tip may be a crucial factor which determines the influence of hydrogen on the fatigue crack growth rate. Figure 15 gives a comparison of amount of martensitic transformation on fracture surface close to the specimen surface and at the edge of fracture surface (see Figure 15b) in SUS304. The difference in the amount of martensitic transformation is clearly detected on fracture surface for the hydrogen-charged specimen and the uncharged specimen, though the difference is not detected at the specimen surface due to the distance of measured points from fracture surface.

Comparing the fatigue crack growth properties in Figures 12, 14 and 15, we can conclude that the martensitic transformation of the microstructure in the neighbourhood of crack tip during fatigue is a crucial factor for the effect of hydrogen on fatigue crack growth. Particularly SUS304 has more martensitic transformation at the same level of stress, though SUS304 has a higher static strength than SUS316 and SUS316L. Since even without hydrogen charge, SUS304, SUS316 and SUS316L contain hydrogen of the level of 2–3 ppm, there should be a possibility of synergetic or interaction effect between the hydrogen and the microstructure. Therefore,

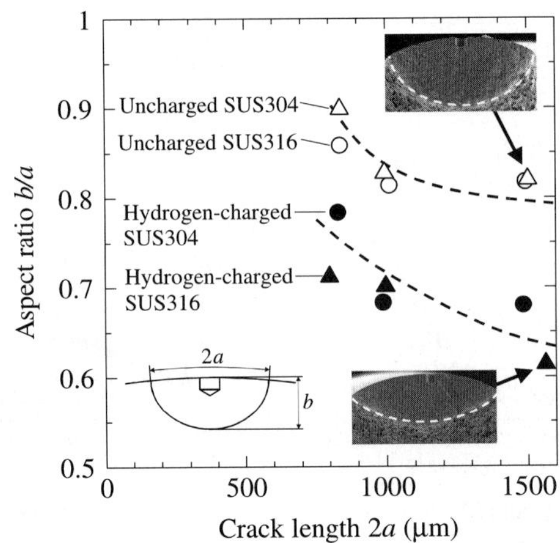

Figure 13. Relationship between crack length $2a$ and aspect ratio b/a measured for uncharged and hydrogen-charged specimen of SUS304 and SUS316 (Kanezaki et al., 2004).

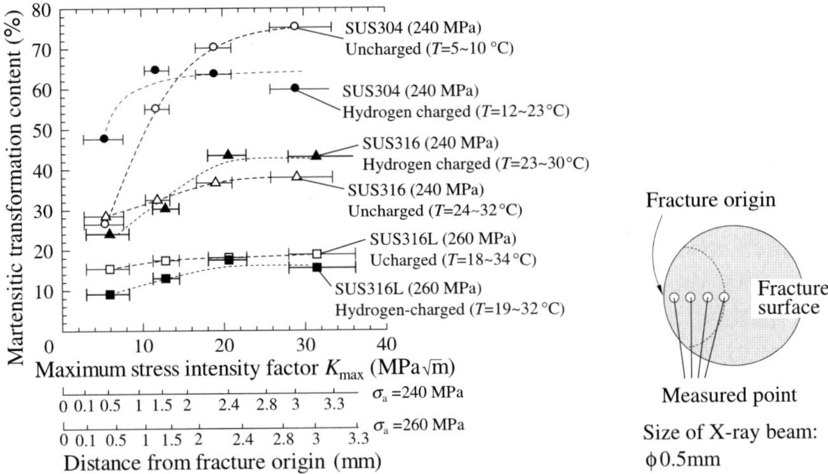

Figure 14. Amount of martensitic transformation at the crack tip in three austenitic stainless steels without hydrogen charging (Kanezaki et al., under preparation). (T: Ambient temperature during fatigue test.)

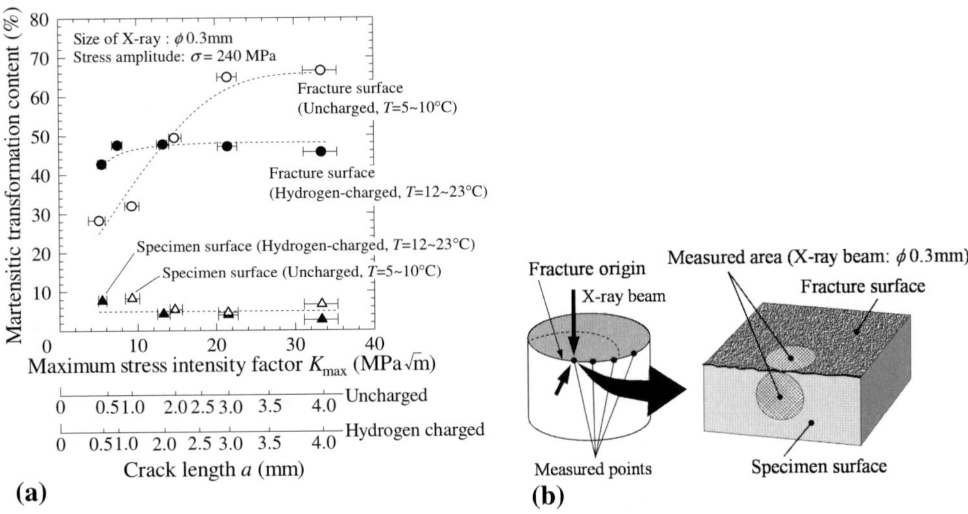

(a) **(b)**

Figure 15. Comparison of amount of martensitic transformation on fracture surface close to the specimen surface and at the edge of fracture surface (Kanezaki et al., under preparation). (T: Ambient temperature during fatigue test.) (a) Amount of martensitic formation (b) Detail of measured points.

considering the fact that the effect of hydrogen on crack growth of SUS316L which has less martensitic transformation probably due to the stable austenitic structure is very small, it is presumed that hydrogen influences the crack growth through a synergetic mechanism with martensite.

Considering that hydrogen diffuses only into a thin surface layer of specimen, it may look strange in Figures 14 and 15 that there is a difference of the amount of martensitic transformation at deep subsurface zone for SUS304 specimens with and without hydrogen charge. The reason is as follows. It has been known that the amount of martensitic transformation under plastic deformation depends strongly on the temperature during experiment (Fiedler et al. 1954). In the present study, the experiment of SUS304 with hydrogen charge was conducted at an ambient temperature of 12–23°C and that without hydrogen charge was conducted at 5–10°C. In general, more amount of martensitic transformation is produced at lower temperature (Fiedler et al., 1954). Modifying the data of Figures 14 and 15 based on the temperature effect, we can draw Figures 16 and 17.

Figures 16 and 17 reveal the following aspects of hydrogen effect. SUS304 even without hydrogen charge produces more martensitic transformation than SUS316 and SUS316L. At the thin surface layer of SUS304 specimens with high hydrogen content, further more martensitic transformation can be produced. However, the mechanism of hydrogen enhanced martensitic transformation is still unclear. The martensitic transformation of SUS316L is very small, though more than 2 ppm hydrogen is contained in the microstructure without hydrogen charge. As will be explained in Section 3.3, in case of a Cr-Mo steel whose microstructure is totally martensitic, a drastic decrease in fatigue strength occurs even in the hydrogen content of 1 ppm. Therefore, we can conclude that the synergetic or coupling effect between hydrogen and martensite is a crucially important factor for fatigue behaviours of alloys.

Figures 18–21 show fatigue cracks emanating from the small hole in uncharged and hydrogen-charged specimens (Kanezaki et al., under preparation). In SUS304 and SUS316, the crack growth path in the hydrogen-charged specimens was relatively linear, whereas the crack growth path in the uncharged specimens was more zigzag in nature. It was also noted that in the case of uncharged specimens of SUS304 and SUS316, large numbers of slip bands were observed in the surface after cyclic loading, whereas when these alloys were charged, relatively few slip bands (localization of slip bands) were observed. On the other hand, in SUS316L and SUS405, any

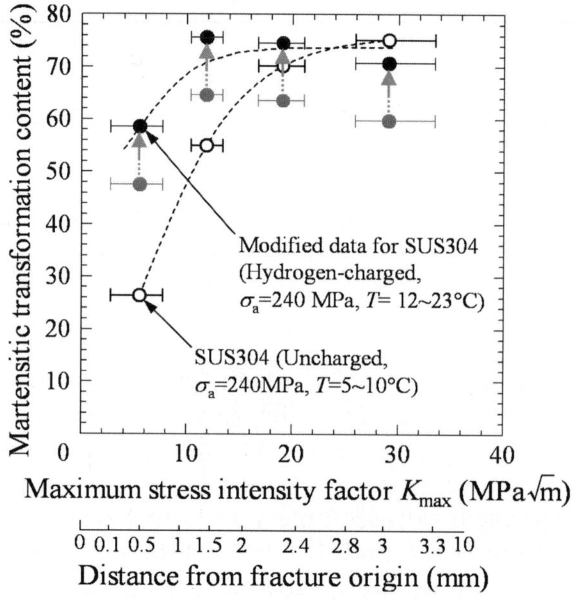

Figure 16. Martensitic transformation data for SUS304 modified from Figure 14 by considering the temperature effect. *T*: Ambient temperature during fatigue test.

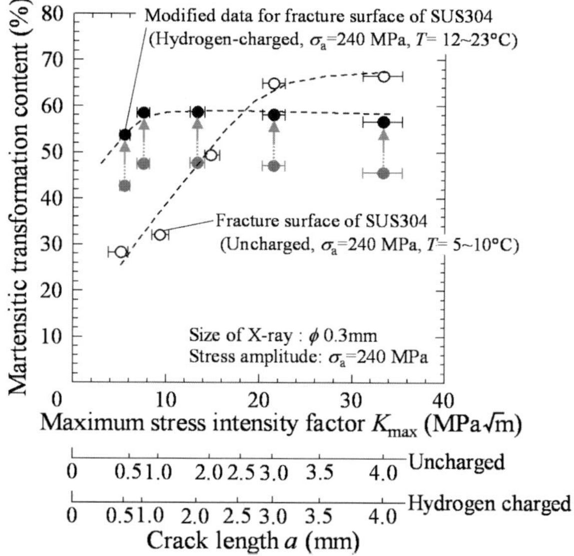

Figure 17. Martensitic transformation data for SUS304 modified from Figure 15a by considering the temperature effect. T: Ambient temperature during fatigue test.

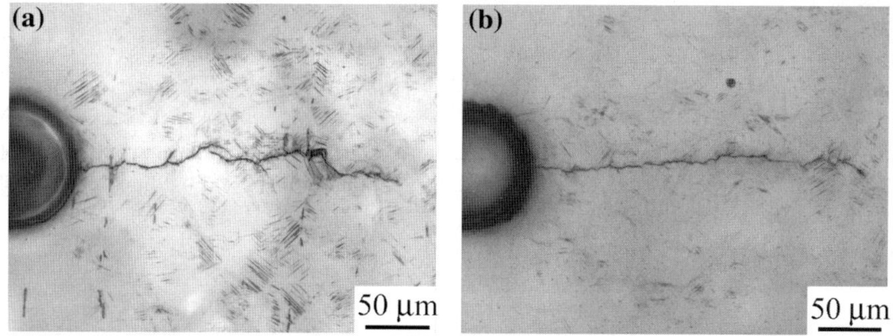

Figure 18. Crack emanating from an artificial hole in the uncharged and hydrogen-charged specimen of SUS304 (Kanezaki et al., under preparation). (a) Uncharged ($C_H = 2.2$ ppm, $\sigma_a = 280$ MPa, $N = 10500$) (b) Hydrogen-charged ($C_H = 3.7$ ppm, $\sigma_a = 280$ MPa, $N = 5500$).

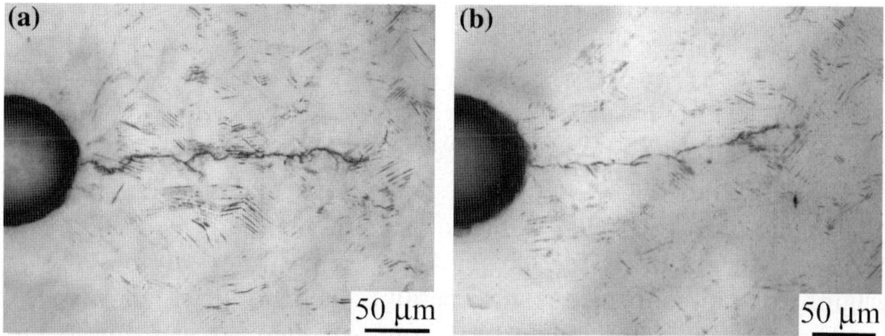

Figure 19. Crack emanating from an artificial hole in the uncharged and hydrogen-charged specimen of SUS316 (Kanezaki et al., under preparation). (a) Uncharged ($C_H = 3.4$ ppm, $\sigma_a = 260$ MPa, $N = 14000$) (b) Uncharged ($C_H = 5.5$ ppm, $\sigma_a = 260$ MPa, $N = 7000$).

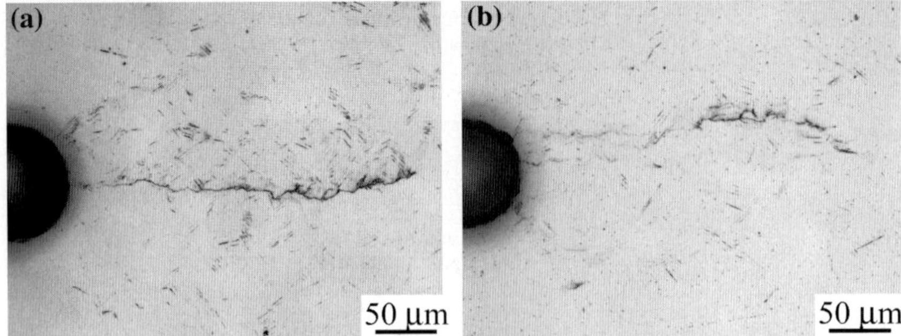

Figure 20. Crack emanating from an artificial hole in the uncharged and hydrogen-charged specimen of SUS316L (Kanezaki et al., under preparation). (a) Uncharged ($C_H = 2.7$ ppm, $\sigma_a = 260$ MPa, $N = 2.35 \times 10^5$) (b) Hydrogen-charged ($C_H = 11.4$ ppm, $\sigma_a = 260$ MPa, $N = 5.25 \times 10^5$).

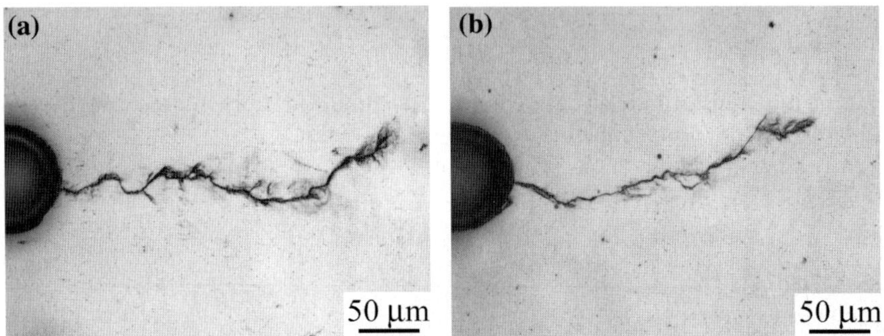

Figure 21. Crack emanating from an artificial hole in the uncharged and hydrogen-charged specimen of SUS405 (Kanezaki et al., under preparation). (a) Uncharged ($C_H = 0.1$ ppm, $\sigma_a = 270$ MPa, $N = 1.3 \times 10^5$) (b) Hydrogen-charged ($C_H = 2.1$ ppm, $\sigma_a = 270$ MPa, $N = 1.35 \times 10^5$).

difference in the number of slip bands formed during cycling between charged and uncharged specimens was not remarkable.

In order to investigate the influence of hydrogen on fatigue properties of materials to be used for FC systems for a long period (15 years), it is necessary to increase the hydrogen content in specimens by improving the hydrogen charging method as well as acquiring the data on actual hydrogen contents developed during a long time use.

3.2. 0.7–13Cr MARTENSITIC STAINLESS STEEL

This material is a candidate material for bearing in hydrogen environment. Figure 22 shows *S-N* data for the 0.7C–13Cr steel (Kanezaki et al., under preparation). It is evident that in case of hydrogen charged specimens a marked increase in fatigue strength occurred. In the uncharged specimens (open circles) almost all of the fatigue fracture origins were at subsurface nonmetallic inclusions which were located near surface (see Figure 23a). This subsurface crack nucleation was found to be due to subsurface tensile residual stresses (see Figure 24). After removal of the surface layer by electropolishing, the fracture origin moved to the core part of specimens (see Figure 25). In contrast, the fracture origins of hydrogen-charged specimens were

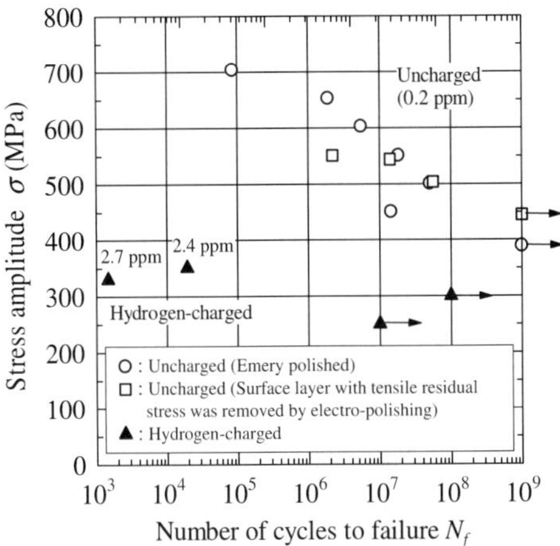

Figure 22. *S-N* data of 0.7C–13Cr steel (Kanezaki et al., under preparation). Hydrogen content of hydrogen-charged specimen was measured 2–4 h after the hydrogen charging.

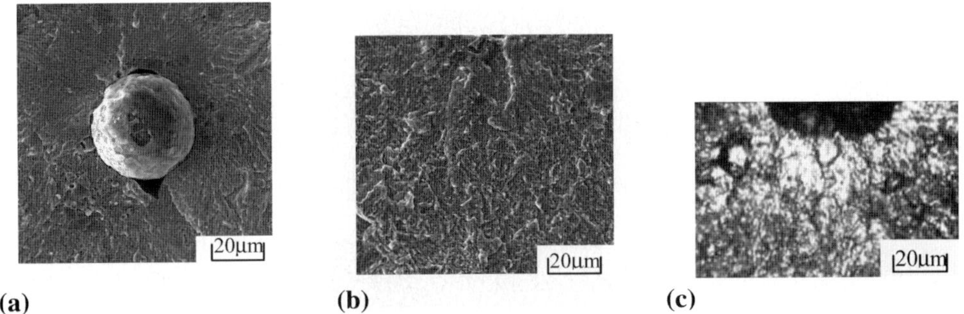

Figure 23. Fracture origins of 0.7C–13Cr steel. (a) Uncharged specimen (SEM) ($\sigma = 600$MPa, $N_F = 5.48 \times 10^6$ Hydrogen content $C_H = 0.2$ppm) (b) Hydrogen-charged specimen (SEM) (c) Optical microscope observation of fracture origin of (b). ($\sigma = 350$MPa, $N_f = 1.96 \times 10^4$ $C_H = 2.1$ppm at 2h after hydrogen charge)

mostly at the specimen surface (Figure 23b, c). This implies that hydrogen easily diffused into the martensitic microstructure and strongly influenced the strength properties of the hard martensitic structure.

3.3. SCM435 STEEL

This material is a candidate material for hydrogen accumulator. Figure 26 shows that the hydrogen content of SCM435 steel decreases with time after charging (Murakami and Nagata, 2005). This suggests that the hydrogen content in specimens also decreases during fatigue testing. Therefore, the hydrogen content before and after a fatigue test was always checked, with the hydrogen content before a test being obtained from Figure 26.

Figure 27 shows *S-N* data obtained in tension-compression fatigue test of specimens with and without hydrogen charging (Murakami and Nagata, 2005). Hydrogen charging drastically decreased the fatigue life and the fatigue strength of SCM435.

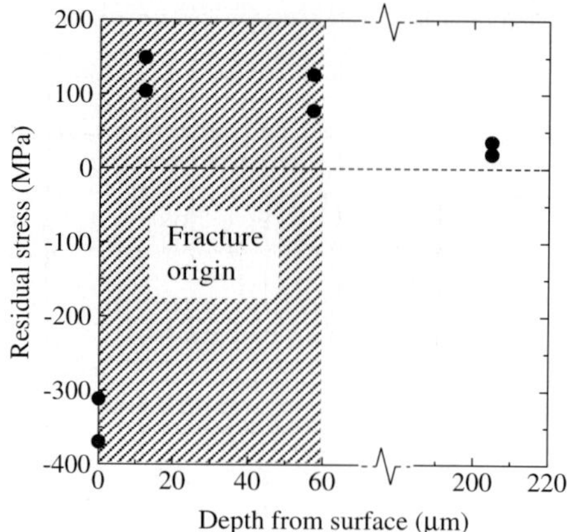

Figure 24. Residual stress at specimen surface after successive electro-polishing for 0.7C-13Cr steel.

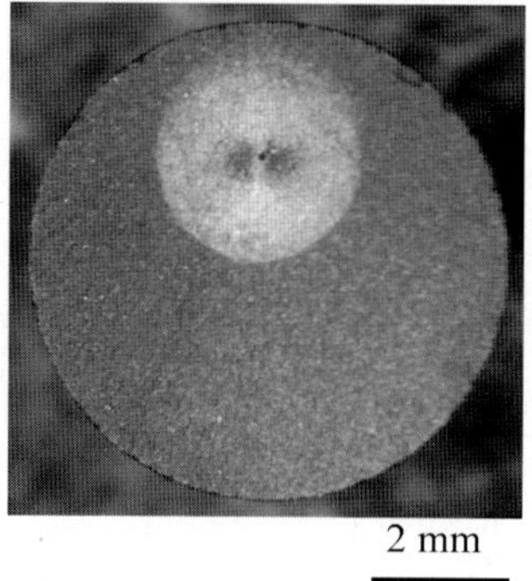

Figure 25. Fracture surface of a 0.7C-13Cr steel specimen whose surface layer with tensile residual stress was removed by electropolishing.

It was observed that in these tests the fatigue fracture origins were all located at sub-surface nonmetallic inclusions. Figures 28a–d show optical micrographs of the fracture origins. Those specimens which had been hydrogen-charged exhibited smaller ODAs (Murakami et al., 1999, 2000, 2001; Nagata and Murakami, 2003; Murakami and Nagata, 2005) than the heat-treated specimens. This finding implies that the fatigue threshold of the microstructure which contains high hydrogen content is much lower than that of the heat-treated microstructure. This trend is confirmed by the arrangement of *S-N* data by the $\sqrt{area}$ parameter model (Murakami and Endo, 1986a, b; Murakami, 2002) (see Figure 29). To investigate this problem in greater detail, hydrogen desorption profiles were determined by TDS.

Figure 30 shows the variation of several hydrogen desorption profiles with time after hydrogen charging as a function of the temperature at which the hydrogen desorption was measured (Murakami and Nagata, 2005). There are two peaks in the profiles. The peak at the lower temperature is due to desorption of hydrogen from

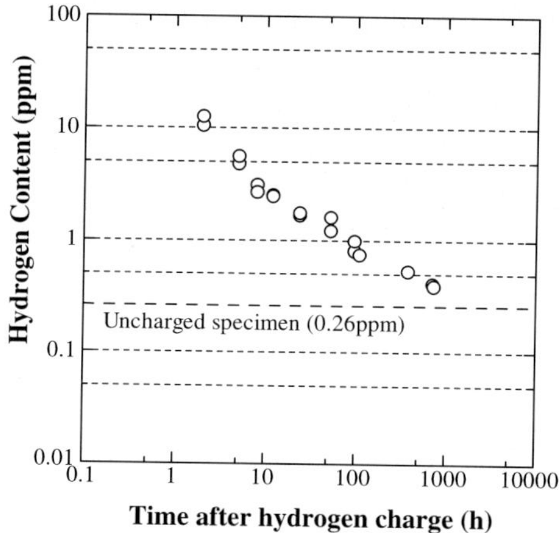

Figure 26. Variation of hydrogen content after hydrogen charge (SCM435(B)) (Murakami and Nagata, 2005).

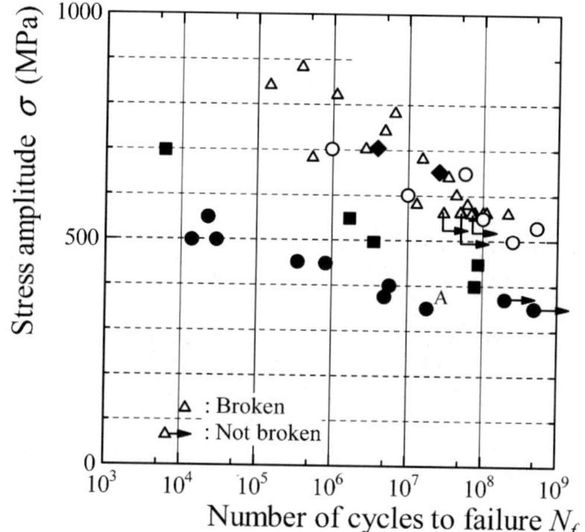

Figure 27. Effect of hydrogen content on *S-N* diagram (Murakami and Nagata, 2005). △: SCM435(A) ○●■◆:SCM435(B) △○:As heat treated specimen (C_H =0.3 ppm) ●■◆ : Hydrogen-charged specimen ●:1.5 h after hydrogen charge ($C_H = 10$ ppm) ■:100 h after hydrogen charge ($C_H = 0.8$ ppm) ◆:4300 h after hydrogen charge ($C_H = 0.3$ ppm) * Hydrogen content of specimens at the beginning of fatigue test was estimated by Fig.26. **Mark "A" indicates the specimen into which hydrogen was recharged every $N = 8.5 \times 10^6$ cycles (25.5 hours).

the microstructure (dislocations, vacancies and grain boundaries). The other, higher-temperature peak, is produced by desorption from hydrogen trapped at nonmetal-lic inclusions (Takai et al. 1996). Figure 30 indicates that hydrogen trapped by the microstructure desorbs and disappears with time from the specimen in air at room temperature, so that the amount of hydrogen trapped approaches that of the heat-treated specimens. To confirm this phenomenon, SIMS was performed on a fracture surface of the hydrogen-charged specimen shown in Figure 28b-1 ($N_f = 8.34 \times 10^5, \sigma = 448$ MPa, $f = 100$ Hz). The fracture surface was immediately buff-polished after the fatigue test to smooth the surface, and then the specimen was transported to the SIMS facility within 30 h. The specimen was kept in Dry Ice during transportation

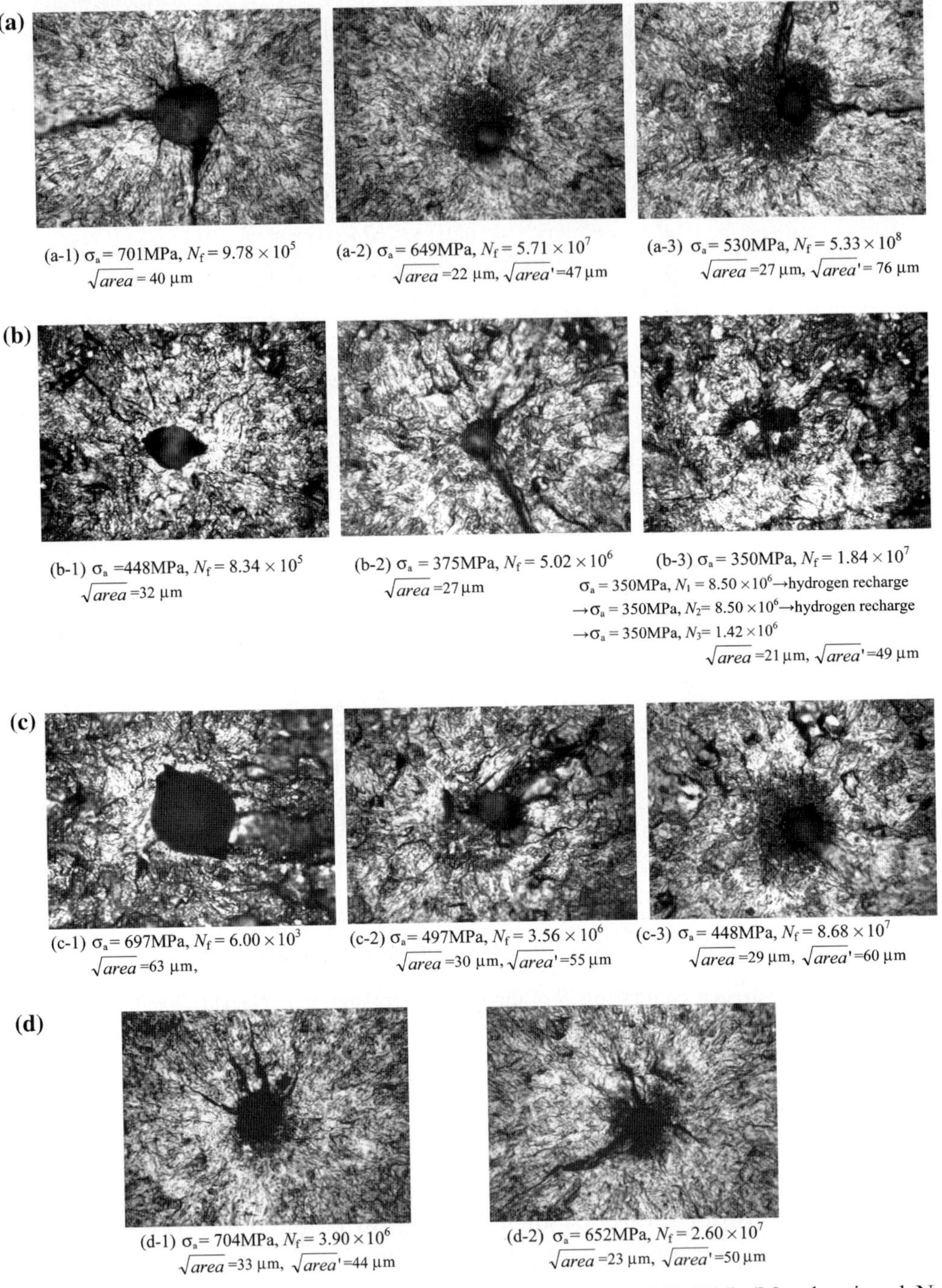

Figure 28. Optical micrographs of the inclusion at fracture origin. (SCM435) (Murakami and Nagata, 2005). ($\sqrt{area}$: Inclusion size, $\sqrt{area'}$: Inclusion size + ODA size). (a) As heat treated specimens (C_H = 0.3 ppm); (b) Fatigue tests were started 1.5 h after hydrogen charge (C_H = 10 ppm); (c) Fatigue tests were started 100 h after hydrogen charge (C_H = 0.8 ppm); (d) Fatigue tests were started 4300 h after hydrogen charge (C_H = 0.3 ppm).

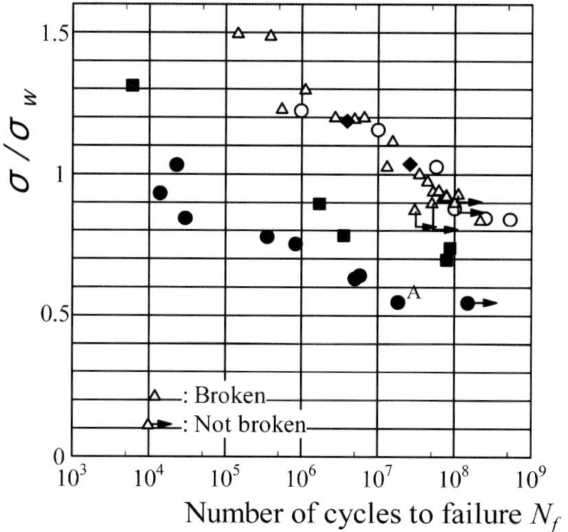

Figure 29. Effect of hydrogen content on the modified *S-N* diagram (Murakami and Nagata, 2005); △:SCM435(A) ○•■♦:SCM435(B); △○:As heat treated specimen ($C_H = 0.3$ ppm); •■♦ : Hydrogen-charged specimen; •:1.5 h after hydrogen charge (C_H =10 ppm); ■:100 h after hydrogen charge (C_H =0.8 ppm); ♦:4300 h after hydrogen charge ($C_H = 0.3$ ppm); σ : Stress amplitude; σ_w : Fatigue limit calculated by the $\sqrt{area}$ parameter model. $\sigma_w = 1.56(HV + 120)/(\sqrt{area})^{1/6}$.

Table 5. Relationship between number of hydrogen ion counts and holding time in the SIMS chamber (Murakami and Nagata, 2005).

	Near the inclusion at fracture origin	Remote microstructure
1 h	20–40	100–150
24 h	15–35	5–35

to prevent desorption and diffusion of hydrogen. Figure 31a and b show the variation of the hydrogen desorption image with time in the vicinity of an inclusion (Murakami and Nagata, 2005). Table 5 shows the relationship between the number of hydrogen ion counts and the holding time in the SIMS chamber at two locations, one in the vicinity of the inclusion and the other in the remote microstructure. Figure 31 and Table 5 demonstrate that although the hydrogen trapped by the inclusion is nondiffusive, the hydrogen in the remote microstructure is diffusive. It may be presumed that the hydrogen trapped by an inclusion contributes to crack initiation as well as to the discrete crack growth process in very early stage of fatigue life, and the diffusive hydrogen above a critical content in the remote microstructure contributes to decrease the fatigue crack growth threshold, resulting in a small ODA size.

Figure 32 shows the relationship between the ODA size, $\sqrt{area}_{ODA}$, and the stress intensity factor range at the periphery of the ODA, ΔK_{ODA}, for uncharged specimens of SCM435(A). The ΔK_{ODA} was calculated from the following equation proposed by Murakami et al. (2002).

$$\Delta K_{ODA} = 0.5\Delta\sigma\sqrt{\pi\sqrt{area}_{ODA}} \text{ (for an internal crack)}$$

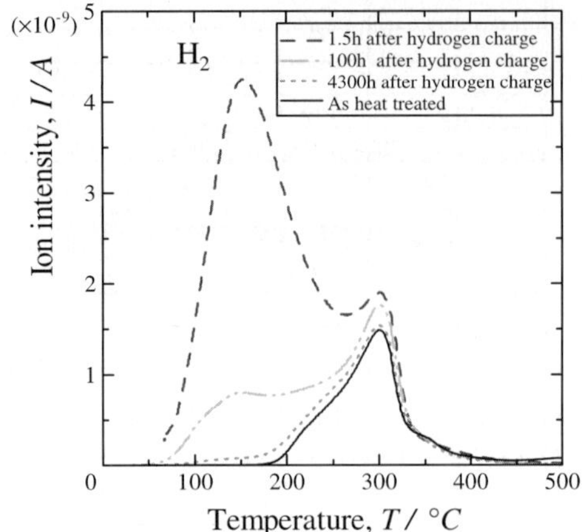

Figure 30. Variation of hydrogen desorption profiles with time after hydrogen charge (SCM435(B)) (Murakami and Nagata, 2005).

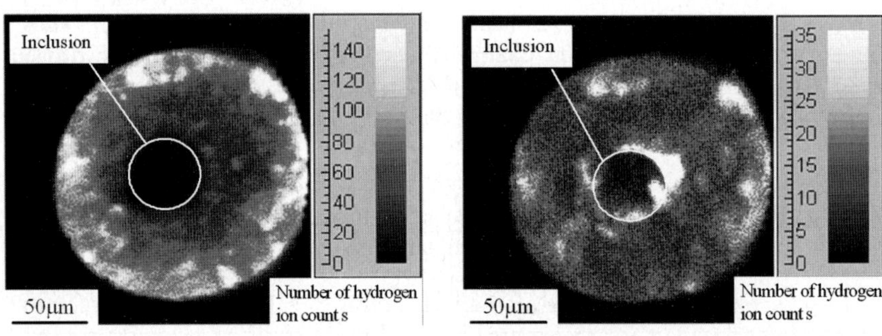

(a) After 1 hour hold in SIMS chamber   **(b)** After 24 hour hold in SIMS chamber

Figure 31. Secondary ion image of hydrogen in the vicinity of inclusion ($Al_2O_3(CaO)_x$) at fracture origin of hydrogen-charged specimen (SCM435(B) (Murakami and Nagata, 2005)) (a) After 1 hour hold in SIMS chambel (b) After 24 hours hold in SIMS chamber.

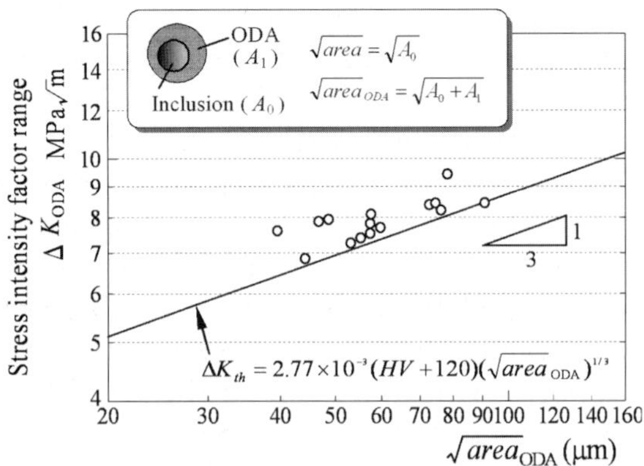

Figure 32. Relationship between $\sqrt{area}_{ODA}$ and ΔK_{ODA} for SCM435(A) without hydrogen charging.

The values of ΔK_{ODA} are proportional to $\left(\sqrt{area}_{ODA}\right)^{1/3}$ in the range of $\sqrt{area}_{ODA} = 40\text{--}100\mu$m and are in good agreement with the prediction by the $\sqrt{area}$ parameter model, which is indicated by the solid line in Figure 32. It must be noted that the values of ΔK_{ODA} are not constant and depend on crack size for small cracks.

3.4. 0.47%C AND 0.45%C MEDIUM CARBON STEELS

An understanding of the influence of hydrogen on slip behaviour is a key factor in making clear the influence of hydrogen on the fatigue mechanism. Annealed carbon steels are more appropriate for observing slip bands formation morphology under cyclic loading than high strength steels. Therefore, annealed 0.47%C steel was used in tension-compression fatigue testing and 0.45%C steel was used in rotating bending fatigue testing to observe the process of slip bands nucleation and crack initiation by the replica method.

The specimens contain a shallow circumferential notch [for 0.47%C (Figure 6c), $K_t = 1.42$] or a shallow notch (0.45%C, Figure 6d) which limits the location of slip band initiation and permits easy observation of the slip bands. The specimen surface of 0.47%C was finished by buffing after the hydrogen charging to remove corrosion pits generated by the charging. The specimen surface of 0.45%C was finished by electropolishing after the charging.

Figure 33 shows the *S-N* data for the uncharged and hydrogen-charged 0.47%C steel (Uyama et al., 2004). Figures 34a and b show that there is a clear difference in slip band morphology between the uncharged and hydrogen charged specimen (Uyama et al., 2004). Slip bands in the uncharged specimens (Figure 34a) gradually spread throughout and finally completely covered the ferrite grains with increasing number of cycles. On the other hand, the slip bands in the hydrogen-charged specimens (Figure 34b) are very discrete, and remain so from the early stages of cycling up to final fracture. Thus, the existence of hydrogen above a critical content causes localization of slip bands formed under cyclic loading. However, in the present tests these differences in slip band morphology appeared to have little effect on fatigue behaviour, as indicated by the results shown in Figure 33.

Figure 35 shows the number of crack initiations at the bottom area ($S = 0.65\,\text{mm}^2$) of shallow notch in three series of 0.45%C specimens having different hydrogen content at the fatigue limit ($\sigma_a = 230\,\text{MPa}$). Figure 36 shows examples of crack initiation in uncharged and hydrogen-charged 0.45%C steel at the fatigue limit observed with plastic replica method. Hydrogen-charging resulted in a marked increase in the number of crack initiations both along grain boundary and in ferrite grain compared with the uncharged specimen and the specimen aged for 270 h after the charging.

3.5. EFFECTS OF HYDROGEN AND PRE-CRACK ON THE REDUCTION OF AREA

So far, loss of ductility during fatigue cycle has been attributed to degradation of material due to "fatigue damage", though the reality of "fatigue damage" has not been revealed explicitly. Similarly, the loss of ductility in slow strain rate tensile test (SSRT) has been attributed to "hydrogen embrittlement", though the reality of "hydrogen embrittlement" has not been revealed explicitly. Murakami et al. (1989)

and Murakami and Miller (2005) made clear the loss of ductility during low-cycle fatigue is not produced by material degradation or so-called fatigue damage (the reality of this term had not been defined clearly) but it is due to the existence of a fatigue crack larger than a critical size. In this study, the effect of hydrogen on the loss of ductility is discussed from the same viewpoint.

A decrease in reduction of area (ductility) was obtained in tensile test carried out by interrupting fatigue test. Figure 37 shows the relationship between crack length and ductility loss. There is a critical crack length (Murakami et al. 1989) L_c for reduction of ductility: $L_c \approx 500\mu$m for SUS304; $L_c \approx 700\mu$m for SUS316.

In the uncharged specimens of SUS304 and SUS316, the similar fracture behaviour was observed. Figure 38 shows fracture appearances of the uncharged specimens with different pre-crack lengths of SUS304. The specimens with pre-crack length $L < L_c$, which did not affect ductility loss, essentially exhibited the so-called cup-and-corn type fracture as shown in Figures 38a and b. In the uncharged specimens with 400μm pre-crack, fracture did not originate from the pre-crack, and cup-and-corn type fractures were obtained at the minimum diameter sections as shown in Figure 38b. This implies that the presence of the crack with $L < L_c$ does not influence the ductility loss. On the other hand, in the specimens with pre-crack length $L > L_c$, shear mode fractures by slips at the crack tip were observed (Figure 38c, d). Thus, the crack with $L > L_c$ causes a shear type fracture and decreases the ductility.

There were no definite differences in fracture morphology and ductility loss between the hydrogen-charged specimens and the uncharged specimens. However, hydrogen-charge into the specimens of SUS304 produced the connection of many small cracks that were nucleated on high hydrogen content surface resulting in a large shear type crack. Figure 39 shows fracture appearances of the hydrogen-charged specimens of SUS304. In the plain specimen with hydrogen charge, a shear type fracture partially appeared due to the cracks produced by tensile stress, which could be observed by the unaided eye (see Figure 39a). In the hydrogen-charged specimen with 1000μm pre-crack, the shear type fracture was observed at a section of specimen

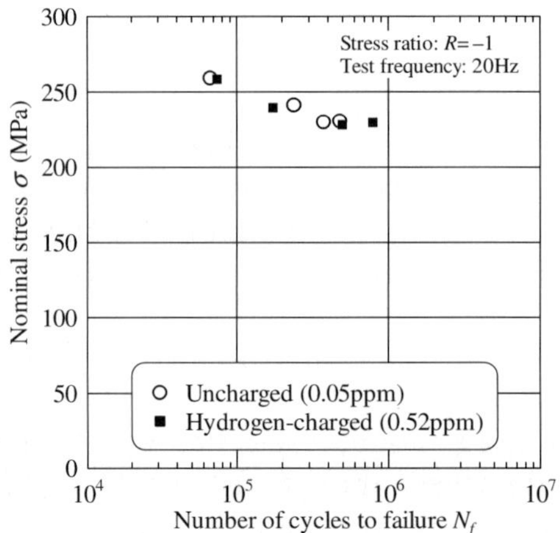

Figure 33. Tension-compression *S-N* data of the unnotched specimens of an annealed 0.47%C steel with and without hydrogen charging (Uyama et al., 2004). (Hydrogen content was measured 2 h after the hydrogen charging.)

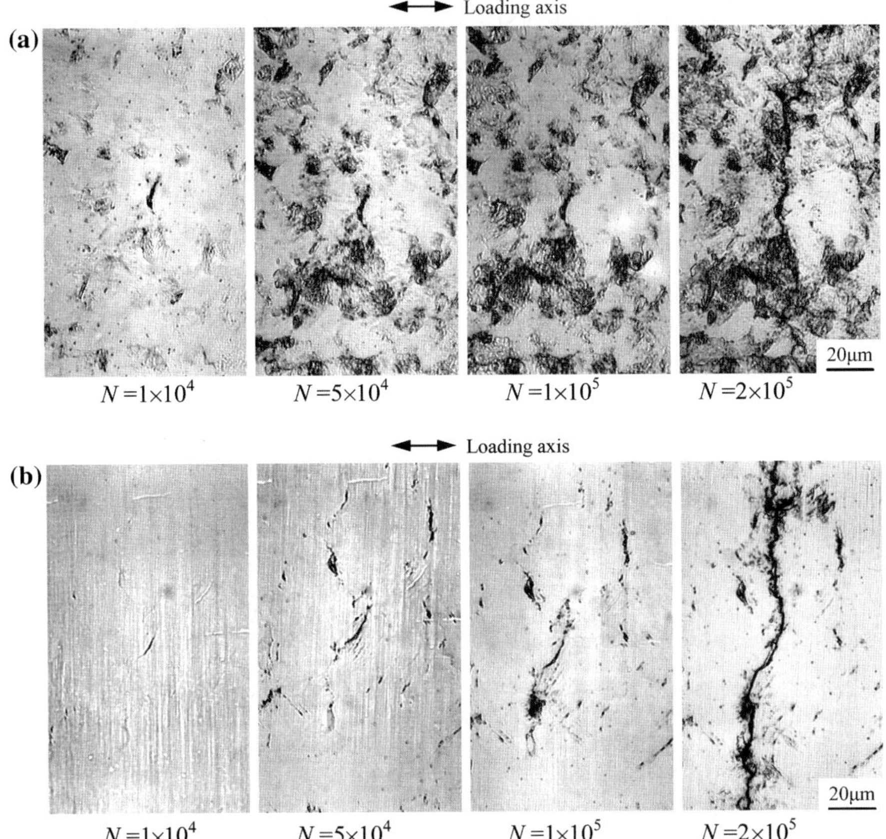

Figure 34. Crack and slip bands of the uncharged and hydrogen-charged specimen of 0.47%C steel under tension-compression (Uyama et al., 2004). (a) Uncharged specimen ($\sigma = 230\,\mathrm{MPa}$, $N_\mathrm{f} = 2.53 \times 10^5$, $C_\mathrm{H} = 0.05\,\mathrm{ppm}$); (b) Hydrogen-charged specimen ($\sigma = 230\,\mathrm{MPa}$, $N_\mathrm{f} = 2.16 \times 10^5$, $C_\mathrm{H} = 0.52\,\mathrm{ppm}$).

other than at the pre-crack as shown in Figure 39c. Many cracks were observed on the surface near the fracture, and especially, at the crack tip. In the case of the hydrogen-charged specimen with 800μm pre-crack, the shear type fracture started from the pre-crack and the ductility decreased more than that of uncharged specimen not only due to the pre-crack but also due to the cracks produced during tensile loading as shown in Figure 39b. However, such cracks did not appear in the hydrogen-charged specimen with 1500μm pre-crack, because the pre-crack was sufficiently large to grow alone in shear mode (see Figure 39d).

The results obtained in the present study suggest that hydrogen causes coalescence of many small cracks, which were nucleated on high hydrogen content surface. However, there is no definite difference in ductility loss between the hydrogen-charged and uncharged specimens. On the other hand, Benson et al. (1968) reported ductility loss on stainless steels in tensile test in high-pressure hydrogen environment. This may be due to the difference in strain rate in two test conditions. In the present study, the tensile tests were conducted in air and the strain rate was $\sim 2.5 \times 10^{-3}\,\mathrm{s}^{-1}$. On the other hand, Benson et al.'s test was conducted in high-pressure hydrogen at a slow strain rate of $\sim 2.9 \times 10^{-4}\,\mathrm{s}^{-1}$. In the condition of slow strain rates in high-pressure hydrogen environments, there is sufficient time for hydrogen diffusion into material (but still only

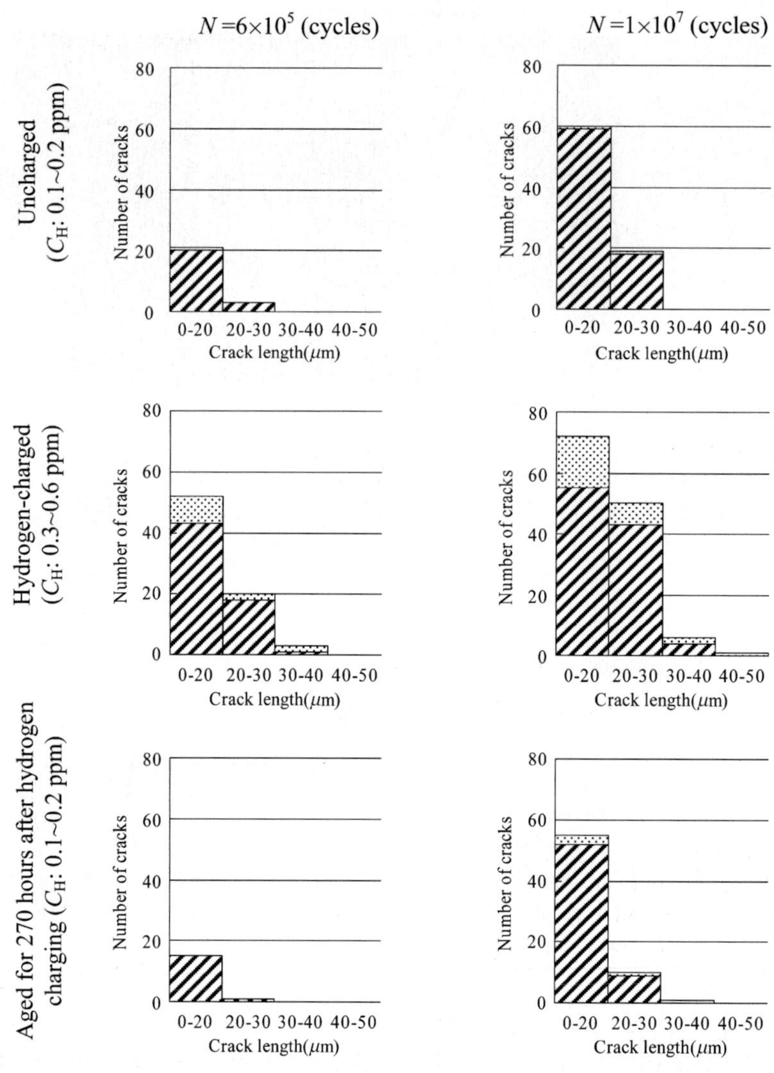

Figure 35. Number of crack initiations in three series of 0.45%C specimens at the fatigue limit ($\sigma_a = 230$ MPa, Rotating bending).

into surface layer) and many cracks initiate and grow during test until a main crack grows up to the critical length after coalescence. Thus, the ductility loss in the present study (strain rate $\sim 10^{-3}\,\mathrm{s}^{-1}$) is not directly caused by prior fatigue history or hydrogen embrittlement but depends on the crack length produced during fatigue cycles.

4. Conclusions

The influence of hydrogen on the fatigue properties of five different stainless steels (SUS304, SUS316, SUS316L, SUS405 and 0.7C-13Cr steel), a Cr-Mo steel (SCM435) and an annealed 0.47%C steel and an annealed 0.45%C steel was investigated by

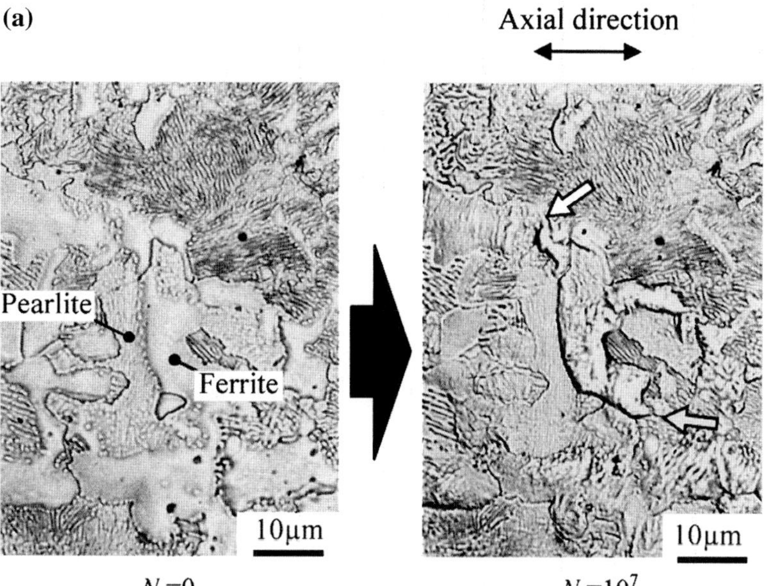

Crack initiation at grain boundary
(Uncharged specimen, R=−1, σ_a=230MPa)

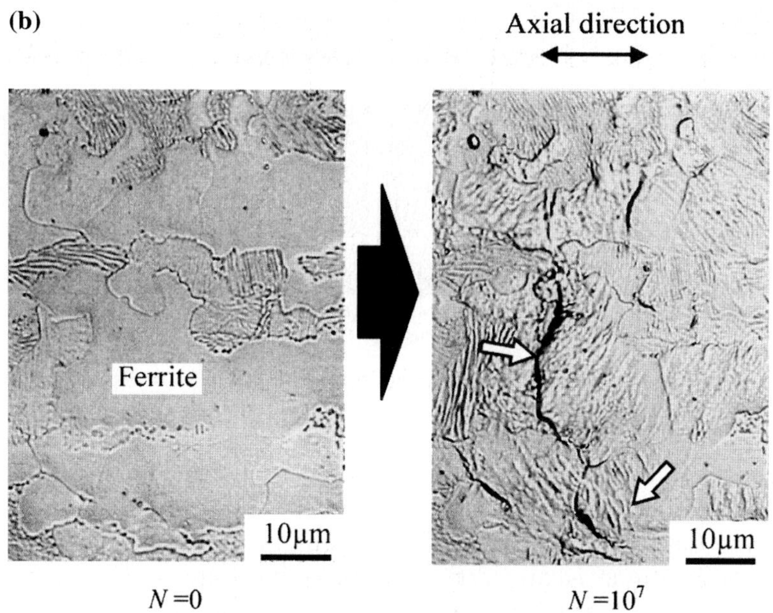

Crack initiation in ferrite grain
(Hydrogen-charged specimen, R=−1, σ_a=230MPa)

Figure 36. Examples of crack initiation in uncharged and hydrogen-charged 0.45%C steel at the fatigue limit observed with plastic replica (Rotating bending). (a) Crack initiation at grain boundary (Uncharged specimen, $R=-1, \sigma_a=230\,\text{MPa}$); (b) Crack initiation in ferrite grain (Hydrogen-charged specimen, $R=-1, \sigma_a=230\,\text{MPa}$).

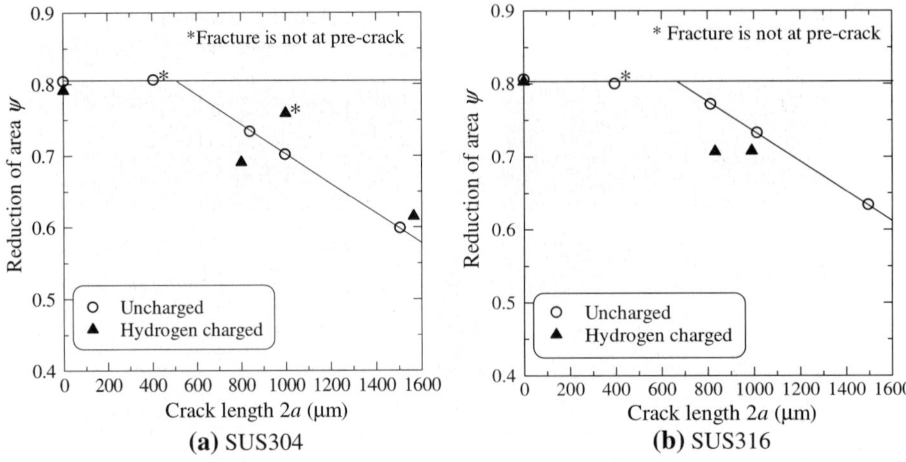

Figure 37. Relationship between crack length and ductility loss (Kanezaki et al., 2004).

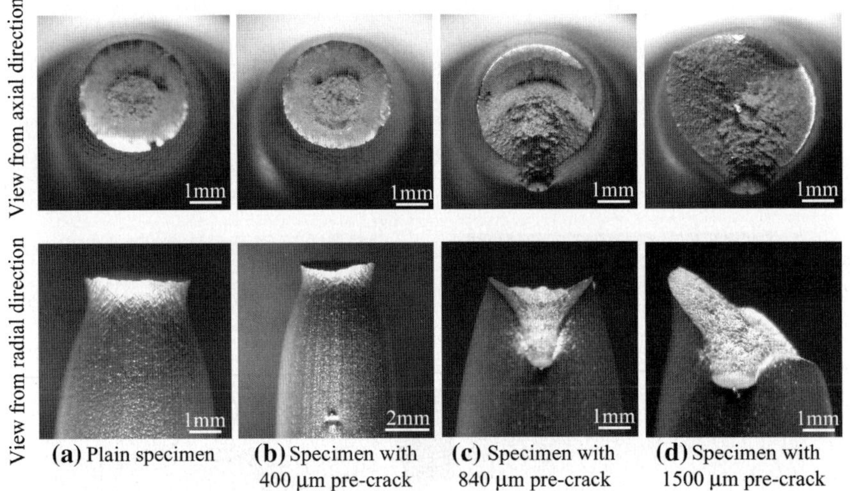

Figure 38. Fracture morphologies of uncharged specimen of SUS304 (Kanezaki et al., 2004).

tension-compression fatigue tests with specimens into which hydrogen was artificially charged.

Reviewing the data we obtained so far, at present we cannot help saying "Role of hydrogen in metal fatigue is mystery!" We have not fully understood the behaviour of hydrogen at an atomic scale. However, we at least have made clear some phenomenological aspects of the effect of hydrogen in metal fatigue.

(1) The hydrogen charging into SUS304 and SUS316 specimens led to a marked increase in the fatigue crack growth rate. The crack growth path in hydrogen-charged specimens was relatively linear, whereas the crack growth path in the uncharged specimen was more zigzag. In addition, in SUS304 and SUS316, numerous slip bands were observed at the surface of uncharged specimens which had been cyclically loaded, whereas relatively few slip bands were observed at the surface of the hydrogen-charged specimens. In contrast, in

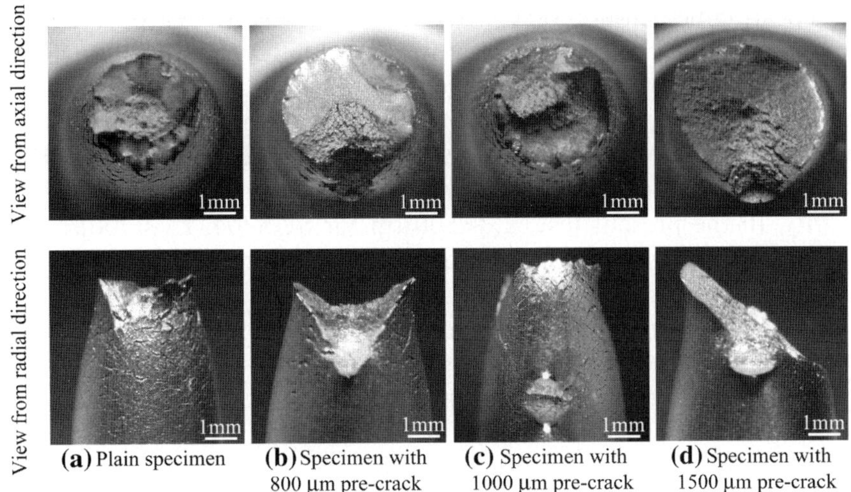

Figure 39. Fracture morphologies of hydrogen-charged specimen of SUS304 (Kanezaki et al., 2004).

SUS316L and SUS405, any differences in crack growth path and slip bands formation between uncharged specimens and hydrogen-charged specimens were not pronounced. In austenitic stainless steels, it appears that one of the crucial factors determining the influence of hydrogen on fatigue crack growth rate is the amount of martensitic transformation generated by cyclic plastic deformation at the vicinity of crack tip.

(2) In 0.7C–13Cr stainless steel (martensitic), the fatigue strength and fatigue life of smooth specimen were markedly decreased by hydrogen charging. Almost all the fracture origins of uncharged specimens were at subsurface nonmetallic inclusions located near the surface. On the other hand, the fracture origin of hydrogen-charged specimens was at specimen surface. This finding implies that hydrogen charging strongly influences the strength properties of hard martensitic structures.

(3) In a Cr–Mo steel (SCM435), fatigue strength and fatigue life decreased with increasing hydrogen content. The fracture origin of hydrogen-charged specimens showed smaller ODAs than the as-heat-treated specimens. This implies that the fatigue threshold of the microstructure which contains high hydrogen content is much lower than that of the as heat-treated microstructure. Observations with TDS and SIMS revealed that hydrogen trapped by the microstructure is diffusive, and the hydrogen trapped by inclusions is non-diffusive. It is assumed that the hydrogen trapped by inclusions contributes to crack initiation as well as discrete crack growth process in very early stage of fatigue life, and that a diffusive hydrogen content above a critical level in the remote microstructure contributes to decrease the fatigue crack growth threshold.

(4) To make clear the influence of hydrogen on the fatigue mechanism, the slip behaviour of grains and the crack initiation behaviour in microstructure was investigated on annealed 0.47%C steel and 0.45%C steel. In 0.45%C steel, slip bands in uncharged specimens gradually spread throughout the ferrite grains with increasing number of cycles. On the other hand, slip bands in hydrogen-

charged specimens were very discrete from early stages of cycling until final fracture, although in the present study this difference had little effect on fatigue strength. In 0.47%C steel, Hydrogen-charging resulted in a marked increase in the number of crack initiations both along grain boundary and in ferrite grain

(5) Reduction of area (ductility) in tensile test carried out by interrupting fatigue test decreased with fatigue cycles. There was a critical crack length for loss of ductility. In the present test, $L_c \approx 500\mu$m for SUS304, $L_c \approx 700\mu$m for SUS316.

(6) In hydrogen-charged SUS304 and SUS316, hydrogen did not directly affect the ductility loss. However, hydrogen causes coalescence of many small cracks, which were nucleated on high hydrogen content surface. The ductility loss is not directly caused by prior fatigue history or hydrogen embrittlement but depends on the length of cracks produced during fatigue cycles of strain rate of $\sim 2.5 \times 10^{-3}\,\mathrm{s}^{-1}$.

Acknowledgements

This research was supported by the Ministry of Education, Science, Sports and Culture, Grant-in-Aid for Specially Promoted Research, 2002–2006, No.1400102 and the NEDO project: Development of Basic Technology for the Safe Use of Hydrogen, 2003–2007.

References

Au, J.J. and Birnbaum, H.K. (1973). Magnetic relaxation studies of hydrogen in iron: relaxation spectra. *Scripta Metallurgica* **7**, 595–604.

Benson, Jr. R.B., Dann, R.K. and Robert, Jr. L.W. (1968). Hydrogen embrittlement of stainless steel. *Transformation of the Metallurgical Society of AIME* **242**, 2199–2205.

Birnbaum, H.K., Robertson, I.M. and Sofronis, P. (2000). Hydrogen effects on plasticity. In: *Multiscale Phenomena in Plasticity*, (edited by Lepinoux, J.), Kluwer Academic Publishers, Dordrecht.

Birnbaum, H.K. and Sofronis, P. (1994). Hydrogen-enhanced localized plasticity-a mechanism for hydrogen-related fracture. *Material Science and Engineering* **A176**, 191–202.

Brass, A.M. and Chene, J. (1998). Influence of deformation on the hydrogen behavior in iron and nickel base alloys: a review of experimental data. *Material Science and Engineering*, **A242**, 210–221.

Clum, J.A. (1975). The role of hydrogen in dislocation generation in iron alloys. *Scripta Metallurgica* **9**, 51–58.

Dufresne, J.F. and Seeger, A. (1976). Hydrogen relaxation in α-iron. Groh P. and Moser P., *Physics Statistics Solutions (a)* **36**, 579–589.

Farrell, K. and Quarrell, A.G. (1964). Hydrogen embrittlement of an ultra-high-tensile steel. *Journal of the Iron and Steel Institute* December, 1002–1011.

Fiedler, H.C., Averbach, B.L. and Cohen M. (1954). The effect of deformation on the martensitic transformation in austenitic stainless steels. *Transaction of the ASME* **47**, 267–290.

Heller, W.R. (1961). Quantum effects in diffusion: internal friction due to hydrogen and deuterium dissolved in α iron. *Acta Metallurgica* **9**, 600–613.

Herms, E., Olive, J.M., Puiggali, M. (1999). Hydrogen embrittlement of 316L type stainless steel. *Materials Science and Engineering* **A272**, 279–283.

Hirth, J.P. (1980). Effect of hydrogen on the properties of iron and steel. *Metallurgical Transactions A* **11A**, 861–890.

Kanezaki, T., Mine, Y., Fukushima, Y. and Murakami, Y. (2004). Effects of hydrogen on fatigue crack growth behaviour and ductility loss of austenitic stainless steels. Proc. of ECF15, CD-ROM.

Kimura, H. and Matsui, H. (1979). Reply to "further discussion on the lattice hardening due to dissolved hydrogen in iron and steel" by Asano and Otsuka. *Scripta Metallurgica* **13**, 221–223.

Magnin, T., Bosch, C., Wolski, K. and Delafosse, D. (2001). Cyclic plastic deformation behaviour of Ni single crystals oriented for single slip as a function of hydrogen content. *Material Science and Engineering* **A314**, 7–11.

Matsui, H. and Kimura, H. (1979). The effect of hydrogen on the mechanical properties of high purity iron III. The dependence of softening on specimen size and charging current density. *Material Science and Engineering* **40**, 227–234.

Mignot, F., Doquet, V. and Sarrazin-Baudoux, C. (2004). Contribution of internal hydrogen and room-temperature creep to the abnormal fatigue cracking of Ti6246 at high K_{max}. *Material Science Engineering* **A380**, 308–319.

Murakami, Y. (2002). *Metal Fatigue: Effect of Small Defects and Nonmetallic Inclusions*, Elsevier, Oxford.

Murakami, Y. (2004). Keynote lecture, The Kick-off Conference for the Fukuoka Hydrogen Energy Project Alliance, **August**, 3.

Murakami, Y. and Endo, M. (1986a). Effects of hardness and crack geometry on ΔK_{th} of small cracks. *Journal of the Society of Material Science Japan* **35**, 911–917.

Murakami, Y. and Endo, M. (1986b). Effects of hardness and crack geometries on ΔK_{th} of small cracks emanating from small defects. In: *The Behaviour of Short Fatigue Cracks, EGF Pub.1*, (edited by Miller, K.J. and delos Rios), Mechanical Engineering Publications, London, 275–293.

Murakami, Y. (Yukitaka), Konishi, H., Takai, K., Murakami Y. (Yasuo) (2000). Acceleration of super-long fatigue failure by hydrogen trapped by inclusions and elimination of conventional fatigue limit. *Tetsu-to-Hagane* **86**, 777–783.

Murakami, Y., Makabe, C. and Nisitani, H. (1989). Effects of small surface cracks on ductility loss in low cycle fatigue of 70 / 30 brass. *Journal of Testing and Evaluation*. JTEVA **17**, 20–27.

Murakami, Y. and Miller, K.J. (2005). What is fatigue damage? A viewpoint from the observation of low cycle fatigue process, *International Journal of Fatigue* **27**, 991–1005.

Murakami, Y. and Nagata J. (2005). Effect of hydrogen on high cycle fatigue failure of high strength steel, SCM435. *Journal of the Society of Material Science Japan* **54**, 420–427.

Murakami, Y., Nomoto, T., Ueda, T. (1999). Factors influencing the mechanism of superlong fatigue failure in steels. *Fatigue Fracture Engineering Material Structures* **22**, 581–590.

Murakami, Y. (Yukitaka), Ueda, T., Nomoto, T., Murakami, Y. (Yasuo) (1998). Analysis of the mechanism of superlong fatigue failure by optical microscope and SEM/AFM observations. Proceedings of the 24th Symposium on Fatigue, 47–50.

Murakami, Y. (Yukitaka), Ueda, T., Nomoto, T., Murakami, Y. (Yasuo) (2000). Mechanism of superlong fatigue failure in the regime of $N > 10^7$ cycles and fractography of the fracture surface. *Transactions of the Japan Society of Mechanical Engineers* **A66**, 311–319.

Murakami, Y., Yokoyama, N.N. and Takai, K. (2001). Effect of hydrogen trapped by inclusions on ultra-long life fatigue failure of bearing steel. *Journal of the Society of Material Science Japan* **50**, 1068–1073.

Nagata, J. and Murakami, Y. (2003). Factors influencing the formation of ODA in ultralong fatigue regime. *Journal of the Society of Material Science Japan* **52**, 966–973.

Sakamoto, Y. and Katayama, H. (1982). The electrochemical determination of diffusivity and solubility of hydrogen in an austenitic type 304 steel. *Journal of the Japan Institute Metals* **46**, 805–814.

Senkov, O.N. and Jonas, J.J. (1996). Dynamic strain aging and hydrogen-induced softening in alpha titanium. *Metallurgica and Materials Transactions A* **27A**, 1877–1887.

Shih, D.S., Robertson, I.M. and Birnbaum, H.K. (1988). Hydrogen embrittlement of α titanium: in situ TEM studies. *Acta Metallurgica* **36**, 111–124.

Shiina, T., Nakamura, T. and Noguchi, T. (2004). Effect of stress ratio on surface- and interior-originating fatigue properties of high strength steel. *Transactions of the Japan Society of Mechanical Engineers* **A70**, 1042–1049.

Takai, K., Homma, Y., Izutsu, K. and Nagumo, M. (1996). Identification of trapping sites in high-strength steels by secondary ion mass spectrometry for thermally desorbed hydrogen. *Journal of the Japan Institute Metals* **60**, 1155–1162.

Uyama, H., Mine, Y. and Murakami, Y. (2004). Effects of hydrogen charge on cyclic stress-strain properties and fatigue behaviour. Proceedings of ECF15 (2004) CD-ROM.

International Journal of Fracture (2006) 138:197–209
DOI 10.1007/s10704-006-0034-2

© Springer 2006

A cohesive zone global energy analysis of an impact loaded bi-material strip in shear

J.G. WILLIAMS[1,*] and H. HADAVINIA[2]
[1]*Mechanical Engineering Department, Imperial College London, SW7 2AZ London, UK*
[2]*Faculty of Engineering, Kingston University, SW15 3DW London, UK*
**Author for Correspondence. (g.williams@ic.ac.uk)*

Received 1 March 2005; accepted 1 December 2005

1. Introduction

The study of impact loaded strips in shear has been extensive in recent years. This was prompted by a major experimental programme by Rosakis and coworkers (e.g. Coker et al., 2001; Rosakis, 2002; Saundrula and Rosakis, 2002) who noted many interesting phenomena in polymer-metal strips and in particular the fact that shear, Mode II, cracks running along a weak interface achieved steady state crack speeds greater than the shear wave velocity and approaching the dilation wave speed. Such failures have been termed "inter-sonic" or "supershear" and generated much interest. The work has also been of considerable interest in geophysics since the experiments can be used to model earthquakes if an unbonded interface is loaded laterally to generate friction forces and a disturbance propagated (e.g. Coker et al., 2005). In such experiments supershear failures have also been observed and indeed there is evidence of their occurrence in real earthquakes (Rosakis, 2002).

There has also been considerable effort devoted to modelling these experiments numerically. Needleman (1999) and Needleman and Rosakis (1999) used hyperelastic cohesive zone models in FE solutions and demonstrated transitions from low speed cracks to speeds approaching the dilatational wave speed by varying both the cohesive strength (Needleman, 1999) and the impact speed (Needleman and Rosakis, 1999). Similar calculations using FV and a Dugdale, constant stress, cohesive zone were reported by Williams, et al. (to be published) in which the cohesive stress and the fracture energy were varied. A general pattern of high crack speeds at low stresses was observed with abrupt transitions at critical stress values. A rather modest effect of fracture energy was reported.

A global energy balance solution of the problem was given in Willams (2005) and was compared with the experimental results. It showed that the high speed crack propagation could be predicted from such an analytical solution which agreed well with the experiments. It also showed that transients observed in the experiments, and in particular the accelerations, could be predicted. This paper is an extension of this solution to include a constant stress cohesive zone and will be compared with the numerical solutions.

2. Global analysis

2.1. STRESSES AND DISPLACEMENTS

The axially loaded strip is shown in Figure 1 divided into three sections. In Section 1 we have the initial crack length and the faces at $y = 0$ and h are unconstrained and stress free. This gives uniaxial tension (or compression) such that,

$$\sigma_{x_1} = \sigma_0, \quad \tau = \sigma_y = 0 \tag{1}$$

and

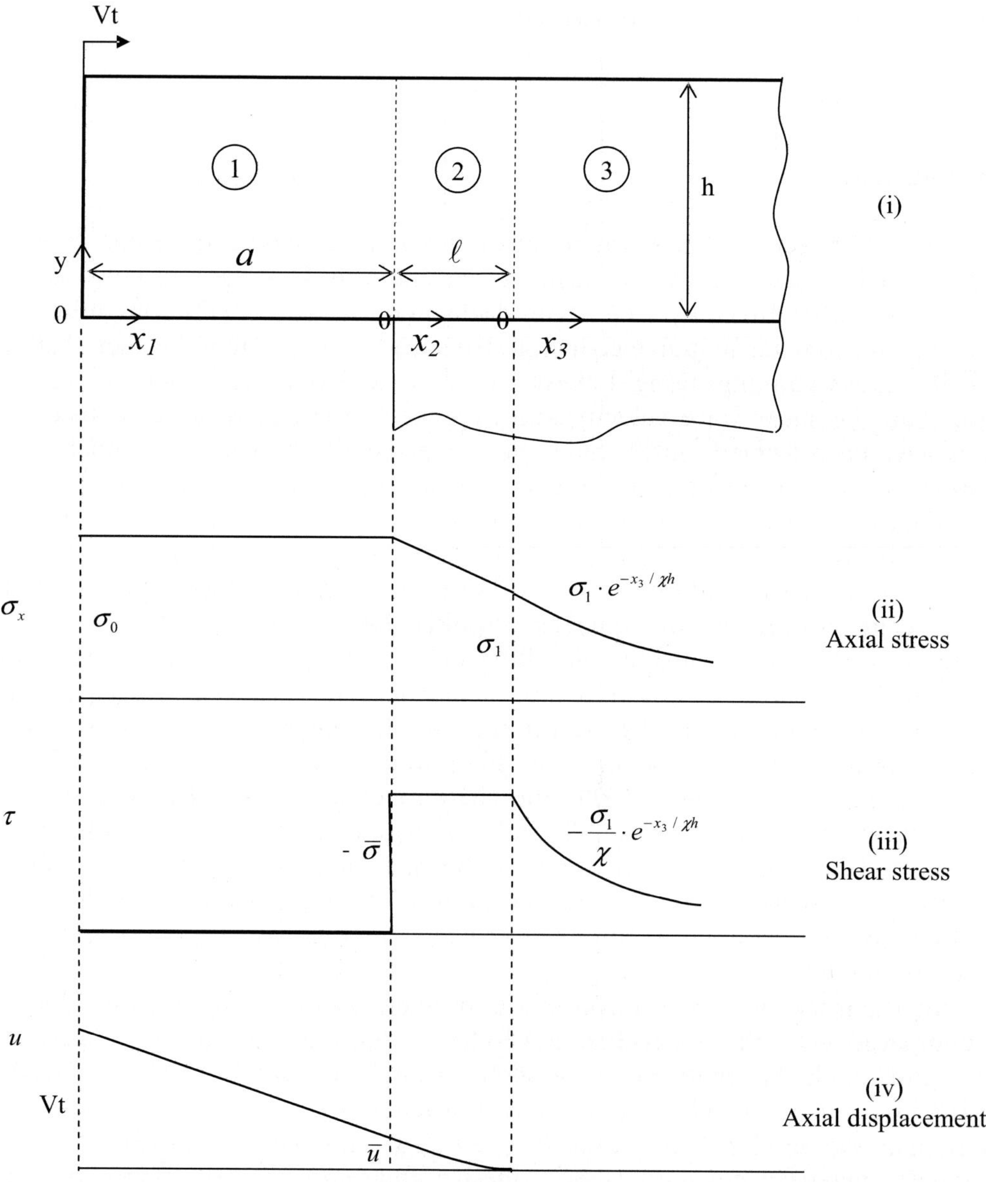

Figure 1. Stress and displacement distributions in the three zones.

$$u_1 = -u_0 \left(1 - \frac{x_1}{a}\right) + \bar{u} \tag{2}$$

where $\bar{u}$ is the displacement at the crack tip and,

$$u_0 = \frac{\sigma_0}{E} \cdot a = Vt \tag{3}$$

for loading at constant velocity V.

In Section 2 we have the cohesive zone for which the shear stress $\tau = -\bar{\sigma}$ at $y = 0$ and at $y = h$, $\tau = \sigma_y = 0$. From equilibrium,

$$\frac{\partial \sigma_{x_2}}{\partial x_2} = -\frac{\bar{\sigma}}{h}$$

i.e.

$$\sigma_{x_2} = \sigma_0 - \frac{\bar{\sigma}}{h} \cdot x_2 \tag{3}$$

and

$$\frac{\partial \tau}{\partial y} = -\frac{\partial \sigma_{x_2}}{\partial x_2} = \frac{\bar{\sigma}}{h}$$

i.e.

$$\tau = -\bar{\sigma}\,(1 - \varsigma), \quad \varsigma = y/h \quad \text{and} \quad \sigma_y = 0 \tag{4}$$

Axial displacements are derived from,

$$\frac{\partial u_2}{\partial x_2} = \frac{\sigma_{x_2}}{E}, \quad \text{i.e. } u_2 = \frac{-1}{E}\left[\sigma_0\,(\ell - x_2) - \frac{\bar{\sigma}}{2h}\left(\ell^2 - x_2^2\right)\right] \tag{5}$$

since $u = 0$ at $x_2 = \ell$, the zone length. At $x_2 = 0$, $u_1 = u_2 = \bar{u}$, i.e.,

$$\bar{u} = \frac{-\ell}{E}\left(\sigma_0 - \frac{\bar{\sigma}\,\ell}{2h}\right) \tag{6}$$

ℓ is determined by the axial stress in region 2 at $x_3 = 0$, i.e.

$$\sigma_1 = \sigma_0 - \frac{\bar{\sigma}}{h}\ell \tag{7}$$

and

$$\bar{u} = \frac{-h}{2E\bar{\sigma}}\left(\sigma_0^2 - \sigma_1^2\right) \tag{8}$$

σ_1 is determined by the constraint in region 3 and this may be modelled approximately by assuming that the axial stress has the form,

$$\sigma_{x_3} = \sigma_1(2\varsigma) \cdot e^{-x_3/\chi h} \tag{9}$$

and that $\sigma_y = 0$ giving zero axial strain along the lower edge, $\varsigma = 0$. χ is a characteristic length determined by the boundary conditions.

From equilibrium we have,

$$\frac{1}{h}\frac{\partial \tau}{\partial \varsigma} = -\frac{\partial \sigma_{x_3}}{\partial x_3}$$

and

$$\tau = -\frac{\sigma_1}{\chi}(1 - \varsigma^2)\,e^{-x_3/\chi h} \tag{10}$$

by putting $\tau = 0$ at $\varsigma = 1$, the upper edge. At $\varsigma = 0$ and $x = 0$

$$\tau = -\overline{\sigma} = \frac{-\sigma_1}{\chi}, \quad \text{i.e., } \sigma_1 = \chi\overline{\sigma}$$

χ may be determined from the compatibility condition since,

$$\frac{\partial v}{h\partial \varsigma} = \frac{-v}{E}\sigma_{x_3} = -\frac{v\sigma_1}{E}(2\varsigma)e^{-x_3/\chi h}$$

and

$$\frac{\partial^2 v}{\partial x_3^2} = -\frac{v}{\chi}\left(\frac{\sigma_1}{Eh}\right)\varsigma^2 e^{-x_3/\chi h}$$

with $v = 0$ at $\varsigma = 0$,
similarly

$$\frac{\partial^2 v}{\partial x_3^2} = \frac{2(1+v)}{E}\frac{\partial \tau}{\partial x_3} - \frac{\partial^2 u}{h\partial \varsigma \partial x_3} = \left(\frac{\sigma_1}{Eh}\right)e^{-x_3/\chi h}\left[\frac{2(1+v)}{\chi^2}(1 - \varsigma^2) - 2\right]$$

Equating these terms and averaging over h, we have:

$$\frac{2(1+v)}{\chi^2}\int_0^1 (1 - \varsigma^2)d\varsigma - 2 = \frac{-v}{\chi^2}\int_0^1 \varsigma^2 d\varsigma$$

i.e.

$$\chi^2 = \tfrac{1}{6}(4 + 5v) \tag{11}$$

An alternative approach is to minimise the strain energy. Thus,

$$U_s = \frac{h}{2E}\int_0^1 \int_0^\infty \left[\sigma_{x_3}^2 + 2(1+v)\tau^2\right]dx_3 d\varsigma = \left(\frac{\sigma_1^2 h^2}{2E}\right)\left[\frac{2}{3}\chi + \frac{8}{15}\frac{(1+v)}{\chi}\right]$$

and

$$\frac{dU_s}{d\chi} = \left(\frac{\sigma_1^2 h^2}{2E}\right)\left[\frac{2}{3} - \frac{8}{15}\frac{(1+v)}{\chi^2}\right] = 0$$

i.e.

$$\chi^2 = \tfrac{4}{5}(1 + v) \tag{12}$$

For $v = 0.3$, as used here, these solutions give $\chi = 0.96$ from Equation (11) and $\chi = 1.02$ from Equation (12).

Some FE results were obtained using ABAQUS to confirm these results. The dimensions of $h = 50$ mm were used and τ determined for σ_1 values of 20, 50, 100 and 200 MPa. The τ values are shown in Figure 2 plotted as $\ln \tau$ vs. x_3. The linear form confirms that exponential dependence assumed in the approximate solution and the numerical values of χ from the intercepts, 1.34, and the slopes, 1.19, are reasonably consistent though somewhat higher than the approximate solution. The increasing values for $x_3 < 20$ mm reflect the singularity. Figure 3 shows the results for σ_{x_3} at $\varsigma = 1/2$ the central section, and a χ value of 1.03 is given by the slopes. The distribution of σ_{x_3} and τ are shown in Figure 4 together with the assumed form for $\sigma_{x_3}, 2\varsigma$, and that derived for $\tau, 1 - \varsigma^2$, again confirming the analysis. σ_y values were very much less than σ_{x_3} and τ as assumed. It is clear that the approximate value of χ is unity and this will be used in the subsequent analysis.

2.2. ENERGY RELEASE RATE (G)

The simplest derivation of G is via the expression for $\bar{u}$, Equation (8).

$$G = -\bar{\sigma}\,\bar{u} = \frac{h}{2E}\left(\sigma_0^2 - \sigma_1^2\right) \tag{13}$$

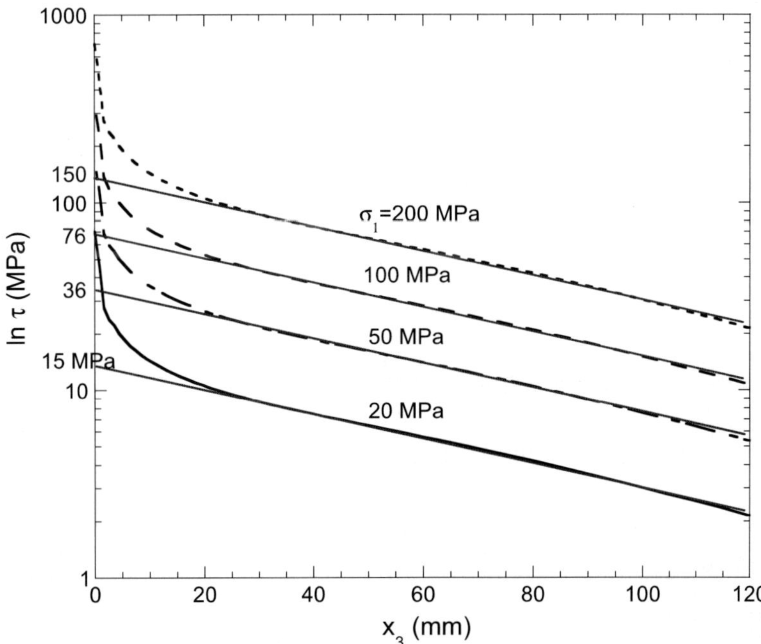

Figure 2. FE results for shear stress, $\varsigma = 0$ the lower edge.

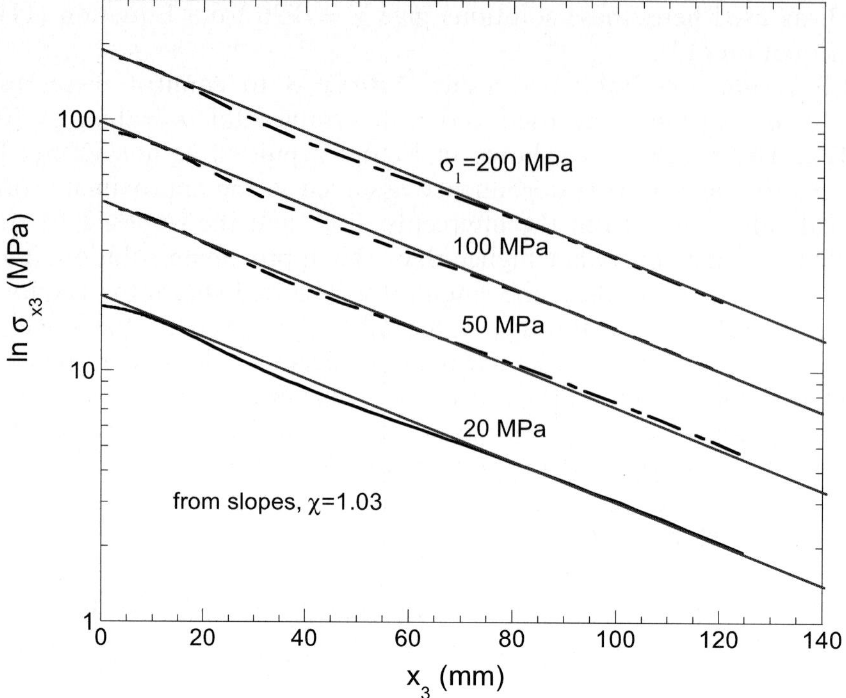

Figure 3. FE results for axial stress at $\varsigma = 1/2$.

i.e. the change in strain energy across the zone. For $\sigma_1 = 0$ we have the usual result for the axially loaded strip (Williams), while for the constrained base, $\sigma_1 = \chi\bar{\sigma}$ and using $\chi = 1$

$$G = \frac{h}{2E}\left(\sigma_0^2 - \bar{\sigma}^2\right) \tag{14}$$

2.3. DYNAMIC EFFECTS

For dynamic crack propagation the solutions for zones 2 and 3 may be adapted by noting that for steady state crack propagation,

$$\frac{\partial u}{\partial t} = \dot{a}\frac{\partial u}{\partial x} \quad \text{and} \quad \frac{\partial^2 u}{\partial t^2} = \dot{a}^2\frac{\partial^2 u}{\partial x^2}$$

and the axial equilibrium equation becomes

$$\frac{\partial \sigma_x}{\partial x} + \frac{\partial \tau}{\partial y} = \rho\frac{\partial^2 u}{\partial t^2} = \alpha^2 E \cdot \frac{\partial^2 u}{\partial x^2}$$

where $\alpha = \dot{a}/C$ and $C = \sqrt{E/\rho}$, the axial wave speed. In zone 2, $\sigma_y = 0$ and hence,

$$E\frac{\partial u}{\partial x} = \sigma_x$$

and

$$\left(1 - \alpha^2\right)\frac{\partial \sigma_{x_2}}{\partial x_2} + \frac{\partial \tau}{\partial y} = 0 \tag{15}$$

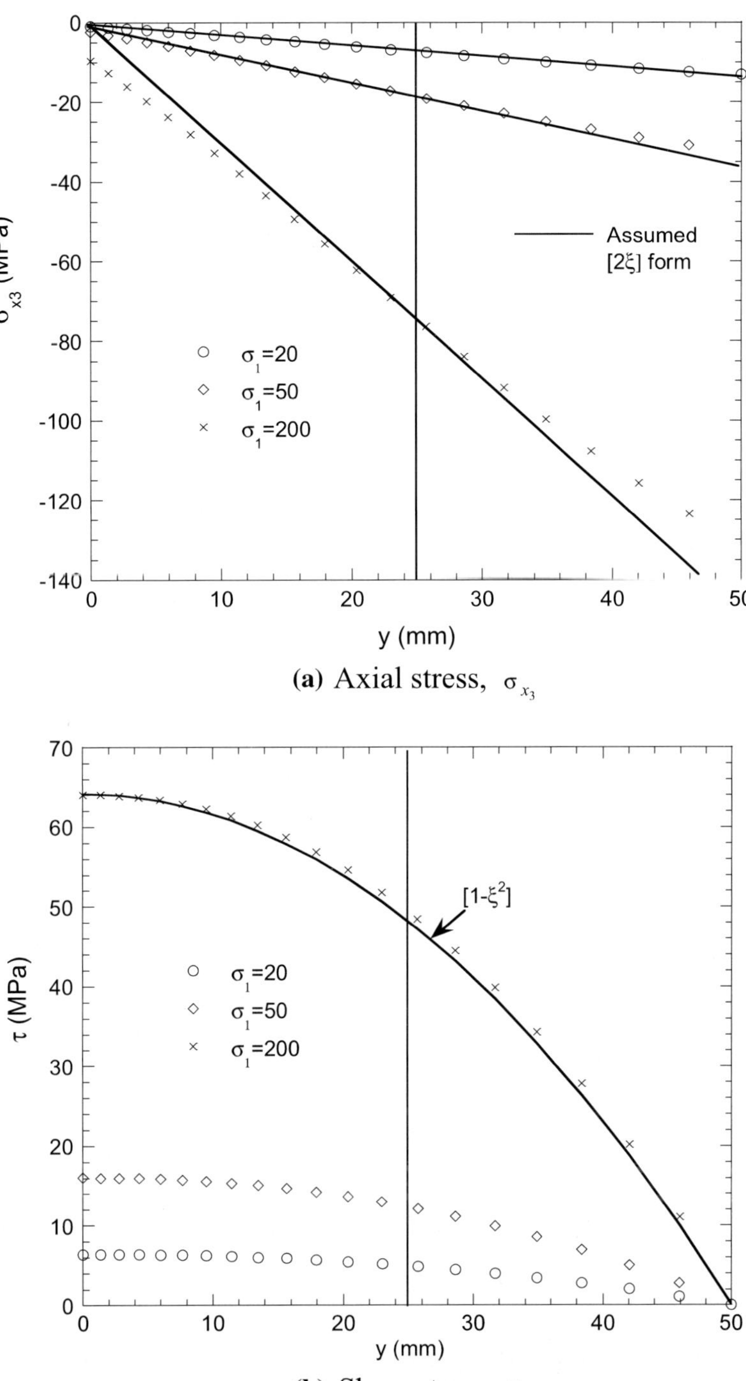

(a) Axial stress, σ_{x_3}

(b) Shear stress, τ

Figure 4. FE results for stress distribution at $x = 50\,\text{mm}$.

Using the same assumption as before, i.e. $\dfrac{\partial \tau}{\partial y} = \dfrac{\overline{\sigma}}{h}$,

$$\sigma_{x_2} = \sigma_0 - \frac{\overline{\sigma}}{1-\alpha^2} \cdot \frac{x_2}{h}$$

and hence

$$\overline{u} = -\frac{\ell}{E}\left(\sigma_0 - \frac{\overline{\sigma}}{1-\alpha^2} \cdot \frac{\ell}{2h}\right)$$

and

$$G = -\overline{\sigma}\,\overline{u} = \frac{\overline{\sigma}\ell}{E}\left(\sigma_0 - \frac{\overline{\sigma}}{1-\alpha^2} \cdot \frac{\ell}{2h}\right)$$

Also $\sigma_{x_2} = \sigma_1$ at $x_2 = \ell$ giving

$$\frac{\overline{\sigma}\,\ell}{h} = (1-\alpha^2)(\sigma_0 - \sigma_1)$$

i.e.

$$G = \frac{h}{2E}(\sigma_0^2 - \sigma_1^2)(1-\alpha^2)$$

If we assume that, as a first approximation, static conditions prevail in the elastic zone 3 then $\chi = 1$ and we have

$$G = \frac{h}{2E}(\sigma_0^2 - \overline{\sigma}^2)(1-\alpha^2) \tag{16}$$

For steady state crack propagation we may approximate σ_0 by,

$$\sigma_0 = E\frac{V}{\dot{a}} = E\frac{V}{C}\frac{1}{\alpha} \tag{17}$$

and if we assume that propagation occurs at a constant G value, G_0, then the equation of motion of the crack is,

$$A = \frac{2G_0}{Eh}\cdot\left(\frac{C}{V}\right)^2 = (1-\alpha^2)\left[\frac{1}{\alpha^2} - \left(\frac{\overline{\sigma}}{E}\frac{C}{V}\right)^2\right] \tag{18}$$

A dynamic form for χ may be found by using Equation (15) from which,

$$\tau = -(1-\alpha^2)\frac{\sigma_1}{\chi}(1-\varsigma^2)e^{-x_3/\chi h}$$

and proceeding via the compatibility condition in $\dfrac{\partial^2 V}{\partial x_3^2}$ as before we have,

$$\chi^2 = (1-\alpha^2)\frac{2}{3}(1+v) + \frac{v}{6} \tag{19}$$

For $v = 0.3$ this may approximated to,

$$\chi = \left(1 - 0\cdot92\alpha^2\right)^{1/2}$$

and Equation (18) becomes

$$A = (1 - \alpha^2)\left[\frac{1}{\alpha^2} - (1 - 0\cdot 92\alpha^2)\left(\frac{\overline{\sigma}}{E}\cdot\frac{C}{V}\right)^2\right] \tag{20}$$

It is of interest to consider α as a function of $\overline{\sigma}$ for fixed A values and Equation (20) is shown in Figure 5a. For low speeds, α is proportional to $\left(\frac{\overline{\sigma}}{E}\frac{C}{V}\right)^{-1}$ but as α increases, there is a very steep change with $\overline{\sigma}$ until α becomes constant at $(1 + A)^{-1/2}$ for finite values of A. This would appear in numerical results as constant speeds for low $\overline{\sigma}$ values with a sharp transition to low values when $\left(\frac{\overline{\sigma}}{E}\frac{C}{V}\right) \approx 1.5$–2. A similar transition is shown in Figure 5b where α is shown as a function of $\left(\frac{E}{\overline{\sigma}}\frac{V}{C}\right)$ for various values of $\frac{2EG_0}{\overline{\sigma}^2 h}$ which describes the case of varying V for fixed G_0 and $\overline{\sigma}$. For low values of $\frac{2EG_0}{\overline{\sigma}^2 h}$ there is a rather sharp transition in α for $\left(\frac{E}{\overline{\sigma}}\frac{V}{C}\right) \approx 0.5$–0.6 i.e. the inverse of the previous case.

3. Comparison with numerical results

Three numerical simulations of the Rosakis experiments (Needleman, 1999; Needleman and Rosakis, 1999; Williams et al., to be published) using polymer and metal specimens have been reported. In both the dimensions were taken from tests; i.e. $h = 50$ mm and $a_0 = 25$ mm. In Williams et al. (to be published) four loading conditions were examined with $G_0 = 1000$ and 3000 J/m^2 and using a constant stress zone and $V = 10$ and 20 m/s. A modulus for the polymer of 6 GPa was assumed and with $\rho = 1,500$ kg/m^3 the wave speed was 2000 m/s. Thus for the four cases the A values were, 0.80, 0.27, 0.20 and 0.07 and for each the velocity history was evaluated for $\overline{\sigma}$ values of 10, 20, 50, 100, 150 and 200 MPa using a FV numerical code. In all cases the lower $\overline{\sigma}$ values gave constant high crack velocities which approached C while with the higher values much lower speeds occurred. In three cases there was an abrupt transition in speed during the propagation. Figure 6 shows the speeds plotted as $\alpha = \dot{a}/C$ versus $\left(\frac{\overline{\sigma}}{E}\frac{C}{V}\right)$. The three transition cases are shown as double points with vertical lines. The curves for Equation (20) are also shown for $A = 0, 0.05, 0.2$ and 0.8 which predicts the transition at $\left(\frac{\overline{\sigma}}{E}\frac{C}{V}\right)$ at about 2 while the numerical results give values of 0.7–1.5. The low stress level speeds are reasonably well predicted but at higher stress values the speeds are rather higher than these predictions.

A similar analysis was conducted in Needleman (1999) in which a hyperelastic cohesive zone model was used with $G_0 = 75$ J/m^2, and $V = 26$ m/s. An elastic modulus of 5.3 GPa was used giving, $C = 2600$ m/s and $A = 6 \times 10^{-3} \cdot \overline{\sigma}$ was varied up to about 100 MPa and the very sharp transition identified at 69 MPa, i.e. $\left(\frac{\overline{\sigma}}{E}\frac{C}{V}\right) = 1.3$. The values taken from Needleman (1999) are also shown in Figure 6 and for $\overline{\sigma} > 69$ MPa the crack speed was measured as the Rayleigh wave speed, i.e. 1170 m/s, $\alpha = 0.45$ in this case. This speed was not found in the data reported in Williams et al. (to be published) and the analysis given here does not predict such a speed. The transition predicted by Equation (20) is dependant on χ and unity was used here. For low A values this is given when $\frac{d}{d\alpha}\left(\frac{\overline{\sigma}}{E}\cdot\frac{C}{V}\right) = 0$, i.e. $\alpha^2 = 1/1.84$ and,

$$\frac{\overline{\sigma}}{E}\cdot\frac{C}{V} = \frac{1.9}{\chi}$$

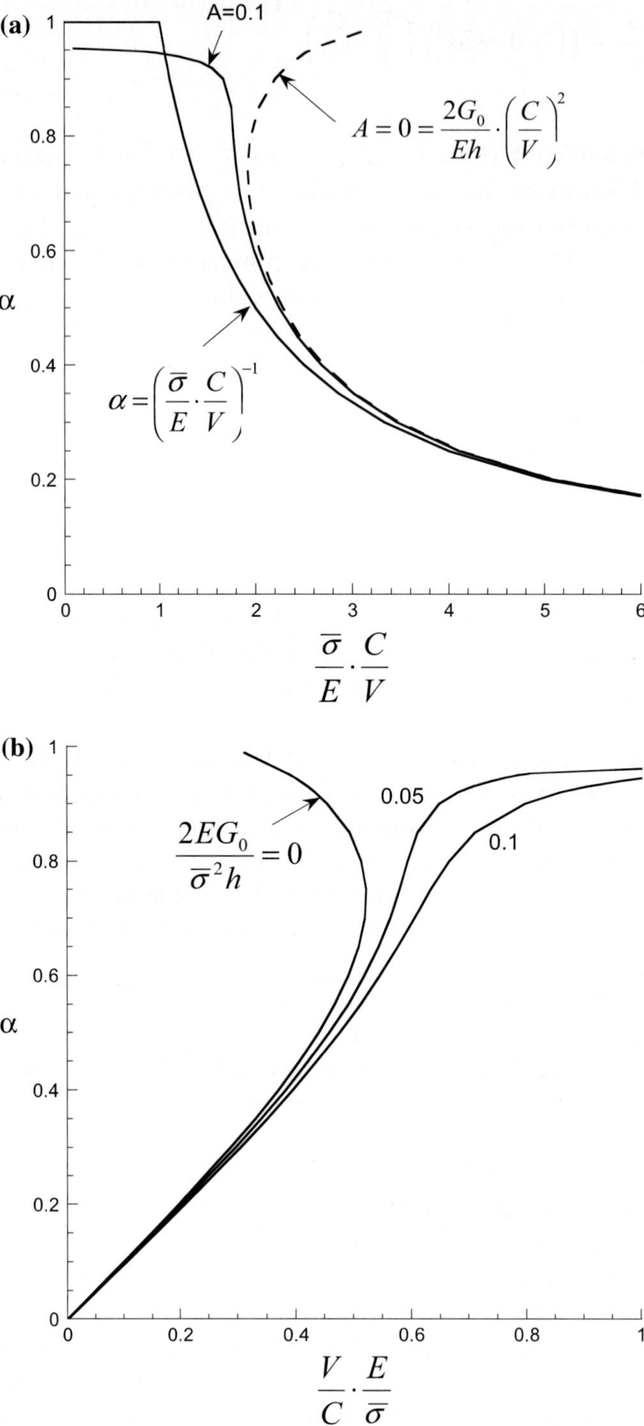

Figure 5. (a) Variation of α as a function of $\overline{\sigma}$ for fixed A. (b) Variation of α as a function of $\dfrac{V}{C}\dfrac{E}{\overline{\sigma}}$ for fixed $\dfrac{2EG_0}{\overline{\sigma}^2 h}$.

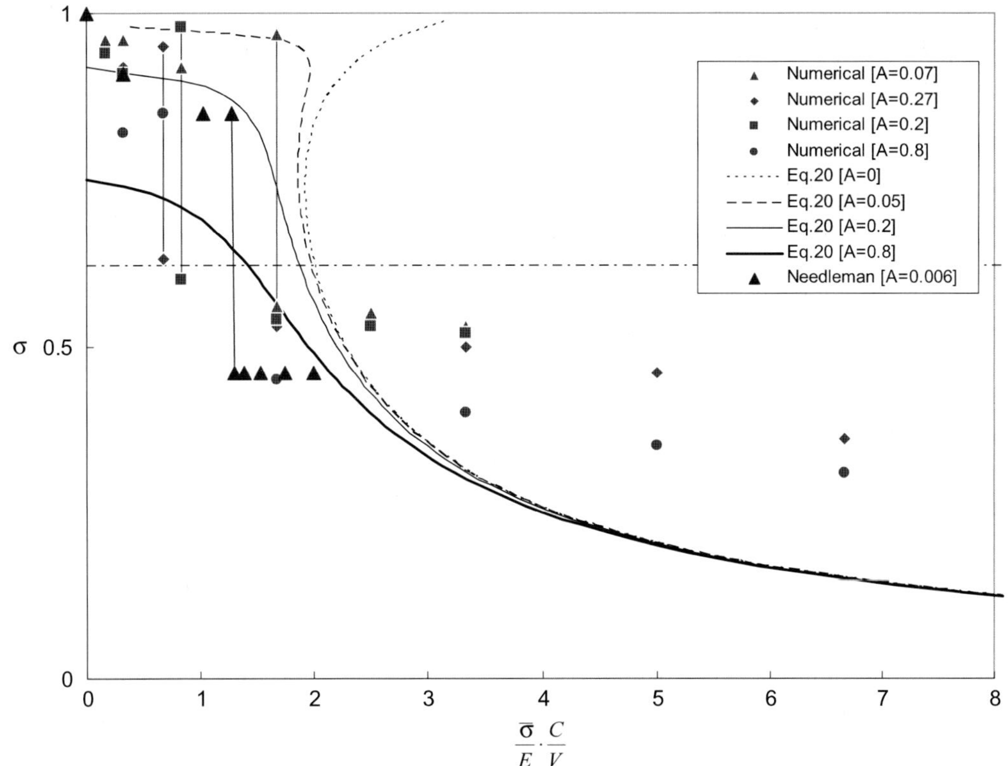

$$\frac{\overline{\sigma}}{E}\cdot\frac{C}{V}$$

Figure 6. Crack speed dependence on the cohesive stress $\overline{\sigma}$. Data from Needleman (1999) and Needleman and Rosakis (1999).

so the transition at 1.3 would suggest $\chi = 1.5$. The transitions in Williams et al. (to be published) were not defined as precisely but were at values of $\left(\frac{\overline{\sigma}}{E}\frac{C}{V}\right)$ values of 0.7, 0.8 and 1.7 giving χ values of 2.7, 2.4 and 1.1.

A rather different experiment is modelled in Needleman and Rosakis (1999) in which a PMMA-steel specimen is impacted on the steel side on the end opposite to the crack. Crack propagation occurs when the wave is reflected back to the crack tip. A range of $\overline{\sigma}$ values was used and V varied to determine the transition impact speed to give high crack speeds, i.e. $\approx 0.9C$. In this case the energy is almost entirely in the steel and,

$$\sigma_0 = E_{st}\left(\frac{V}{C_{st}}\right)$$

where E_{st} is the modulus of the steel and C_{st} the wave speed. The crack propagation is determined by the polymer wave speed and thus we have,

$$G_0 = (1-\alpha^2)\left[\frac{E_{st}h}{2}\left(\frac{V}{C_{st}}\right)^2 - (1-0.9\alpha^2)\frac{\overline{\sigma}^2h}{2E}\right] \tag{21}$$

This case has a rather different dependence of α on $\overline{\sigma}$ at low speeds since σ_0 is not a function of α. However, the transition at higher α values does occur and if we find the condition $\frac{dV^2}{d\alpha^2} = 0$ we have,

$$\alpha^2 = 1 - 1.47 \left(\frac{G_0 E}{\overline{\sigma}^2 h} \right)^{1/2}$$

and

$$\left(\frac{V}{C_{st}} \right)^2 = 0.08 \left(\frac{\overline{\sigma}^2}{E\,E_{st}} \right) + 1.92 \left(\frac{2G_0}{E_{st}h} \right)^{1/2} \left(\frac{\overline{\sigma}^2}{E\,E_{st}} \right)^{1/2} \tag{22}$$

For this case $C_{st} = 6000\,\text{m/s}$, $E = 3.2\,\text{GPa}$ and $E_{st} = 208\,\text{GPa}$ and values of $\overline{\sigma}$ of 16.2, 81, 162 and 243 MPa were used and gave crack speed transitions to α values of about 0.9 for V values of 2–3, 8.75–10, 12.5–15 and 19.75–20 m/s, respectively. These values of V are shown as a function of $\overline{\sigma}$ in Figure 7 together with Equation (22) and the variation of α. The agreement is excellent and α increases very rapidly and is close to unity for $\overline{\sigma} > 50\,\text{MPa}$. There is no valid solution for $\overline{\sigma} < 13\,\text{MPa}$ and the lowest computed values were at 16.2 MPa for which $\alpha = 0.58$.

4. Conclusions

The numerical results show clearly that at low cohesive stresses inter-sonic crack speeds approaching the axial wave speed are predicted. At higher cohesive stresses the crack speeds are much lower and Needleman (1999) reported that they ran at

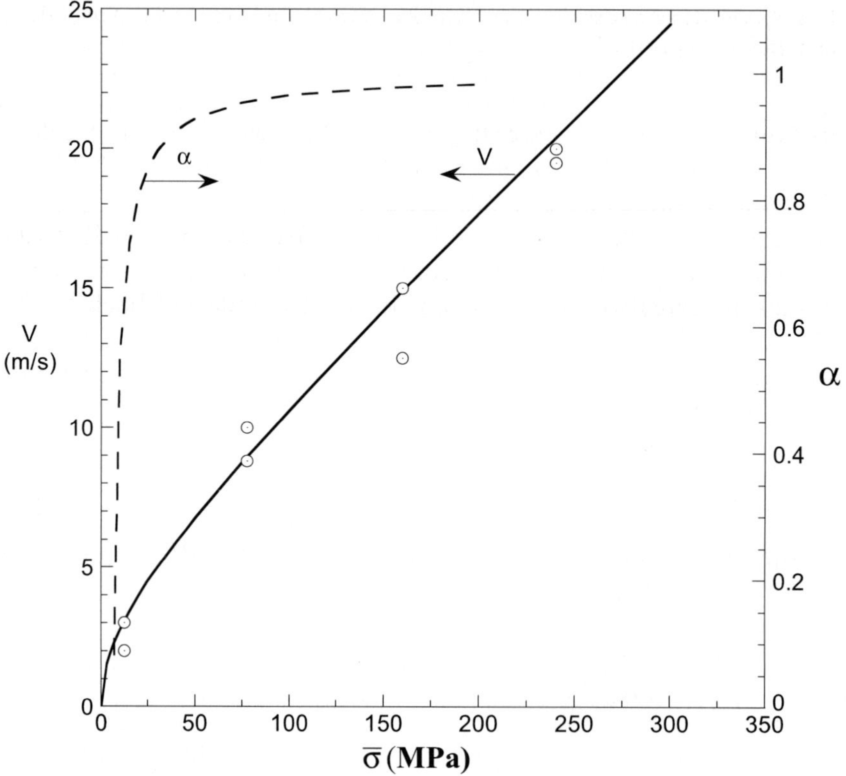

Figure 7. Transition impact velocity V as a function of $\overline{\sigma}$. Data points are from Williams et al. (to be published) and lines are from Equation (22).

the Rayleigh wave velocity. In Williams et al. (to be published), however, a range of speeds was observed from about the Rayleigh velocity to significantly lower values. All the numerical studies showed transitions from the lower speeds to the higher under certain conditions. Such transitions are observed experimentally though it is difficult to prescribe the cohesive stress, or indeed G_0, in the tests and the numerical results are able only to infer the values.

The analytical solution gives a sensible description of the numerical values. The most basic of solutions predicts the intersonic speeds and when a cohesive zone is involved these speeds occur for low stress values. The crack speed dependency on the cohesive stress shows a region of rapid change which coincides well with the transitions observed numerically. This arises from the elastic constraint in the region beyond the cohesive zone.

References

Coker, D., Lykotrafitis, G., Needleman, A. and Rosakis, A.J. (2005). Frictional sliding modes along an interface between identical elastic plates subject to shear impact loading. *JMPS* **53**, 884–922.

Coker, D., Rosakis, A.J. and Needleman, A. (2001). Dynamic crack growth along a polymer composite-homalite interface. *JMPS* **51**, 425–460.

Needleman, A. (1999). An analysis of intersonic crack growth under shear loading. *Journal of Applied Mechanics* **66**, 847–857.

Needleman, A. and Rosakis, A.J. (1999). The effect of bond line strength and loading rate on the conditions governing the attainment of intersonic crack growth along interfaces. *JMPS* **47**, 2411–2449.

Rosakis, A.J. (2002). Intersonic shear cracks in fault ruptures. *Advances in Physics* **51**(4), 1198–1257.

Saundrula, O. and Rosakis A.J. (2002). Effects of loading and geometry in the sub sonic–intersonic transition of a bimaterial interface crack. *Engineering Fracture Mechanics* **70**, 309–337.

Williams, J.G. (2005). A global energy analysis of impact loaded bi-material strips. *Engineering Fracture Mechanics* **72**, 813–826.

Williams, J.G. A review of the determination of energy release rates for strips in tension and bending – Part II: Dynamic Solutions. *Journal Strain Analysis for Engineering Design* **28**, 247–256.

Williams, J.G., Tropsa, V. and Ivankovic, A. Impact shear loading of strips – same numerical results. (to be published).

International Journal of Fracture (2006) 138:211–218
DOI 10.1007/s10704-006-0030-6

© Springer 2006

Laboratory earthquakes

ARES J. ROSAKIS[1,*], HIROO KANAMORI[2] and KAIWEN XIA[3]

[1]*Graduate Aeronautical Laboratories, California Institute of Technology, Pasadena, CA 91125, USA*
[2]*Seismological Laboratory, California Institute of Technology, Pasadena, CA 91125, USA*
[3]*Department of Civil Engineering, University of Toronto, 35 St. George Street, Toronto, ON M5S 1A4, Canada*
Author for correspondence (E-mail: rosakis@aero.caltech.edu)

Received 1 March 2005; accepted 1 December 2005

Abstract. We report on the experimental observation of the phenomenon of, spontaneously nucleated, supershear rupture and on the visualization of the mechanism of subRayleigh to supershear rupture transition in frictionally-held interfaces. The laboratory experiments mimic natural earthquakes. The results suggest that under certain conditions supershear rupture propagation can be facilitated during large earthquakes.

Key words: Earthquake rupture, supershear, subRayleigh, transition.

1. Introduction

Vertically dipping crustal faults are long pre-existing weak planes that extend tens of kilometers perpendicularly to the earth's surface and often host catastrophic earthquake rupture events. The geometry (planarity and length) of such faults is often simple enough to apply appropriately modified concepts of dynamic fracture mechanics to the study of the physics underlying their rupture process. Due to the nature of earthquakes however, direct full field and real time observations of the rupture process are prohibited while even strong motion data have limitations of spatial resolution. As a result, most efforts to date have focused on complicated analytical studies and on extensive numerical modeling of dynamic rupture processes using finite element, finite difference, and boundary element methods. As clearly elucidated by Rice et al. (2001), the nature and stability of the predicted rupture process depends very strongly on the choice of cohesive or frictional laws employed in the modeling and, as a result, experimental validation of the fidelity of such calculations becomes of primary importance.

Despite continuous efforts starting from the early 1970's (Dieterich, 1972; Brune, 1973; Dieterich and Kilgore, 1994; Anooshehpoor and Brune, 1999), there are still many mysteries regarding earthquake rupture dynamics. One of the pressing questions relevant to Seismic hazard is how fast real earthquake ruptures can propagate. As shown by Rosakis (Rosakis et al., 1999), shear cracks in coherent, adhesive, engineering interfaces can propagate at a supershear velocity (faster than the shear wave speed of the material) in various bonded bimaterials subjected to impact loading. However, questions remain about the possibility of supershear growth of a frictional, earthquake type, ruptures whose nucleation is spontaneous in nature (absence of stress wave loading) and whose propagation takes place on a mostly frictional, incoherent, interface. In this paper we describe work in progress related to the scientific questions posed above. This work capitalizes heavily on scientific knowledge on shear fracture processes in heterogeneous, engineering material systems and layered structures of the type studied in the past 10 years at Caltech under our engineering composites program. The scientific principles obtained through our past work have been extended to the study of a dynamic rupture problem that occurs over length scales that are 5–6 orders or magnitude larger than the equivalent engineering applications.

2. Recent reports of supershear earthquake fault rupture

The M_s 8.1 (M_w 7.8) central Kunlunshan earthquake that occurred on 14 November, 2001, was an extraordinary event from the point of view of dynamic rupture mechanics. The rupture occurred over a long, near-vertical, strike-slip fault segment of the active Kunlunshan fault and featured an exceptionally long (400 km) surface rupture zone and large surface slip displacements (Lin et al., 2002). Modeling of the rupture speed history (Bouchon and Vallee, 2003) suggests rupture speeds that are slower than the Rayleigh wave speed, c_R, for the first 100 km, transitioning to supershear for the remaining 300 km of rupture growth. Other events, such as the 1979 Imperial Valley earthquake (Archuleta, 1984; Spudich and Cranswick, 1984), the 1992 Landers earthquake (Olsen et al., 1997), the 1999 Izmit earthquake (Bouchon et al., 2001), and the 2002 Denali earthquake (Ellsworth et al., 2004) may also have featured supershear speeds. Supershear was also predicted theoretically (Burridge, 1973; Burridge et al., 1979) and numerically (Andrews, 1976; Das and Aki, 1977). Even with these estimates and predictions at hand, the question of whether natural earthquake ruptures can propagate at supershear speeds is still a subject of active debate. In addition, the exact mechanism for transition from subRayleigh (speed earthquake-type rupture starts with) to supershear rupture speed is not clear. One answer to this question was provided by the 2-D Burridge–Andrews Mechanism (BAM) (Andrews, 1976) which is a mechanism introduced to circumvent restrictions imposed by classical fracture mechanics theories. Classical dynamic fracture theories of growing shear cracks have many similarities to the earthquake rupture processes. Such theories treat the rupture front as a distinct point (sharp tip crack) of stress singularity. Such conditions are closer to reality in cases that feature coherent interfaces of finite intrinsic strength and toughness. The singular approach ultimately predicts that dynamic shear fracture cannot propagate in the small velocity interval between C_R and C_S, the shear wave speed of the material, and thus excludes the possibility of a smooth transition from subRayleigh to supershear. The introduction of a distributed rupture process zone has allowed fracture mechanics to better approximate the conditions that exist during real earthquake events (Rosakis, 2002) and to describe mechanisms for a subRayleigh rupture to enter the supershear regime. According to the two-dimensional BAM, a shear rupture accelerates to a speed very close to c_R soon after its initiation. A peak in shear stress is found to propagate at the shear wave front and is observed to increase its magnitude as the main rupture speed approaches c_R. At that point, the shear stress peak may become strong enough to promote the nucleation of a secondary micro-rupture whose leading edge propagates at c_P, the P wave speed of the material. Shortly thereafter, the two ruptures join up and the combination propagates at c_P (Rosakis, 2002). Recent numerical investigations of frictional rupture have suggested alternative, asperity based, three dimensional mechanisms (Day, 1982; Madariaga and Olsen, 2000; Dunham et al., 2003). Whether and how supershear rupture occurs during earthquakes has important implications for seismic hazards because the rupture speed influences the character of near-field ground motions.

3. The experiments

To answer the above stated questions, we conducted experiments that mimic the earthquake rupture processes. Our goal was to examine the physical plausibility and conditions under which supershear ruptures can be generated in a controlled laboratory environment. We studied spontaneously nucleated dynamic rupture events in incoherent, frictional interfaces held together by far-field tectonic loads. Thus we departed from experimental work that addresses the dynamic shear fracture of coherent interfaces loaded by stress waves (Rosakis et al., 1999; Rosakis, 2002) which was of direct relevance to the dynamic failure of Naval structures. In this study, a fault is simulated using two photoelastic plates (Homalite) held together by friction and the far-field tectonic loading is simulated by far-field pre-compression (Figure 1a–c). A unique device that triggers the rupture in a highly controlled manner is used to nucleate the dynamic rupture while preserving the spontaneous nature of the rupturing. This triggering is achieved by an exploding wire technique. The fault forms an acute angle with the compression axis to provide the shear driving force for continued rupturing.

The triggering mechanism is inspired by recent numerical work on rupture along frictional interfaces (Rice et al., 2001). Experimentally, it is a convenient way of triggering the system's full-field, high-speed diagnostics (Figure 1a) that would otherwise be unable to capture an event with total duration of $\sim 50\mu$s.

More than 50 experiments featuring a range of α and far-field pressure P were performed and the symmetric bilateral rupture process histories were visualized in intervals of 2μs. Depending on P and α, rupture speeds that are either subRayleigh or supershear were observed. The maximum shear stress field for an experiment with $\alpha = 25°$ and $P = 7\,$MPa (Figure 2a) shows that the speed of the rupture tip is very close to c_R and follows closely behind the circular shear wave front which is emitted at the time of rupture nucleation. The same was found to be true for smaller angles and lower pressures. For an experiment with $\alpha = 25°$ and $P = 15\,$MPa (Figure 2b), the circular trace of the shear wave is also visible and is at the same location as in Figure 2a. However, in front of this circle a supershear disturbance, featuring a Mach cone (pair of shear shock waves) is clearly visible. For this case, the sequences of images before 28μs have a similar form to the image displayed in Figure 2b, and reveal a disturbance that was nucleated as supershear. Its speed history $v(t)$ is determined independently by either the rupture length record or by measuring the inclination angle, β, of the shear shocks with respect to the fault plane and using the relation $v = c_S/\sin\beta$. Its speed was 1970 m/s, which is close to the longitudinal wave speed c_P. In previous experiments involving strong, coherent, interfaces and stress wave loading, stable rupture speeds near $\sqrt{2}c_S$ were observed (Rosakis et al., 1999). This apparent discrepancy can be explained by referring to the rupture

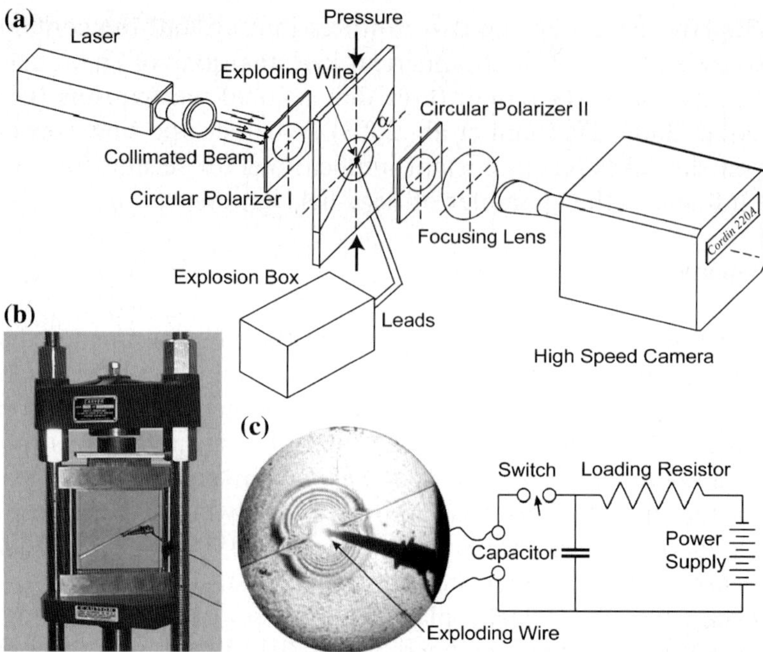

Figure 1. The diagnostics is photoelasticity combined with high-speed photography (up to 10^8 frames/s). The fault system is simulated by using two photoelastic plates (Homalite-100, shear modulus $G = 1.4\,GPa$, Poisson's ration $\nu = 0.35$, density $\rho = 1200\,kg/m^3$) held together by friction. The interface (fault) is inclined at an angle α to the horizontal promoting strike-slip rupture events (a). The carefully prepared interface has a measured static coefficient of friction $\mu^s = 0.6$; the dynamic coefficient of friction μ^d is estimated by finding the critical α of triggered events, which is between $10°$ and $15°$, and hence $\mu^d = 0.2$ is estimated. The far-field tectonic loading is simulated by uniaxial compression exerted at the top and bottom ends of the system by a hydraulic press (b). The dynamic rupture is nucleated at the center of the simulated fault by producing a local pressure pulse in a small area of the interface. A thin wire of $0.1\,mm$ in diameter is inserted in a small hole of the same size. An electronic condenser is then discharged turning the metal into expanding plasma to provide the controllable pressure pulse (c).

velocity dependence on the available energy per unit crack advance within the supershear regime (Rosakis, 2002). This energy attains a maximum value at speeds closer to $\sqrt{2}\,c_S$ for strong interfaces while for weaker interfaces, this maximum moves towards c_P.

To visualize a transition within our field of view (100 mm), we kept $\alpha = 25°$ but reduced P to $9\,MPa$ (Figure 3a–c). The circular traces of P and S waves are visible followed by a rupture propagating initially at c_R (Figure 3a). A small secondary rupture appears in front of the main rupture and propagates slightly ahead of the S wave front (Figure 3b). The two ruptures coalesce and the leading edge of the resulting rupture grows at a speed close to c_P. The transition length L here is $\sim 20\,mm$ (Figure 3d).

4. Modeling and conclusions

The above transition phenomenon is comparable with BAM, which was described by Andrews (1976). Andrews investigated this transition in a parameter space spanned

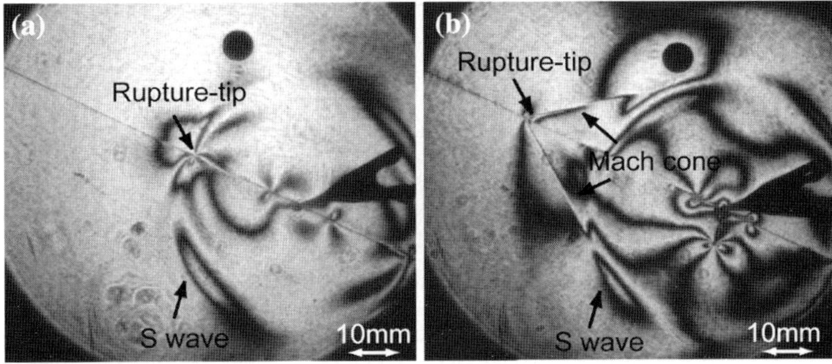

Figure 2. Purely subRayleigh ($\alpha = 25°$, $P = 7$ MPa) (a) and purely supershear ($\alpha = 25°$, $P = 15$ MPa) (b) rupture at the same time (28μs) after triggering.

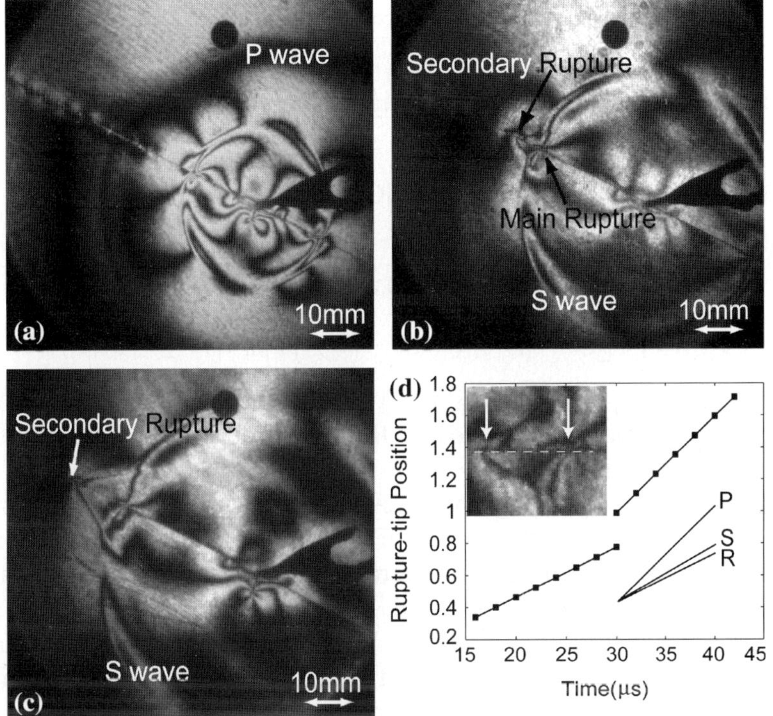

Figure 3. Visualization of the subRayleigh to supershear rupture transition ($\alpha = 25°$ $P = 9$ MPa). (a–c) were taken at 18μs, 30μs and 38μs respectively. In the rupture-tip history plot (d), we included lines corresponding to P, S and Rayleigh waves as reference.

by a normalized supershear transition length L/L_c and a non-dimensional driving stress parameter s ($s = (\tau^y - \tau)/(\tau - \tau^f)$). The parameters τ, τ^y and τ^f are the resolved shear stress on the fault, the static and the dynamic strength of the fault, respectively, which describe the linear slip-weakening frictional law. In our experiment, s, can be expressed as $s = (\mu^s \cos\alpha - \sin\alpha)/(\sin\alpha - \mu^d \cos\alpha)$. Andrews' result can be written as $L = L_c f(s)$, where the function $f(s)$ has been given numerically and can be approximated by $f(s) = 9.8(1.77 - s)^{-3}$. The normalizing length L_c is the critical length for unstable rupture nucleation and is proportional to the rigidity G and

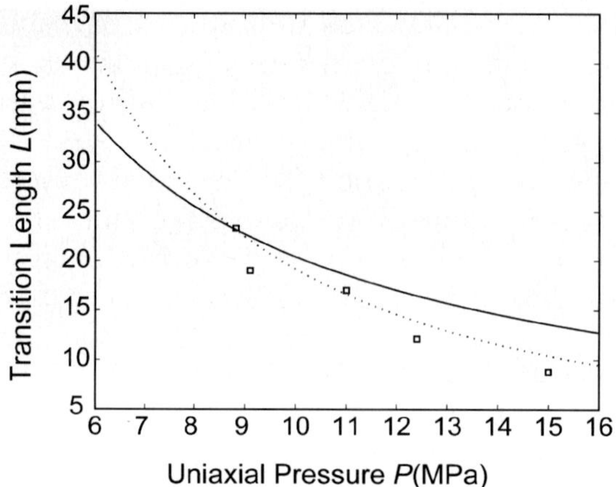

Figure 4. Transition length as a function of far-field load. Solid curve is Andrew's theory, dashed curve is modified theory and squares are experimental data.

to d_0, which is the critical or breakdown slip of the slip weakening model. L can then be expressed as:

$$L = f(s)[(1+v)/\pi][(\tau^y - \tau^f)/(\tau - \tau^f)^2]Gd_0. \tag{1}$$

Applying Equation 1 to our experiments, the transition length is inversely proportional to the applied uniaxial pressure P as:

$$L = f(s)[(1+v)/\pi]G\left[(\mu^s - \mu^d)/(\tan\alpha - \mu^d)^2\right](d_0/P). \tag{2}$$

We can compare our experiments to Andrews' theory (Figure 4). Although the theory qualitatively captures the trends of the experiments, the data exhibits a dependence on pressure stronger than P^{-1}.

A natural way to modify Andrews' results is to introduce some micro-contact physics, and to thus consider the effect of pressure on d_0. We first note that there exists a linear relation between a characteristic surface length (half-distance between contacting asperities, D) and the critical slip distance d_0($d_0 = c[(\tau^y - \tau^f)/\tau^f]^M D$, where c and M are constants) (Xia et al., 2004). We then denote the normal stress applied on the fault as σ($\sigma = P\cos^2\alpha$) and assume that the average radius of n contacting asperities, a_0, is constant. As the pressure over a macroscopic contact area $A(=n\pi D^2)$ is increased, n, as well as the real contact area $A_r(=n\pi a_0^2)$ increase. By defining the hardness H as the ratio of the normal force N to A_r, N can be expressed as: $N = HA_r = \sigma A = AP\cos^2\alpha$. Substitution of A and A_r in terms of D and a_0 respectively, gives $D = \sqrt{H}a_0\cos\alpha P^{-1/2}$. Using the relation $d_0 \propto D$, d_0 is found to depend on the pressure as $d_0 \propto P^{-1/2}$. By further using Equation 2, a modified expression relating L to P emerges featuring a stronger pressure dependence ($L \propto P^{-3/2}$). This modified relation which agrees well with the experimental data (Figure 4) is given by:

$$L = f(s)\frac{1+v}{\pi}G\frac{\mu^s - \mu^d}{[\sin\alpha - \mu^d\cos\alpha]^2}2c\left(\frac{\mu^s - \mu^d}{\mu^s}\right)^M \sqrt{H}a_0 P^{-\frac{3}{2}}\cos^{-1}\alpha. \tag{3}$$

For seismic applications, we rewrite Equation 1 in terms of the effective stress $\tau_e = \tau - \tau^f$ as $L = f(s)(1+v)(1+s)Gd_0/\pi\tau_e$. Application of this equation to both seismic faulting and to laboratory data allows us to scale L from laboratory to seismological conditions. The stress τ_e in our experiments is chosen to be of the same order as that in seismology. The ratio of rigidity of the Earth's crust to Homalite is 25. We estimate $L = 20$ mm from the experiment where $P = 9$ MPa and $\alpha = 25°$, and for which $d_0 = 10\mu$m (obtained using Equation 2). The values of d_0 for large earthquakes are often estimated as 50 cm to 1 m (Xia et al., 2004). Thus, if s is approximately the same under laboratory and crustal conditions, L for earthquakes can be estimated to be in the range between 25 and 50 km. Because s can be different, and the estimate of d_0 for earthquakes is uncertain, this value should be taken as an order of magnitude estimate. Nevertheless, it is of the same order as that inferred for the Kunlunshan event (Bouchon and Vallee, 2003).

The large L required for supershear is one of the reasons why only a few earthquake events have been observed to feature supershear speeds. It suggests that in such cases the tectonic stress is fairly close to the static fault strength (i.e., small s), which has important implications for evolution of rupture in large earthquakes.

Acknowledgements

Authors appreciate fruitful discussions with T. Heaton and G. Ravichandran from Caltech and J. R. Rice from Harvard. This study is supported by NSF grant EAR-0207873 and Naval Research grant N00014-03-1-0435 (Yapa D. S. Rajapakse program monitor).

References

Andrews, D.J. (1976). Rupture velocity of plane strain shear cracks. *Journal of Geophysical Research* **81**(32), 5679–5687.

Anooshehpoor, A. and Brune, J.N. (1999). Wrinkle-like weertman pulse at the interface between two blocks of foam rubber with different velocities. *Geophysical Research Letters* **26**(13), 2025–2028.

Archuleta, R.J. (1984). A faulting model for the 1979 imperial-valley earthquake. *Journal of Geophysical Research* **89**(NB6), 4559–4585.

Bouchon, M., Bouin, M.P., Karabulut, H., Toksöz, M.N., Dietrich, M. and Rosakis, A.J. (2001). How fast is rupture during an earthquake? New insights from the 1999 Turkey earthquakes. *Geophysical Research Letters* **28**(14), 2723–2726.

Bouchon, M. and Vallee, M. (2003). Observation of long supershear rupture during the magnitude 8.1 Kunlunshan earthquake. *Science* **301**(5634), 824–826.

Brune, J.N. (1973). Earthquake Modeling by stick-slip along precut surfaces in stressed form rubber. *Bulletin of the Seismological Society of America* **63**(6), 2105–2119.

Burridge, R. (1973). Admissible speeds for plane-strain self-similar shear cracks with friction but lacking cohesion. *Geophysical Journal of the Royal Astronomical Society* **35**(4), 439–455.

Burridge, R., Conn, G. and Freund, L.B. (1979). Stability of a rapid mode-ii shear crack with finite cohesive traction. *Journal of Geophysical Research* **84**(NB5), 2210–2222.

Das, S. and Aki, K. (1977). Numerical study of 2-dimensional spontaneous rupture propagation. *Geophysical Journal of the Royal Astronomical Society* **50**(3), 643–668.

Day, S.M. (1982). 3-Dimensional simulation of spontaneous rupture – the effect of nonuniform prestress. *Bulletin of the Seismological Society of America* **72**(6), 1881–1902.

Dieterich, J.H. (1972). Time-dependent friction as a possible mechanism for aftershocks. *Journal of Geophysical Research* **77**(20), 3771–3781.

Dieterich, J.H. and Kilgore, B.D. (1994). Direct observation of frictional contacts – new insights for state-dependent properties. *Pure and Applied Geophysics* **143**(1–3), 283–302.

Dunham, E.M., Favreau, P. and Carlson, J.M. (2003). A supershear transition mechanism for cracks. *Science* **299**(5612), 1557–1559.

Ellsworth, W.L., Çelebi, M., Evans, J.R., Jensen, E.G., Nyman, D.J. and Spudich, P. (2004). Processing and Modeling of the Pump Station 10 Record from the November 3, 2002, Denali Fault, Alaska Earthquake. Eleventh International Conference of Soil Dynamics and Earthquake Engineering, Berkeley, California, January 7–9.

Lin, A.M., Fu, B.H., Guo, J.M., Zeng, Q.L., Dang, G.M., He, W.G. and Zhao, Y. (2002). Co-seismic strike-stip and rupture length produced by the 2001 M-s 8.1 Central Kunlun earthquake. *Science* **296**(5575), 2015–2017.

Madariaga, R. and Olsen, K.B. (2000). Criticality of rupture dynamics in 3-D. *Pure and Applied Geophysics* **157**(11–12), 1981–2001.

Olsen, K.B., Madariaga, R. and Archuleta, R.J. (1997). Three-dimensional dynamic simulation of the 1992 Landers earthquake. *Science* **278**(5339), 834–838.

Rice, J.R., Lapusta, N. and Ranjith, K. (2001). Rate and state dependent friction and the stability of sliding between elastically deformable solids. *Journal of the Mechanics and Physics of Solids* **49**(9), 1865–1898.

Rosakis, A.J. (2002). Intersonic shear cracks and fault ruptures. *Advances in Physics* **51**(4), 1189–1257.

Rosakis, A.J., Samudrala, O. and Coker, D. (1999). Cracks faster than the shear wave speed. *Science* **284**(5418), 1337–1340.

Spudich, P. and Cranswick, E. (1984). Direct observation of rupture propagation during the 1979 imperial valley earthquake using a short baseline accelerometer array. *Bulletin of the Seismological Society of America* **74**(6), 2083–2114.

Xia, K.W., Rosakis, A.J. and Kanamori, H. (2004). Laboratory earthquakes: the sub-rayleigh-to-supershear rupture transition. *Science* **303**, 1859–1861.

International Journal of Fracture (2006) 138:219–240
DOI 10.1007/s10704-006-0059-6

© Springer 2006

Electromigration failure of metal lines

HIROYUKI ABÉ[1], KAZUHIKO SASAGAWA[2,*] and MASUMI SAKA[3]
[1]*Tohoku University, Sendai 980-8579, Japan*
[2]*Department of Intelligent Machines and System Engineering, Hirosaki University, Bunkyo-cho3, Hirosaki 036-8561, Japan*
[3]*Department of Nanomechanics, Tohoku University, Aoba 6-6-01, Aramaki, Aoba-ku, Sendai 980-8579, Japan*
Author for correspondence (E-mail:sasagawa@cc.hirosaki-u.ac.jp)

Received 1 March 2005; accepted 1 December 2005

Abstract. With the scaling down process of microcircuits in semiconductor devices, the density of electric current in interconnecting metal lines increases, and the temperature of the device itself rises. Electromigration is a phenomenon that metallic atoms constructing the line are transported by electron wind. The damage induced by electromigration appears as the formation of voids and hillocks. The growth of voids in the metal lines ultimately results in electrical discontinuity. Our research group has attempted to identify a governing parameter for electromigration damage in metal lines, in order to clarify the electromigration failure and to contribute to circuit design. The governing parameter is formulated based on the divergence of the atomic flux by electromigration, and is denoted by *AFD*. The prediction method for the electromigration failure has been developed by using *AFD*. The *AFD*-based method makes it possible to predict the lifetime and failure site in universal and accurate way. In the actual devices, the metal lines used in the integrated circuit products are covered with a passivation layer, and the ends of the line are connected with large pads or vias for current input and output. Also, the microstructure of metal line distinguishes the so-called bamboo structured line from polycrystalline line depending on the size of metallic grains relative to the line width. Considering the damage mechanisms depending on such line structure, our research group has made a series of studies on the development of the prediction method. This article is dedicated to make a survey of some recent achievements for realizing a reliable circuit design against electromigration failure.

Key words: Electromigration, failure, integrated circuit, threshold current density.

1. Introduction

In semiconductor devices the integration of microcircuits progresses and interconnecting metal lines get finer and finer. With the scaling down process, the density of electric current in the metal line increases, and the temperature of the device itself rises. From the trends of the operating conditions such as high current density and high temperature, it is anticipated that the issue of the metal line failure due to electromigration becomes more serious now and in the future. Electromigration is a phenomenon that metallic atoms constructing the line are transported by electron wind. The damage induced by electromigration appears as the formation of voids and hillocks. The growth of voids in the metal lines ultimately results in electrical discontinuity. Therefore, it is required from the viewpoint of ensuring the reliability of semiconductor integrated circuits (ICs) that the lifetime of metal line is predicted accurately.

Our research group has attempted to identify a governing parameter for electromigration damage in the metal lines, in order to clarify the electromigration failure and to contribute to circuit design. The governing parameter is formulated based on the divergence of the atomic flux by electromigration, and is denoted by *AFD*. The prediction method for the electromigration failure has been developed by using *AFD*. The *AFD*-based method makes it possible to predict the lifetime and possible failure site in universal and accurate way. The width of metal lines ranges from several tens nm to several tens μm at present. In this study, the lines of 980 nm to 9.9 μm in width are treated. Though the reliability of finer lines than that treated in this study is an interesting issue, it remains to be seen in future. Recently, the semiconductor industry begins to employ Cu interconnects. But, Al lines are widely used even now. We focus on the reliability issues of Al lines.

Multi-level interconnections constructed in IC are illustrated in Figure 1(a). The metal lines used in the IC products are covered with a passivation layer as shown in Figure 1(b), and the ends of the line are connected with large pads or vias for current input and output as shown in Figure 1(c) and (d), respectively. Also, the microstructure of metal line distinguishes the so-called bamboo structured line from polycrystalline line depending on the size of metallic grains relative to the line width, see Figure 1(e) and (f). The line width of around 1 μm gives its transition. Considering the damage mechanisms depending on such line structure, our research group has made a series of studies on the development of the prediction method for the metal line failure (Sasagawa et al., 1998, 1999, 2000, 2001, 2002a, b, 2003; Hasegawa et al., 2003; Hasegawa, 2004). This article is dedicated to make a survey of some recent achievements for realizing a reliable circuit design against electromigration failure.

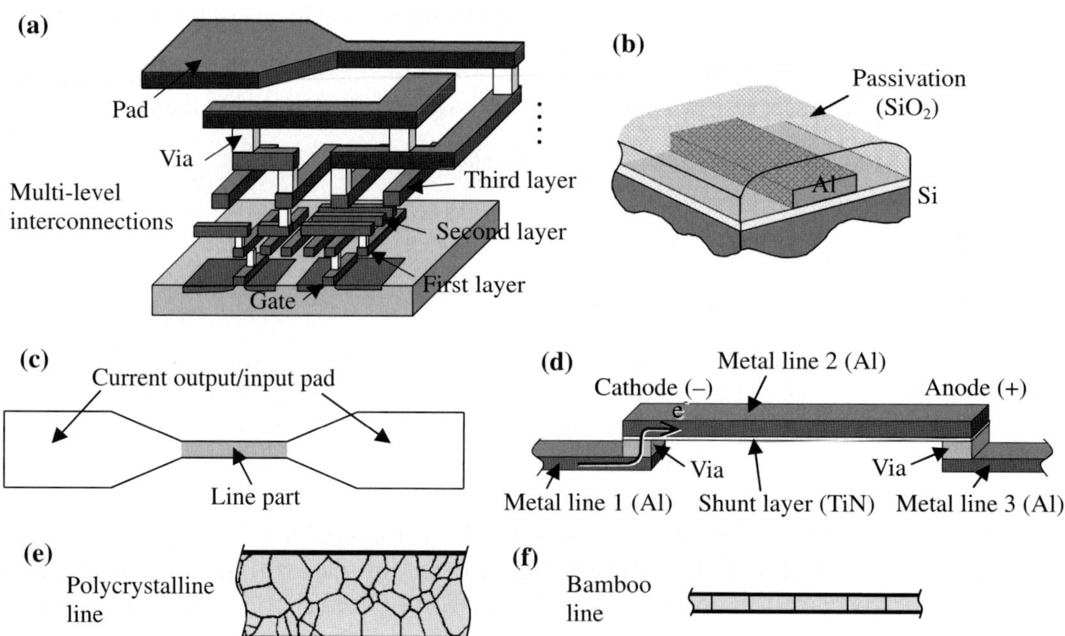

Figure 1. Line structures treated. (**a**) Multi-level interconnections, (**b**) Passivated line, (**c**) Line connected with pads, (**d**) Via-connected line, (**e**) Polycrystalline line, (**f**) Bamboo line.

2. Unified approach based on atomic flux divergence

Up to now, the prediction of the electromigration failure, i.e., estimates of lifetime and failure location have been attempted by using an empirical equation (Black, 1969) and numerical simulations (Nikawa, 1981; Marcoux et al., 1989; Kirchheim and Kaeber, 1991). On the other hand, it is known that there is a threshold current density of electromigration damage, j_{th}, below which no electromigration damage appears in the case of the via-connected metal lines. The evaluation of the threshold of the current density is also of great interest. Conventionally, the threshold value has been evaluated based on the assumption that the product of the threshold and line-length is constant (Blech, 1976; Oates, 1991). The predictions of lifetime and failure site and the evaluation of j_{th} have been attempted individually in those papers. Introduction of the governing parameter for electromigration damage, *AFD*, makes it possible to perform the predictions and the evaluation in a unified way. The *AFD*-based method has resolved some problems inherent in the conventional methods as summarized in Table 1.

So far, we have formulated the governing parameter for electromigration damage, which integrates all the factors affecting the damage, i.e., the line structures, film characteristics, operating conditions such as current density and temperature, and atomic density (Sasagawa et al., 1998, 2000, 2002a), and have approached to the development of the reliability evaluation method by using the governing parameter (Sasagawa et al., 1999, 2001, 2002b, 2003). Since the preliminary stage of the development of the practical evaluation method, where the surface of the treated lines is bared so that there is no passivation, we have attempted to identify the governing parameter for electromigration damage, and have confirmed that the governing parameter is associated well with the volume of the void formed in metal lines (Sasagawa et al., 1998, 2000). On the other hand, the metal lines in ICs are generally covered with passivation layer. The governing parameter for electromigration damage in the passivated line has been formulated by considering the effect of the passivation on electromigration mechanisms (Sasagawa et al., 2002a). We have named the governing parameter for the unpassivated line and that for the passivated line *AFD* and *AFD**, respectively. Hereafter, *AFD* collectively means the both parameters. The usage of *AFD* makes it possible to clarify the damage behavior due to electromigration and to evaluate the reliability of metal lines in a universal, accurate and engineering way, as described in Table 1.

3. Development of *AFD*-based method

3.1. GENERAL EXPRESSION OF *AFD*

In this section, a general expression of the governing parameter for electromigration damage is given. First, let us consider the atoms coming in and going out of a rectangle as shown in Figure 2, where lengths of sides of the rectangle are a and b, thickness is unity, and J_ξ and J_η are the components of atomic flux vector $\boldsymbol{J}$ in Cartesian coordinates ξ and η. The divergence of atomic flux in the $a \times b$ rectangle is obtained by integrating the normal components of atomic flux along the sides of the rectangle.

Table 1. Comparison between conventional methods and the *AFD*-based method.

Conventional methods	Present method using *AFD*
Lifetime prediction	
Empirical equation (Black, 1969)	*AFD-based simulation of failure process* (Sasagawa et al., 1999, 2001, 2002b, 2003)
Simple	
Not universal (Many and long term experiments are necessary for the respective line shapes even if the lines are made of the same metallic films)	Universal (Once the film characteristic constants are obtained, the failure prediction of any shaped line is possible under arbitrary operating conditions)
Inaccurate (It is difficult to predict a phenomenon in operating condition from that in accelerated condition) (McPherson, 1986)	Accurate prediction for not only lifetime but also failure site
Failure location	
Numerical simulation (e.g., Kirchheim and Kaeber, 1991)	*AFD-based simulation of electromigration behavior* (Sasagawa et al., 1998, 2000, 2002a)
Main purpose is to clarify the damage mechanisms	*AFD* corresponds with actual amount of damage
Method for determining the film characteristic constants of the line to be predicted is not developed	Film characteristic constants can be derived by simple experiments to measure the amount of damage
→ Not necessarily suitable for quantitative failure prediction	
Evaluation of threshold current density[*], j_{th}	
Assumption of product of threshold and line-length being constant (e.g., Blech, 1976)	*AFD-based simulation for building-up process of atomic density distribution (incubation period)* (Hasegawa, 2004)
Simple and easy	Universal and accurate (Once the film characteristic constants are obtained, the evaluation of the threshold in any shaped line is possible under arbitrary temperature)
Effect of line shape on j_{th} is not considered (Application is limited to only straight line)	
The constant depends on temperature	

[*] j_{th} does not exist in case of metal lines with pads.

It gives the general expression of the governing parameter for electromigration damage, *AFD*, as

$$AFD = \frac{1}{ab}\left[\int_0^b J_\xi(a,\eta)\mathrm{d}\eta - \int_0^b J_\xi(0,\eta)\mathrm{d}\eta + \int_0^a J_\eta(\xi,b)\mathrm{d}\xi - \int_0^a J_\eta(\xi,0)\mathrm{d}\xi\right]. \quad (1)$$

The parameter *AFD* represents the number of atoms decreasing per unit time and unit volume.

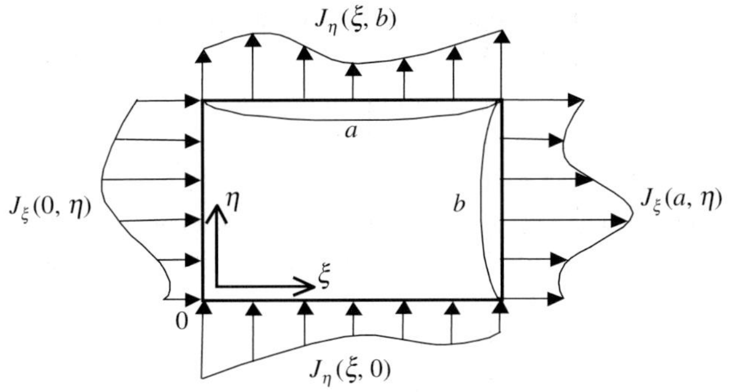

Figure 2. Rectangular region to calculate the atomic flux divergence.

Based on the modification of the Huntington–Grone's equation (Huntington and Grone, 1961), the atomic flux in a passivated metal line is assumed to be represented (Sasagawa et al., 2002a) by

$$|\boldsymbol{J}| = \frac{N D_0}{kT} \exp\left\{ -\frac{Q_+ \kappa\Omega\,(N-N_T)/N_0 - \sigma_T\Omega}{kT} \right\} \left(Z^* e\rho j^* - \frac{\kappa\Omega}{N_0} \frac{\partial N}{\partial l} \right), \qquad (2)$$

where N is the number of atoms per unit volume, that is, atomic density, D_0 a prefactor, k Boltzmann's constant, T the absolute temperature, Q the activation energy for atomic diffusion, κ the effective bulk modulus (Korhonen et al., 1993), N_T the atomic density under tensile thermal stress σ_T, N_0 the atomic density at a reference condition, Ω the atomic volume $(=1/N_0)$, Z^* the effective valence, e the electronic charge, $\rho[=\rho_0\{1+\alpha(T-T_\mathrm{s})\}]$ temperature-dependent resistivity, ρ_0 and α the electrical resistivity and the temperature coefficient at the substrate temperature T_s. Symbols j^* and $\partial N/\partial l$ are the components of the current density vector and atomic density gradient in the direction of $\boldsymbol{J}$, respectively.

The effect of the stress generated in the metal line on diffusivity is given by the term $\kappa\Omega(N-N_T)/N_0 - \sigma_T\Omega$ (Ainslie et al., 1972; Park and Thompson, 1997). On the other hand, the effect of back flow of atoms induced by atomic density gradient is found as $(\kappa\Omega/N_0)\partial N/\partial l$ (Blech, 1976, 1998; Korhonen et al., 1993). These two effects are taken into account in Equation (2).

The derivation of $D_0, Q^*[= Q - \sigma_T\Omega], Z^*, \kappa, \rho_0$ and α can be seen in Sasagawa et al. (2002a), Hasegawa (2004). The quantity N_0 is obtained under the condition of stress-free and at 300 K (Villars, 1997) and N_T can be approximated by N_0 (Sasagawa et al., 2002a).

By utilizing *AFD* the atomic density at any location within the metal line is calculated by

$$N = N_0 - \int_0^t AFD\, \mathrm{d}t, \qquad (3)$$

where t is time. The atomic density N is closely related to the initiation and growth of voids and hillocks.

3.2.   Polycrystalline Line Connected with Current Input and Output Pads

3.2.1.   *AFD for polycrystalline line connected with current input and output pads*

Now, let us derive the governing parameter for electromigration damage in the polycrystalline line by using Equation (1). Lattice diffusion and interface diffusion between metal line and its surroundings can be neglected in the polycrystalline line because the grain boundary becomes the main diffusion path of atoms. When microstructure of the polycrystalline line having an average grain size d is taken into account (see Figure 3(a)), consider a rectangular region including three grain boundaries as illustrated in Figure 3(b) (Sasagawa et al., 1998, 2002a). The value of d is determined by making observation of the microstructure of the test line (Sasagawa et al., 2002a). In Figure 3, a triple point of grain boundaries is designated by using (x, y) global coordinates, and the location along the sides of the rectangle is related to the local coordinate system (ξ, η). Symbol η_i $(i = 1, 2, \ldots, 6)$ is the intersection point of the grain boundary having a certain width and the left or right side of the rectangle, θ the angle between the ξ-axis and x-axis, ℓ the length of the grain boundary, J_{I}, J_{II} and J_{III} are the atomic flux along the each grain boundary and are obtained from Equation (2) as functions of Cartesian coordinates x and y, $J_{\xi\mathrm{I}}$, $J_{\xi\mathrm{II}}$ and $J_{\xi\mathrm{III}}$ the ξ-components of J_{I}, J_{II} and J_{III}, respectively. The relative angle between grain boundaries is denoted by $\pi/3 + \Delta\varphi$, which is assumed to be constant. By considering small value of $\Delta\varphi$, the lengths a and b can simply be expressed as $\sqrt{3}d/4$ and d, respectively. The boundary conditions with regard to the atomic flux, which

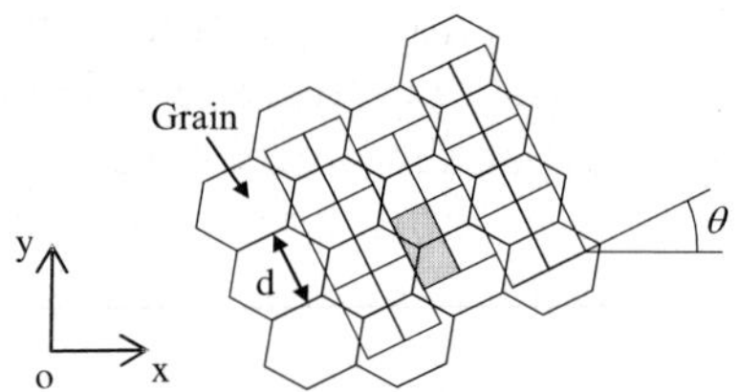

(a) Microstructure assumed for polycrystalline line

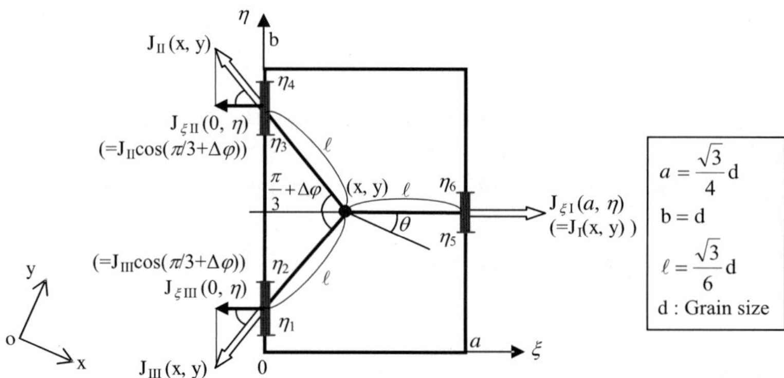

(b) Enlargement of the rectangular region of microstructure

Figure 3. Rectangular region to calculate the atomic flux divergence in polycrystalline line. **(a)** Microstructure assumed for polycrystalline line and **(b)** Enlargement of the rectangular region of microstructure.

are required for *AFD* calculation, are given from Figure 3(b). There are incoming and outgoing of atoms only through two parts of the left side of the rectangle; $\eta_1 < \eta < \eta_2$ and $\eta_3 < \eta < \eta_4$, and a part of the right side; $\eta_5 < \eta < \eta_6$. In the other parts of the sides, there is no incoming or outgoing of atoms. Equation (1) is applied to the rectangle considering such boundary conditions. Then, the parameter is obtained as a function of θ and it is explicitly denoted by AFD_θ. By considering the whole range of θ; from 0 to 2π, we represent the expected value of positive value of AFD_θ, which is closely associated with the growth rate of voids. In this way, the governing parameter *AFD* for polycrystalline line is formulated (Sasagawa et al., 2002a) as follows:

$$AFD = \frac{1}{4\pi} \int_0^{2\pi} (AFD_\theta + |AFD_\theta|)\,\mathrm{d}\theta. \tag{4}$$

The parameter *AFD* was obtained by introducing an averaged polycrystalline microstructure to avoid a complicated and time-consuming calculation. However the lifetime was appropriately predicted even in the simplified method (Sasagawa et al., 2002b). We performed simple acceleration test, and the film characteristic constants, e.g., D_0, $Q^*[= Q - \sigma_T \Omega]$, Z^* and κ were derived based on *AFD*. Through the discussion on the validity of the derived characteristic constants, the use of *AFD* was verified (Sasagawa et al., 2002a). It is noted that the constants are independent of acceleration conditions.

3.2.2. *Lifetime prediction of the polycrystalline line connected with pads*
Next, our effort was dedicated to the development of the prediction method for electromigration failure using *AFD* (Sasagawa et al., 2002b). Lifetime and possible failure location in the passivated polycrystalline line are predicted by means of the numerical simulation of the failure process, covering the building up of atomic density distribution, void initiation, void growth and ultimately line failure in the incubation and damage progress periods as shown in Figure 4. In the simulation to predict the lifetime

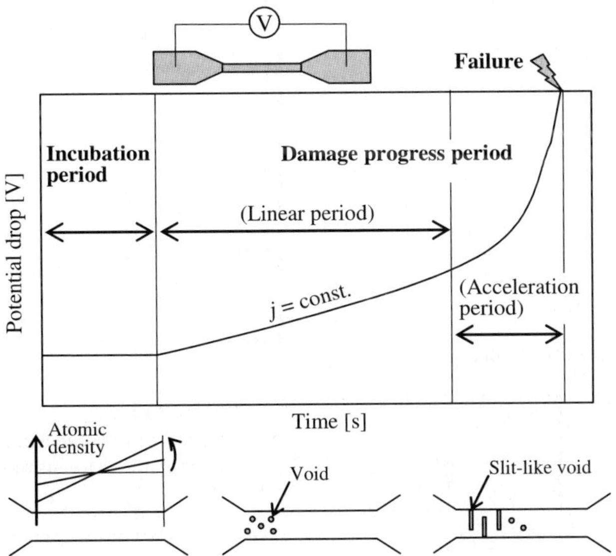

Figure 4. Schematic diagram of the failure process.

and failure location of the metal line, it is divided into elements and the outgoing and incoming of atoms in each element are calculated based on *AFD*. The distributions of current density and temperature are obtained by the two-dimensional finite element (FE) analysis of an electrothermal problem. The calculation of the simulation consists of the part for the incubation period covering the period for the building up of atomic density distribution and the part for the progress period of electromigration damage (see Figure 4). Incubation period means the time period from the start of the current apply to the beginning of the increase in potential drop over the line, which is induced by void formation. The schema of the numerical simulation of the failure process is illustrated in Figure 5.

The former part of the simulation is for the process occurring in the incubation period based on *AFD*. The atomic density in an element depends largely on the orientation of microstructure shown by θ in Figure 3. Then, the atomic density taken in relation to θ is denoted by N^* and all the values of N^* for the whole range of θ are calculated. We can assume that there are the critical atomic density for void initiation, $N^*_{\min}$, and that for hillock initiation, $N^*_{\max}$. The repetitive calculation for

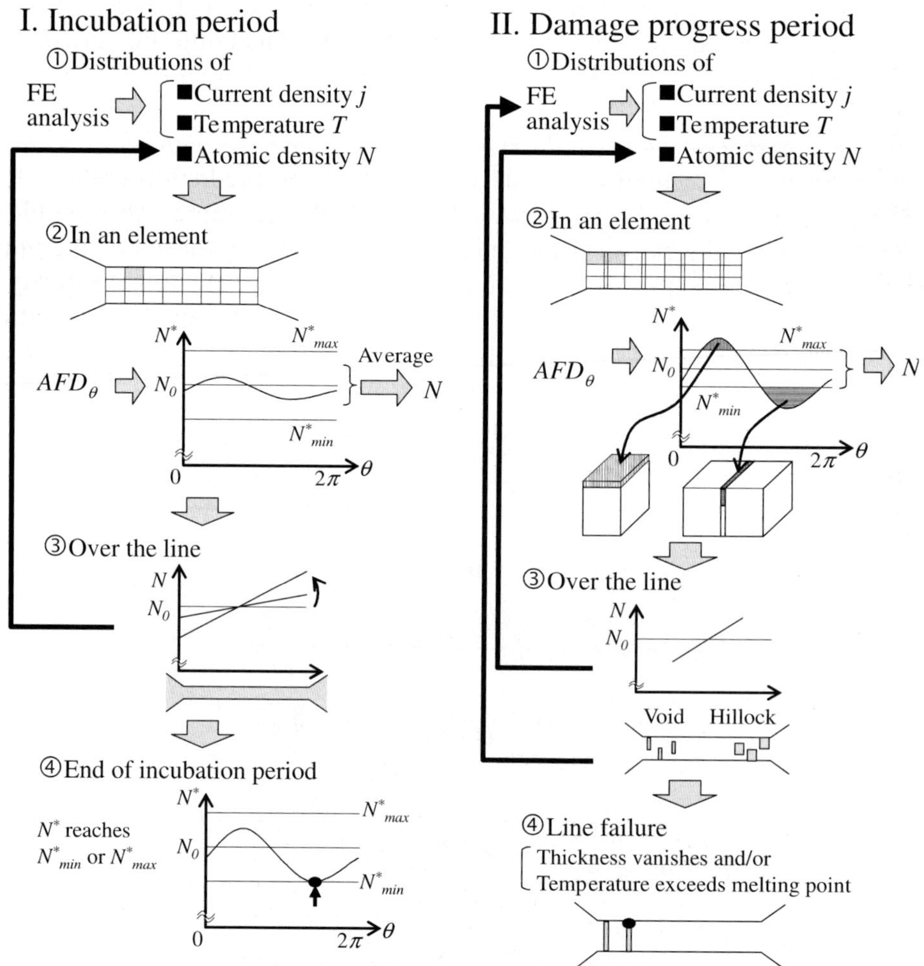

Figure 5. Schema of numerical simulation for failure process of polycrystalline lines connected with pads.

the simulation of the incubation period is carried out until an atomic density N^* in the line reaches $N^*_{\min}$ or $N^*_{\max}$. This means that we find the case in which the initial void or hillock occurs most easily in the line. The initial void or hillock is expected to appear in actual metal lines around the case. In succession, the simulation of the damage-progress period is started by using the atomic density distribution built up in the simulation of the incubation period. The values of $N^*_{\min}$ and $N^*_{\max}$ can be obtained by a simple experiment based on *AFD* (Sasagawa et al., 2002b). It is noted that $N^*_{\min}$ and $N^*_{\max}$ depend on passivation. Therefore for stronger passivation $N^*_{\min}$ becomes smaller, $N^*_{\max}$ becomes larger and the incubation period is extended. In the above simulation, the average of all the values of N^* in the element is obtained as an expected value of the atomic density of the element and it is denoted by N. The distribution of N over the line is used for the calculation of *AFD*.

In the simulation of the damage-progress period, the thickness of elements is changed based on the value of *AFD* to simulate the formation of voids and hillocks. The voids selectively grow along the grain boundary in polycrystalline lines, and they result in slit-like voids extending toward the line width and linking themselves (see Figure 6). By considering this morphology of void growth, finite element meshes are introduced as shown in Figure 7, based on measurements of the average grain size and the effective width of the slit-like void. The changes in current density and temperature distributions due to void growth are taken into account in calculating *AFD*. The FE analysis of current density and temperature in the line is carried out again by considering the change in each element thickness. The increase in thickness of the element used in the simulation is also taken into account because the formation of collapsed hillocks may affect the distributions of current density and temperature. The increase in element thickness is assumed to be conducted only in the neighboring elements of the slim elements for simplicity. The simulation for the period of the damage progress is carried out repeatedly until the line fails, where the line failure is defined as the state that the entire line width is occupied by the elements whose thickness vanishes and/or the elements whose temperature exceeds the melting point.

3.2.3. *Verification and application of AFD-based method to the polycrystalline line with pads*

In order to verify the *AFD*-based method, the electromigration failures in five kinds of straight samples were predicted as shown in Figures 8–10. First, Samples 1 and

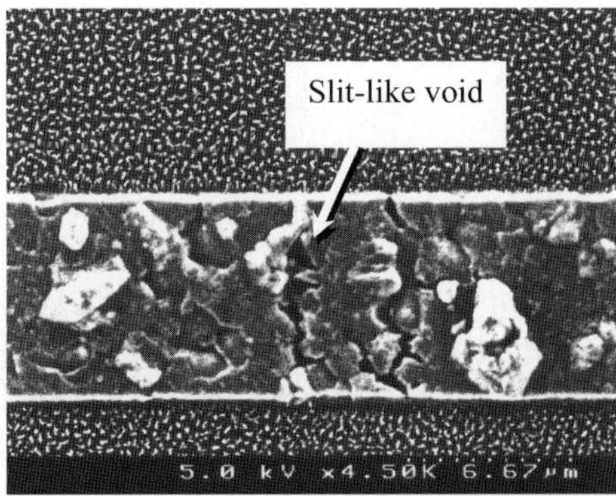

Figure 6. Slit-like void in passivated polycrystalline lines.

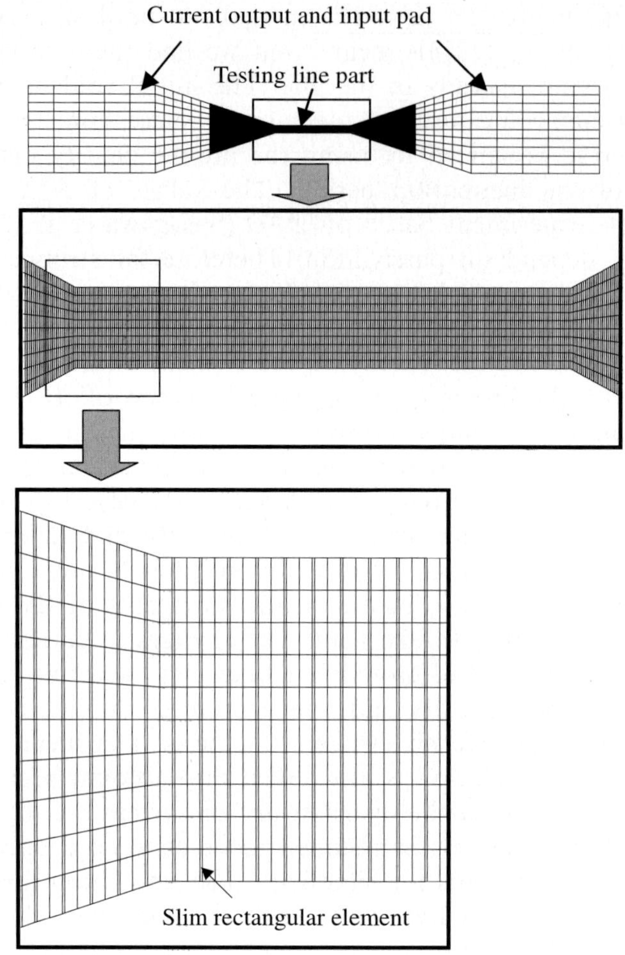

Figure 7. Finite element mesh and its magnification; thickness is decreased only at very slim elements.

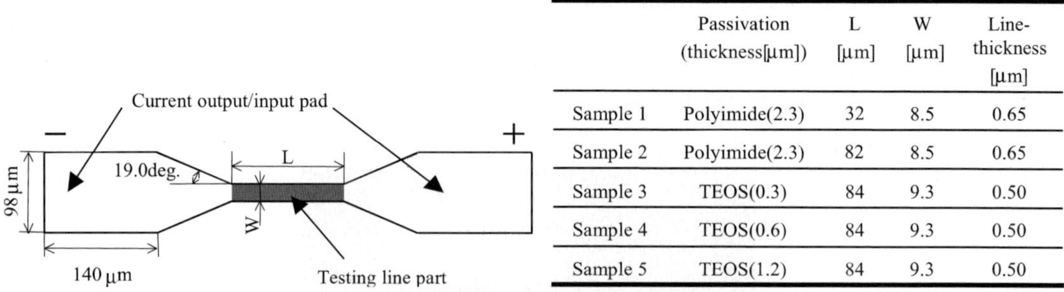

	Passivation (thickness[μm])	L [μm]	W [μm]	Line-thickness [μm]
Sample 1	Polyimide(2.3)	32	8.5	0.65
Sample 2	Polyimide(2.3)	82	8.5	0.65
Sample 3	TEOS(0.3)	84	9.3	0.50
Sample 4	TEOS(0.6)	84	9.3	0.50
Sample 5	TEOS(1.2)	84	9.3	0.50

Figure 8. Dimensions of polycrystalline lines.

2 had different line-lengths and passivation layer was made of polyimide. Through the predicted lifetimes agreed with experimental ones in both lines having different lengths, the usefulness of the *AFD*-based method was shown (Sasagawa et al., 2002b). As the application of our prediction method to some practical problems, the method for predicting electromigration failure considering the passivation thickness was developed by Sasagawa et al. (2003). Dependency of lifetime on the passivation

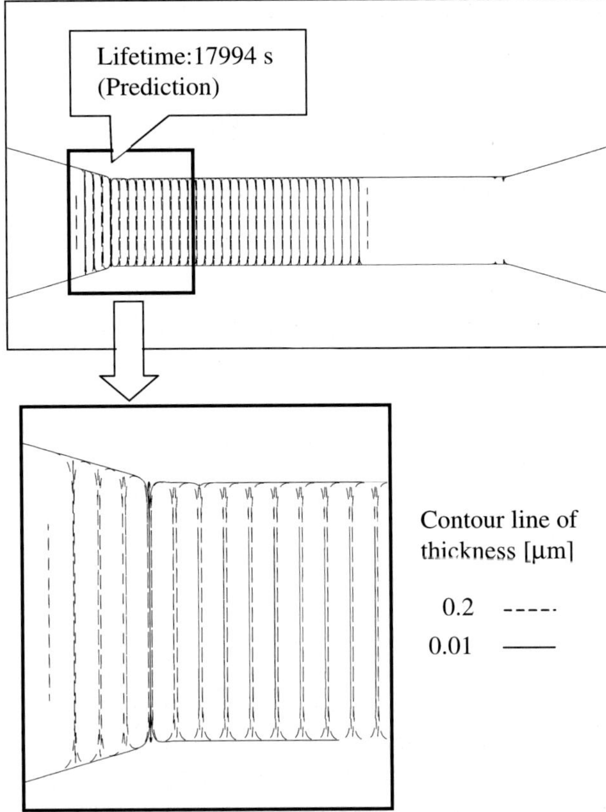

Figure 9. Prediction results in the case of Sample 1.

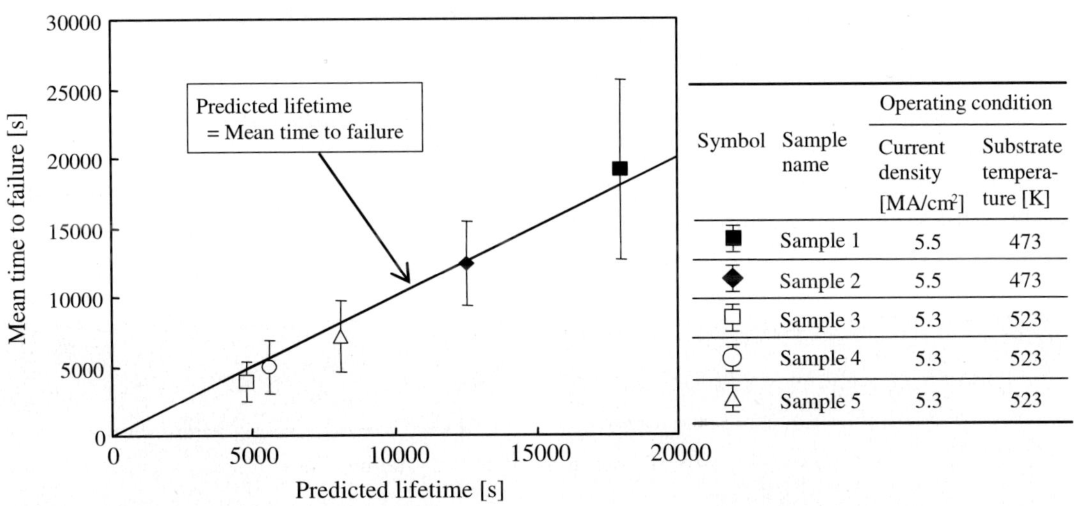

Symbol	Sample name	Operating condition	
		Current density [MA/cm²]	Substrate temperature [K]
■	Sample 1	5.5	473
◆	Sample 2	5.5	473
▯	Sample 3	5.3	523
○	Sample 4	5.3	523
△	Sample 5	5.3	523

Figure 10. Relationship between the predicted lifetime and the mean time to failure.

thickness was obtained by the *AFD*-based method. By utilizing the dependency, it is possible to determine the passivation thickness considering the required lifetime. This fact was proved by treating three kinds of Al lines covered with tetraethyl orthosilicate (TEOS) film having different thickness (Samples 3–5 in Figure 8). The relationship between the predicted lifetime and the mean time to failure experimentally measured is summarized in Figure 10. It was found the lifetimes predicted by our method agreed well with those obtained in experiment in spite of differences in operating conditions, line-length, passivation material and passivation thickness.

3.3.   POLYCRYSTALLINE LINE CONNECTED WITH VIAS

3.3.1.   *AFD for ends of polycrystalline line*

The metal lines in ICs are often connected with vias and multi-level interconnections are constructed. The schematic illustration of a typical interconnection with via is shown in Figure 1(d). The via is made of Al or tungsten (W). The metal lines are often stacked on the shunt layer made of refractory metal such as titanium nitride (TiN), for which the electrical current is able to bypass the void formed in the line. In the metal line structure with via, no atoms are supplied to the cathode end of the line by electromigration because the line is not directly connected with a reservoir of the atoms such as current output pad and the atomic flow is intercepted by vias. The metal line connected with via, therefore, has a failure mode that the cathode edge of the line drifts in the longitudinal direction as a result of electromigration.

The governing parameter *AFD* for electromigration damage at the ends of passivated polycrystalline line was expressed by considering the boundary condition with respect to atomic diffusion, namely, there is no incoming and outgoing of atoms at the cathode end and anode end, respectively (Hasegawa et al., 2003). Furthermore, our research group developed a simple method of determining the film characteristic constants by utilizing *AFD* and the measurement of the drift velocity of line end as shown in Figure 11. The use of *AFD* for the electromigration damage at the ends of passivated polycrystalline line was verified through the discussion on the validity of the characteristic constants (D_0, $Q^*[= Q - \sigma_T \Omega]$, Z^* and κ) derived by the *AFD*-based method with simple experiments (Hasegawa, 2004).

3.3.2.   *Evaluation of threshold current density for polycrystalline line connected with vias*

There is a threshold current density of electromigration damage, j_{th}, in the line connected with vias. The evaluation method of j_{th} was developed for the passivated poly-

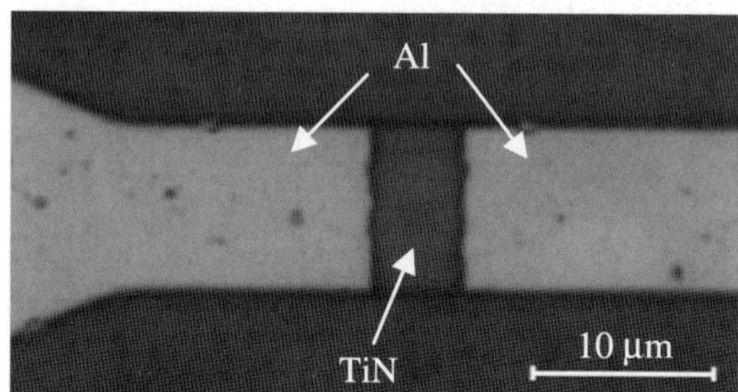

(a) Before current supply

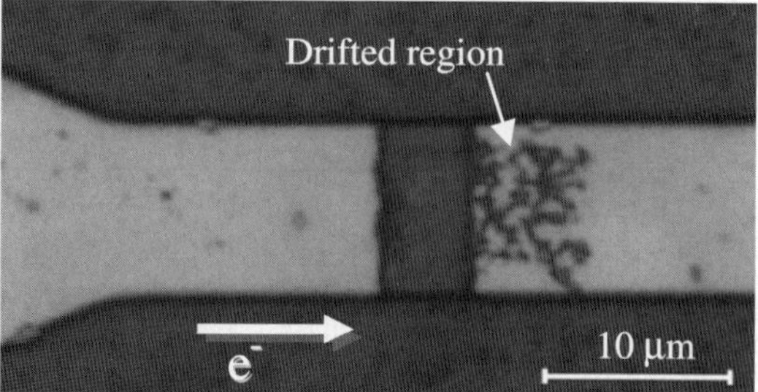

(b) After current supply

Figure 11. Observation of the end of the polycrystalline line modeled on via-connected line before and after current supply. **(a)** Before current supply and **(b)** After current supply.

crystalline line also by using *AFD*. In the evaluation, the line is divided into elements (see Figure 12) and the process occurring in the incubation period is simulated based on the value of *AFD* in each element. Several kinds of input current density, j, are supposed, and then the simulation is performed under each current density at a certain substrate temperature. The value of N^* depending on θ is calculated by using the value of *AFD*. Then, the atomic density in each element, N, is obtained by averaging all the values of N^* in the element. The threshold current density j_{th} is found such that if the supposed j is larger than j_{th}, the value of N^* soon reaches the critical density for void formation N^*_{min}, and if j is smaller than j_{th}, the value of N^* does not reach N^*_{min}, and the atomic density distribution holds the steady state. Figure 13 shows this tendency in terms of N, where N_0 is the value given at the reference condition as noted in Section 3.1.

The result of the threshold evaluation for the test line modeled on the via-connected line (see Figure 14) at 538 K is shown in Figure 15(a). In the figure, the smallest value of N^* in the steady state just discussed concerning θ and the position over the line is plotted against the supposed current density. It is shown that the smallest value of N^* normalized by N_0 approaches N^*_{min}/N_0 with increase in the current density, where N_0 denotes the initial atomic density and N^*_{min} is determined by a simple experiment (Sasagawa et al., 2002b). The smallest value of N^* becomes N^*_{min} when the atomic density gradient in the line reaches a critical value $\partial N/\partial x|_{\text{th}}$. The back flow induced by the gradient can no longer counterbalance the electromigration without damage initiation. Consequently, the threshold current density was evaluated to be $0.18\,\text{MA/cm}^2$ from an intersection of the linear solid line with solid squares and the dashed line indicating the value of N^*_{min}/N_0.

In order to verify the result of evaluation, the experiment was performed by using the same line-shape and under the same substrate temperature as those in the simulation. In the accelerated test, the metal lines were subjected to three kinds of direct current with density of 1.2, 0.9 and $0.6\,\text{MA/cm}^2$ under substrate temperature being 538 K. After the current was applied until 20% up in potential drop, the drift velocity was measured from the observation of the drifted line end (see Figure 11). The drift velocity was defined by the black area of the drifted region in Figure 11 divided

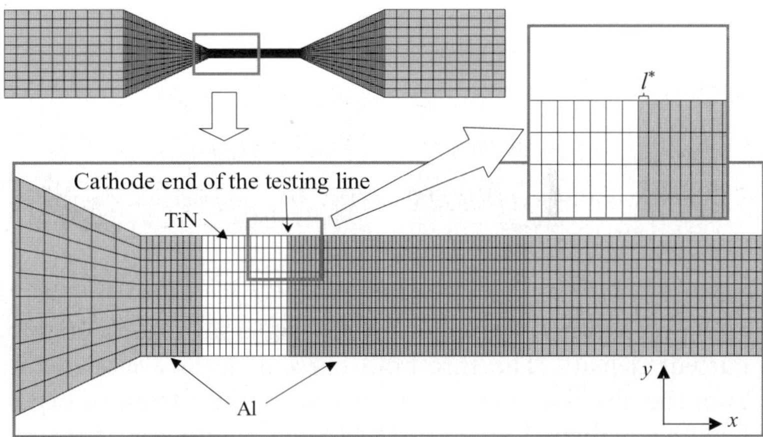

Figure 12. Supposed polycrystalline line and mesh generation of finite elements ($l^* = 0.658d$).

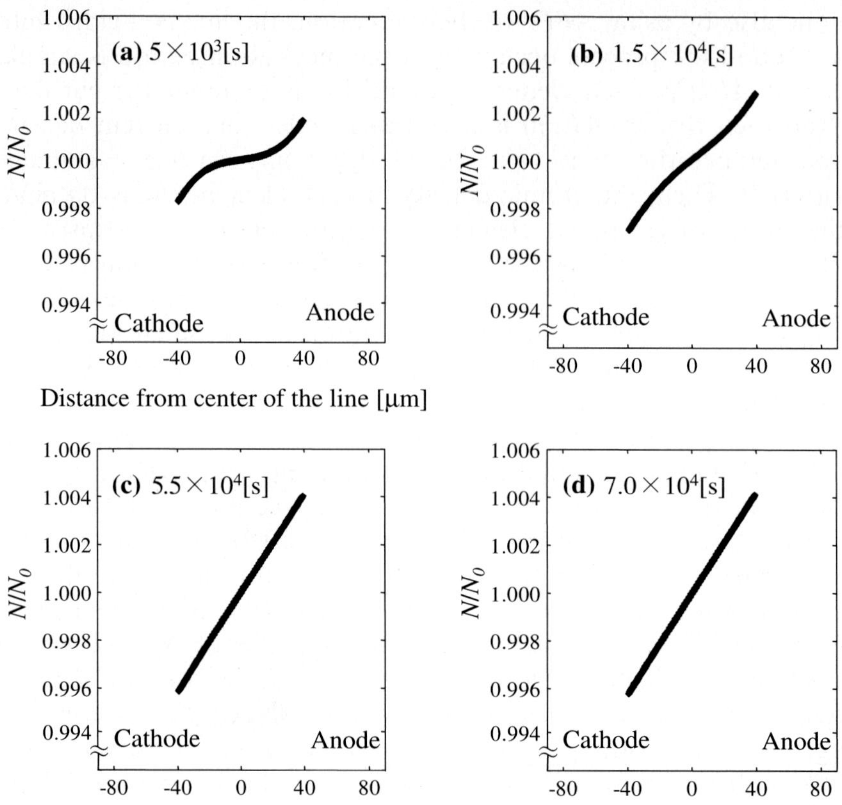

Figure 13. Change in atomic density distribution with time $[j(=0.1) < j\mathrm{th}(=0.18)\,\mathrm{MA/cm^2}]$.

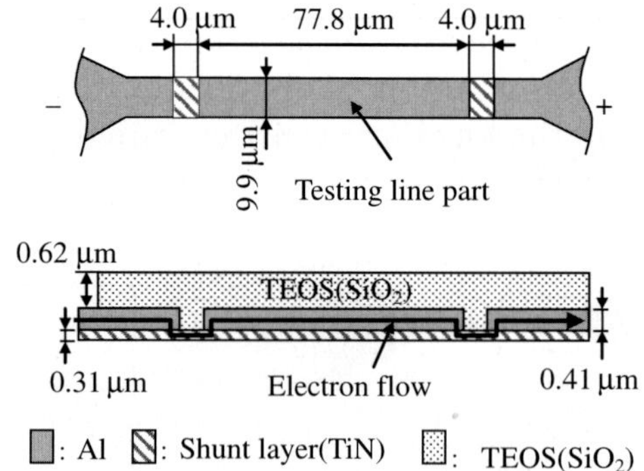

Figure 14. Dimensions of the polycrystalline test line modeled on via-connected line.

by the line-width. Figure 15(b) presents the experimental results of the drift velocity against input current density. The threshold current density was obtained as $0.22 \pm 0.05\,\mathrm{MA/cm^2}$ from the abscissa intercept by linear extrapolation of experimental data. The agreement of the evaluated value ($0.18\,\mathrm{MA/cm^2}$) with the measured one ($0.22 \pm 0.05\,\mathrm{MA/cm^2}$) confirms the usefulness of the evaluation method.

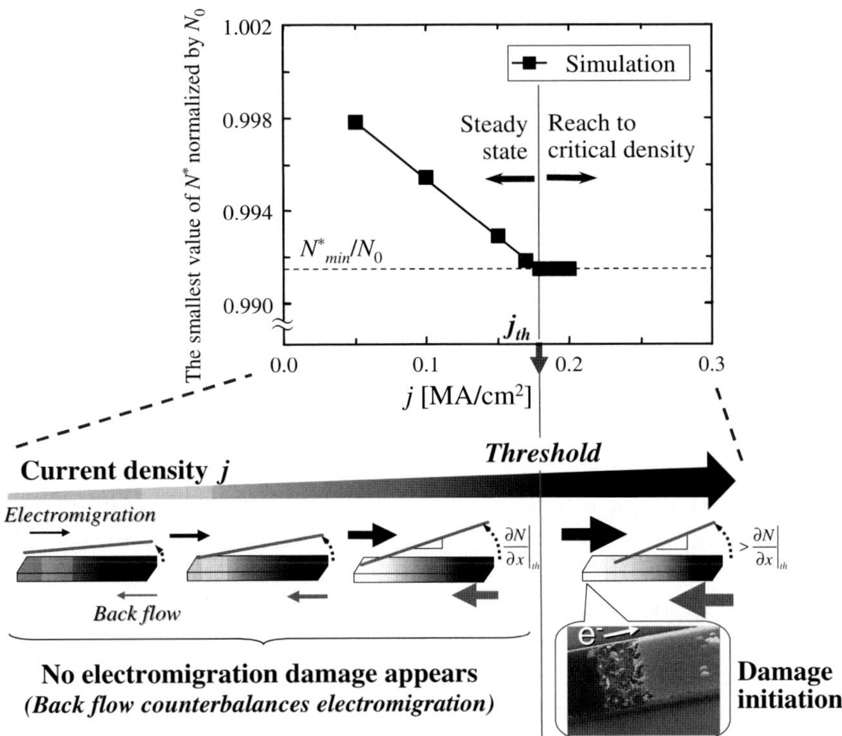

(**a**) Evaluation of the threshold current density

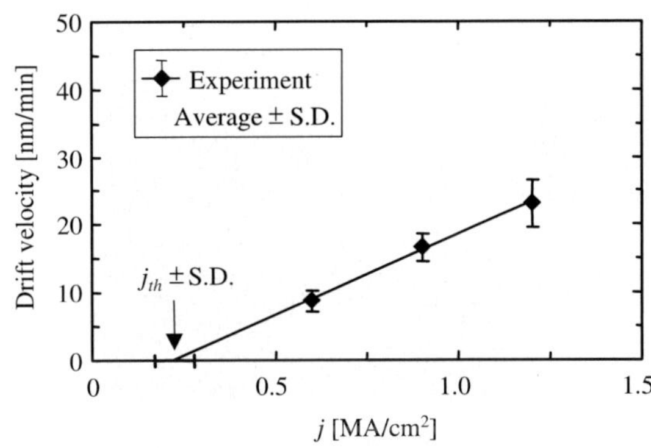

(**b**) Measurement of the threshold current density

Figure 15. Comparison of the evaluation and measurement of j_{th} in the polycrystalline line connected with vias (substrate temperature 538 K). (**a**) Evaluation of the threshold current density and (**b**) Measurement of the threshold current density.

3.4. BAMBOO LINE CONNECTED WITH VIAS

3.4.1. *AFD for bamboo lines*

Now, let us move onto the case of bamboo structured lines. In this case, atomic diffusion along grain boundary is negligible, and the lattice diffusion including interface diffusion is dominant. The consideration of the grain microstructure is not necessary in Figure 2. Here, let us assume that the lengths a and b in Figure 2 are sufficiently small. The atomic flux at the middle point of the sides of the rectangle, $J_\xi(0, b/2)$, $J_\xi(a, b/2)$, $J_\eta(a/2, 0)$ and $J_\eta(a/2, b)$ can stand for the flux on the corresponding sides because of small a and b. And the local coordinate system (ξ, η) is transformed to global Cartesian coordinate system (x, y). Equation (1) is calculated by considering the boundary condition with respect to the atomic flux and then the governing parameter *AFD* for bamboo line can be obtained as:

$$AFD = \mathrm{div}\, J. \tag{5}$$

The way to determine the film characteristic constants was also developed by using *AFD* (Hasegawa, 2004). The method is based on the simple measurement of drift velocity of the line end as shown in Figure 16.

3.4.2. *Evaluation of threshold current density for bamboo line connected with vias*

An evaluation method of j_{th} was developed for the passivated bamboo line connected with vias. The method is based also on *AFD*, and is realized by means of numerical simulation for the process occurring in the incubation period. In the simulation, the line is divided into elements as shown in Figure 17 and, by calculating the atomic density in the element N based on *AFD*, the building-up process of the distribution of atomic density over the line is simulated. Some kinds of input current densities are supposed, and then the simulation is performed under each current density j at an operating temperature. On the other hand, there is a critical atomic density for

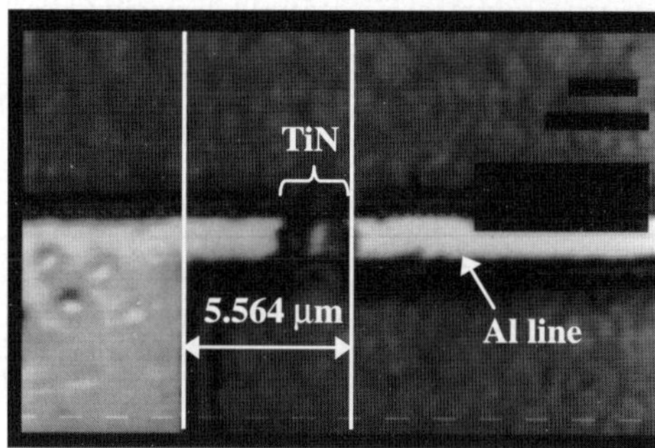

(a) Before current supply

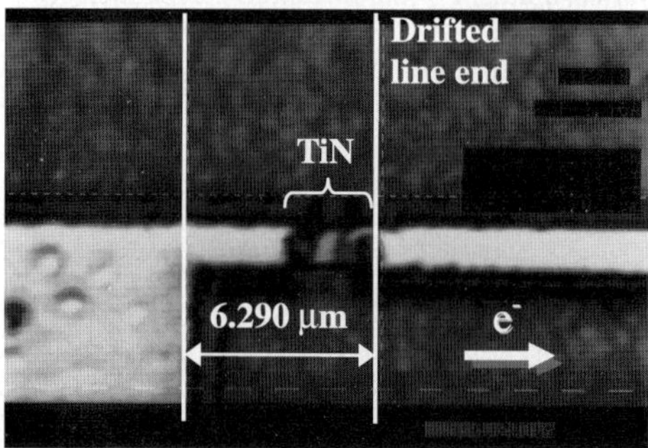

(b) After current supply

Figure 16. The observation of the cathode end of bamboo test line before and after current supply. (a) Before current supply and (b) After current supply.

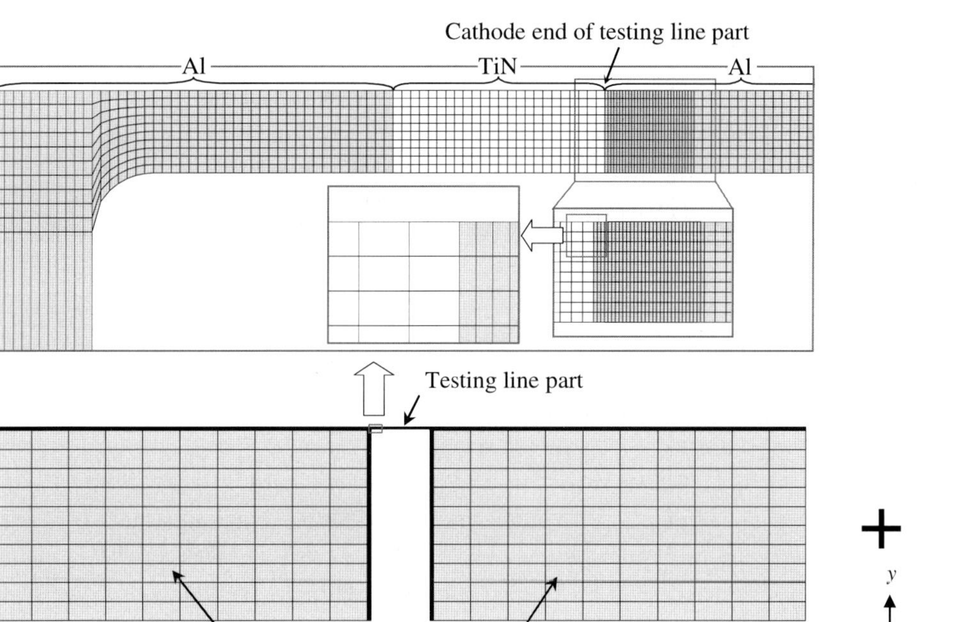

Figure 17. Supposed bamboo line and mesh generation of finite elements.

void formation also in bamboo line, which is defined as N_{min} and determined by a simple experiment (Hasegawa, 2004). The threshold current density j_{th} is found such that if j is larger than j_{th}, the value of N in any element soon reaches N_{min}, and if j is smaller than j_{th}, the distribution of N holds steady state without reaching N_{min}.

The evaluation result for the test line modeled on the via-connected line (see Figure 18) at a temperature of 523 K is shown in Figure 19(a). The figure shows the smallest value of N in the line, which is normalized by N_0, versus input current density j, where the smallest value of N is obtained from the steady distribution of N over

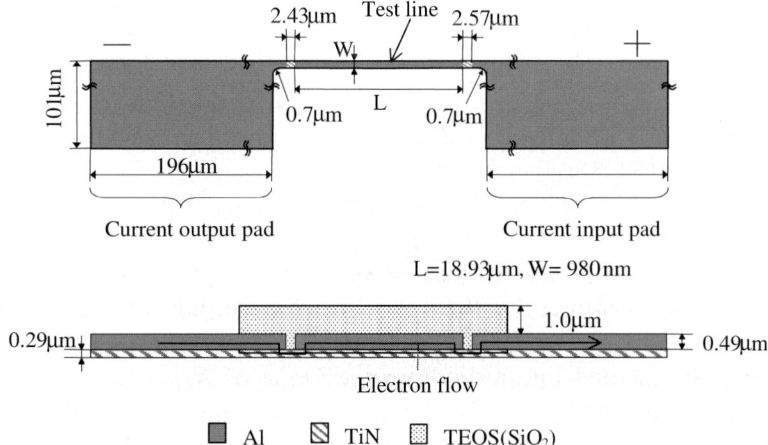

Figure 18. Dimensions of the bamboo test line modeled on the via-connected line.

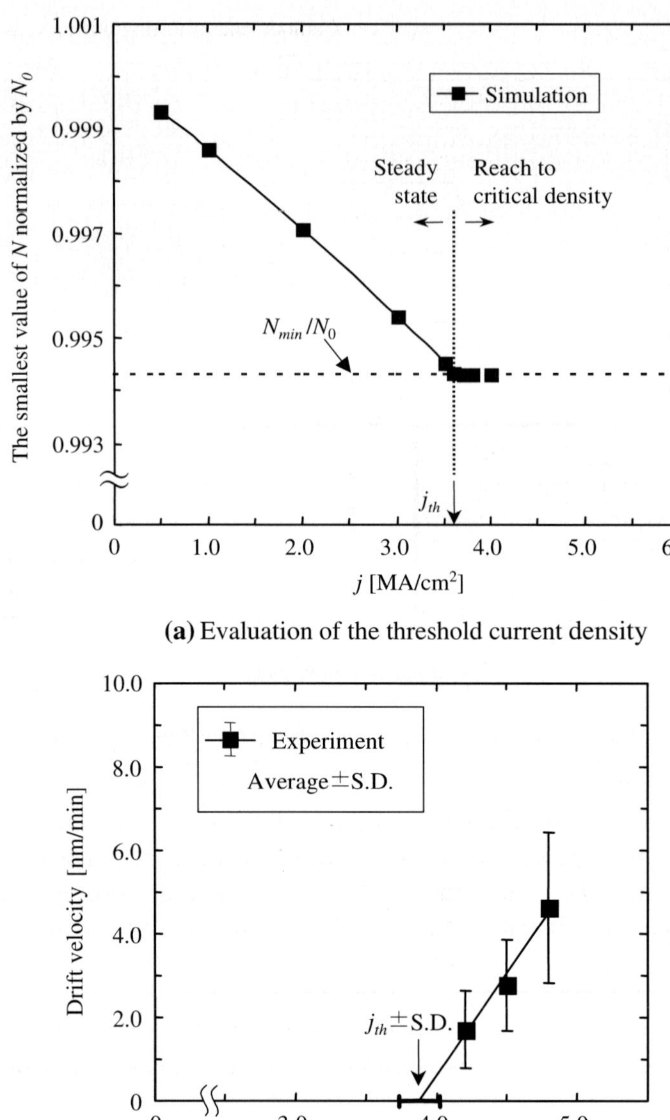

(a) Evaluation of the threshold current density

(b) Measurement of the threshold current density

Figure 19. Comparison of the evaluation and measurement of j_{th} in the bamboo line connected with vias (substrate temperature 523 K). **(a)** Evaluation of the threshold current density and **(b)** Measurement of the threshold current density.

the line. It is shown that the steady smallest value of N approaches N_{min} with increase in the current density, as shown by the solid line. Consequently, the threshold current density was evaluated to be 3.62 MA/cm^2 from the intersection of the solid line with solid squares and the dashed line indicating the value of N_{min}/N_0.

In order to verify the evaluation result, the threshold current density was experimentally obtained concerning the same line-shape and condition as those in the simulation. Three kinds of direct current with density of 4.8, 4.5 and 4.2 MA/cm^2 were supplied under substrate temperature of 523 K. After current was supplied until 20% increase in potential drop, the drift velocity was obtained by observation of the drifted line end (see Figure 16). Figure 19(b) presents the experimental results of drift velocity against the input current density. The threshold current density of 3.84 ± 0.14 MA/cm^2 was got from the abscissa intercept by linear extrapolation of experimental data. It was found that the evaluation method of j_{th} in the bamboo line gave the good agreement between the evaluation (3.62 MA/cm^2) and the measurement (3.84 ± 0.14 MA/cm^2). The evaluation method of the threshold current density was shown to be successfully constructed based on *AFD* also for the passivated bamboo line.

4. Achievements

Our research group has developed the *AFD*-based method for the reliability evaluation of IC metal line against electromigration failure. It is summarized as shown in Figure 20. The governing parameter *AFD* integrates all the factors affecting the damage, i.e., the line structures, film characteristics, operating conditions such as current density and temperature, and atomic density. The parameter *AFD* gives the number of atoms decreasing per unit time and unit volume. By utilizing *AFD* the distribution of atomic density N within the metal line can be calculated. Then it is judged whether the atomic density is beyond a critical value, for the transition from the incubation period to the damage-progress period or for seeking the threshold current density. An excess of the atomic density over the critical value is used for reproducing the damage-progress to line failure, and lifetime and possible failure site are predicted.

Concerning the polycrystalline line connected with pads, the governing parameter *AFD* has been identified in the passivated polycrystalline line. The prediction method of electromigration failure has been developed based on the numerical simulation using *AFD*, and furthermore, the way to determine the thickness of the passivation layer considering the required lifetime of the line has been developed.

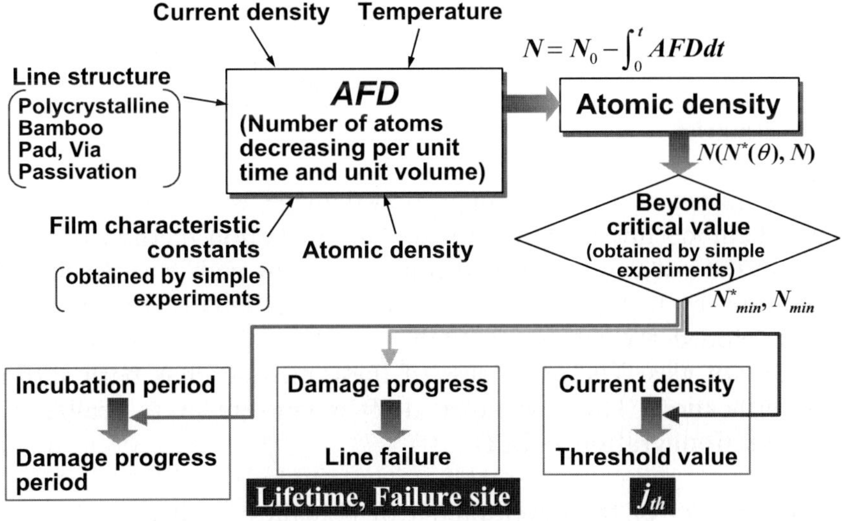

Figure 20. Summary of *AFD*-based method.

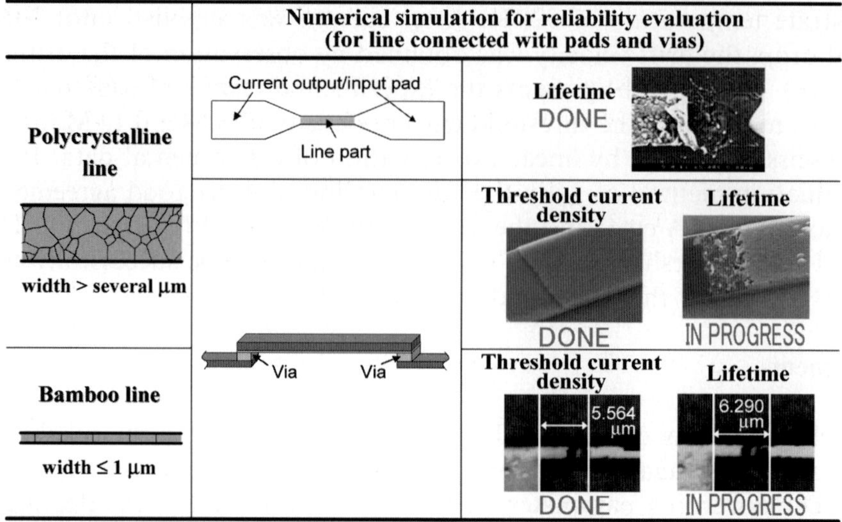

Figure 21. Lifetime prediction for line connected with pads and evaluation of threshold current density j_{th} for via-connected line.

Respecting the polycrystalline line connected with vias, the governing parameter *AFD* at the ends of the passivated polycrystalline line has been identified. The evaluation method of the threshold current density for the passivated polycrystalline line connected with vias has been developed by using *AFD*. It has been shown that the evaluation result of threshold current density in the passivated polycrystalline line agreed with the experimental one, and the usefulness of the *AFD*-based method has been confirmed.

As to the bamboo structured line connected with vias, the governing parameter *AFD* in the passivated bamboo line has also been identified. The evaluation method of the threshold current density for the passivated bamboo line has also been developed by using *AFD*. The usefulness of the *AFD*-based method has been shown experimentally.

These results are described again in Figure 21.

5. Concluding remarks

The governing parameters for electromigration damage, *AFD*, in passivated polycrystalline line and bamboo line were identified. The prediction method of the electromigration failure in the polycrystalline line connected with pads was developed by using *AFD*. Furthermore, by using *AFD*, the evaluation methods of the threshold current density were developed for the polycrystalline and bamboo lines the ends of which were connected with vias. The usefulness of these methods has been shown through experimental verification. The parameter *AFD* is convenient to realize a reliability evaluation for electromigration failure.

As a future prospect of the development of reliability evaluation method, the followings would be pointed:

(1) The reliability evaluation method is expected to be applied to the line structure of Cu metallization.
(2) Bundles of carbon nanotubes (from several tens nm to one μm) are applicable to vias and lines because of the tolerance of higher current density and lower electrical resistance.
(3) The method could be applied to the reliability evaluation of metal lines and electrodes in not only silicon ICs but also electronic devices such a printed circuit board, liquid crystal display and giant magnetoresistive head, etc.
(4) When we consider the line-width from 980 nm treated here down to several tens nm, it is still an open question and interesting problem how our approach including experimental technique can be effective to the finer lines.

References

Ainslie, N.G., d'Heurle, F.M. and Wells, O.C. (1972). Coating, mechanical constraints, and pressure effects on electromigration. *Applied Physics Letters* **20**, 173–174.

Black, J.R. (1969). Electromigration failure modes in aluminum metallization for semiconductor devices. *Proceedings of the IEEE* **57**, 1587–1593.

Blech, I.A. (1976). Electromigration in thin aluminum films on titanium nitride. *Journal of Applied Physics* **47**, 1203–1208.

Blech, I.A. (1998). Diffusional back flows during electromigration. *Acta Materialia* **46**, 3717–3723.

Hasegawa, M., Sasagawa, K., Saka, M. and Abé, H. (2003). Expression of a governing parameter for electromigration damage on metal line ends. *Proceedings of InterPACK'03 (CD-ROM), ASME*, Paper ID IPACK2003-35064.

Hasegawa, M. (2004). Evaluation method of electromigration damage in IC metal lines and its application to practical problems. Ph.D. dissertation, Tohoku University, Sendai, Japan.

Huntington, H.B. and Grone, A.R. (1961). Current-induced marker motion in gold wires. *Journal of Physics and Chemistry of Solids* **20**, 76–87.

Kirchheim, R. and Kaeber, U. (1991). Atomistic and computer modeling of metallization failure of integrated circuits by electromigration. *Journal of Applied Physics* **70**, 172–181.

Korhonen, M.A., Børgesen, P., Tu, K.N. and Li, C.-Y. (1993). Stress evolution due to electromigration in confined metal lines. *Journal of Applied Physics* **73**, 3790–3799.

Marcoux, P.J., Merchant, P.P., Naroditsky, V. and Rehder, W.D. (1989). New 2D simulation model of electromigration. *Hewlett-Packard Journal* June, 79–84.

McPherson, J.W. (1986). Stress dependent activation energy. *Proceedings of the 24th IEEE International Reliability Physics Symposium, IEEE*, 12–18.

Nikawa, K. (1981). Monte Carlo calculations based on the generalized electromigration failure model. *Proceedings of the 19th IEEE International Reliability Physics Symposium*, IEEE, 175–181.

Oates, A.S. (1991). Electromigration in multilayer metallization: drift-controlled degradation and the electromigration threshold of Al–Si–Cu/TiN$_X$O$_Y$/TiSi$_2$ contacts. *Journal of Applied Physics* **70**, 5369–5373.

Park, Y.J. and Thompson, C.V. (1997). The effects of the stress dependence of atomic diffusivity on stress evolution due to electromigration. *Journal of Applied Physics* **82**, 4277–4281.

Sasagawa, K., Nakamura, N., Saka, M. and Abé, H. (1998). A new approach to calculate atomic flux divergence by electromigration. *Transactions of the ASME, Journal of Electronic Packaging* **120**, 360–366.

Sasagawa, K., Naito, K., Saka, M. and Abé, H. (1999). A method to predict electromigration failure of metal lines. *Journal of Applied Physics* **86**, 6043–6051.

Sasagawa, K., Hasegawa, M., Saka, M. and Abé, H. (2000). Atomic flux divergence in bamboo line for predicting initial formation of voids and hillocks. *Theoretical and Applied Fracture Mechanics* **33**, 67–72.

Sasagawa, K., Hasegawa, M., Naito, K., Saka, M. and Abé, H. (2001). Effects of corner position and operating condition on electromigration failure in angled bamboo lines without passivation layer. *Thin Solid Films* **401**, 255–266.

Sasagawa, K., Hasegawa, M., Saka, M. and Abé, H. (2002a). Governing parameter for electromigration damage in passivated polycrystalline line. *Journal of Applied Physics* **91**, 1882–1890.

Sasagawa, K., Hasegawa, M., Saka, M. and Abé, H. (2002b). Prediction of electromigration failure in passivated polycrystalline line. *Journal of Applied Physics* **91**, 9005–9014.

Sasagawa, K., Hasegawa, M., Yoshida, N., Saka, M. and Abé, H. (2003). Prediction of electromigration failure in passivated polycrystalline line considering passivation thickness. *Proceedings of Inter-PACK'03 (CD-ROM), ASME*, Paper ID IPACK 2003-35065.

Villars, P. (1997). *Pearson's Handbook Desk Edition – Crystallographic Data for Intermetallic Phases*, Vol. 1, ASM International, Materials Park, USA.

International Journal of Fracture (2006) 138:241–262
DOI 10.1007/s10704-006-0033-3

© Springer 2006

Modern domain-based discretization methods for damage and fracture

RENÉ DE BORST[1,2]
[1] *Faculty of Aerospace Engineering, Delft University of Technology, NL-2600 GB Delft, the Netherlands (e-mail: R.deBorst@TUDelft.nl)*
[2] *LaMCoS – UMR CNRS 5514, I.N.S.A. de Lyon, F-69621 Villeurbanne, France*

Received 1 March 2005; accepted 1 December 2005

Abstract. Standard domain-based discretization methods that have been developed for continuous media are not well suited for treating propagating (or evolving) discontinuities. Indeed, they are approximation methods for the solution of partial differential equations, which are valid on a domain. Discontinuities divide this domain into two or more parts. Conventionally, special interface elements methods are placed *a priori* between the continuum finite elements to capture discontinuities at locations where they are expected to emerge. More recently, discretization methods have been proposed, which are more flexible than standard finite element methods, while having the potential to capture propagating discontinuities in a robust, efficient and accurate manner. Examples are meshfree methods, finite element methods that exploit the partition-of-unity property of finite element shape functions, and discontinuous Galerkin methods. In this contribution, we shall present an overview of these novel discretization techniques for capturing propagating discontinuities, including a comparison of their similarities and differences.

Key words: Damage, discontinuity, discontinuous Galerkin, fracture, finite elements, interfaces, partition-of-unity method.

1. Introduction

Generally, two types of fracture analyses can be distinguished. The more elementary concerns the computation of fracture properties for a given, stationary crack. Typically, this relates to properties like stress intensity factors or the J-integral. Often, linear elastic fracture mechanics is used as the underlying theory. With the proper knowledge of these quantities, fracture mechanics makes it possible to determine if a crack will propagate, in which direction (although a number of different hypotheses exist) and, for dynamic problems, at which speed the crack will propagate. Since the stress field is singular at the crack tip in linear elastic fracture mechanics, computational methods have been developed for capturing this singularity, especially for coarse discretizations (Henshell and Shaw, 1975; Barsoum, 1976).

More difficult is the simulation of crack propagation. Originally, this was almost exclusively done in a finite element context, either in a discrete, or in a smeared format. In the discrete approach, a stress intensity factor is computed. On basis of this information it is decided if, and if yes, how much, the crack will propagate. The crack is advanced, a new mesh is generated for the new geometry and the process is repeated (Ingraffea and Saouma, 1987). Essentially, this approach consists of a series of computations for a stationary crack. In the smeared approach (Rashid, 1968;

242   *R. de Borst*

de Borst and Nauta, 1985; Rots, 1991), the state of the stress tensor and the internal variables at an integration point are considered to be representative for the tributary area of the finite element belonging to this integration point and the discrete crack is replaced by a damaged area. Approaches like this essentially follow a damage mechanics format (Lemaitre and Chaboche, 1990; de Borst and Gutiérrez, 1999).

The method of simulating the propagation of a discrete crack by a sequence of linear elastic fracture mechanics calculations is possible by virtue of the linear nature of the theory. In nonlinear fracture mechanics, including cohesive-zone models (Dugdale, 1960; Barenblatt, 1962; Hillerborg et al., 1976), this no longer holds true and methods must be developed that allow for tracing crack propagation in a nonlinear sense. Promising avenues are meshfree methods (Nayroles et al., 1992; Belytschko et al., 1994; Liu et al., 1995; Duarte and Oden, 1996; Fleming et al., 1997; Krysl and Belytschko, 1999) and finite element methods that exploit the partition-of-unity property of finite element shape functions (Babuska and Melenk, 1997; Belytschko and Black, 1999; Moes et al., 1999; Wells and Sluys, 2001; Wells et al., 2002; Remmers et al., 2003; Simone, 2004) while, more recently, authors have also applied discontinuous Galerkin methods (Zienkiewicz et al., 2003) to problems of fracture and damage evolution (Mergheim et al., 2004; Wells et al., 2004). Another method to incorporate discontinuities in finite element has recently been proposed in (Hansbo and Hansbo 2004), (see also Mergheim et al., 2005) for its application to cohesive-zone models.

This contribution starts by giving an outline of the domain-based discretization methods that are amenable to the numerical analysis of damage and fracture propagation: conventional interface elements, meshfree methods, where the element-free Galerkin method (Belytschko et al., 1994; Fleming et al., 1997; Krysl and Belytschko, 1999) is taken as a prototypical meshfree method, the partition-of-unity approach and, finally, discontinuous Galerkin methods. Relations between these methods exist, and for some cases, in particular when the discontinuity coincides with the grid lines, correspondences will be shown.

2. Zero-thickness interface elements

The classical way to represent discontinuities in solids is to introduce zero-thickness interface elements between two neighbouring (solid) finite elements, e.g. Figure 1 for a planar interface element. The governing kinematic quantities in interfaces are relative displacements: v_n, v_s, v_t for the normal and the two sliding modes, respectively. When collecting these relative displacements in a relative displacement vector $\mathbf{v}$, they can be related to the displacements at the upper $(+)$ and lower sides $(-)$ of the

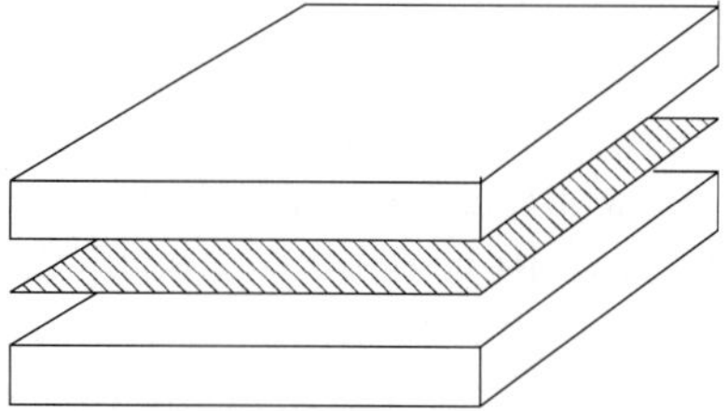

Figure 1. Planar interface element between two three-dimensional finite elements.

interface, $u_n^-, u_n^+, u_s^-, u_s^+, u_t^-, u_t^+$, by

$$\mathbf{v} = \mathbf{L}\mathbf{u} \tag{1}$$

with $\mathbf{u}^{\mathrm{T}} = (u_n^-, \ldots, u_t^+)$ and $\mathbf{L}$ an operator matrix:

$$\mathbf{L} = \begin{bmatrix} -1 & 0 & 0 \\ +1 & 0 & 0 \\ 0 & -1 & 0 \\ 0 & +1 & 0 \\ 0 & 0 & -1 \\ 0 & 0 & +1 \end{bmatrix}. \tag{2}$$

The displacements contained in the array $\mathbf{u}$ are interpolated in a standard manner, as

$$\mathbf{u} = \mathbf{H}\mathbf{a}, \tag{3}$$

where

$$\mathbf{H} = \operatorname{diag}\left[\mathbf{h}\,\mathbf{h}\,\mathbf{h}\,\mathbf{h}\,\mathbf{h}\,\mathbf{h}\right] \tag{4}$$

with $\mathbf{h}$ an $1 \times N$ matrix containing the interpolation polynomials, and $\mathbf{a}$ the element nodal displacement array,

$$\mathbf{a} = \left(a_n^1, \ldots, a_n^N, a_s^1, \ldots, a_s^N, a_t^1, \ldots, a_t^N\right)^{\mathrm{T}} \tag{5}$$

with N the total number of nodes in the interface element. The relation between nodal displacements and relative displacements for interface elements is now derived from Equations (1) and (3) as:

$$\mathbf{v} = \mathbf{L}\mathbf{H}\mathbf{a} = \mathbf{B}_i\mathbf{a}, \tag{6}$$

where the relative displacement-nodal displacement matrix $\mathbf{B}_i$ for the interface element reads:

$$\mathbf{B}_i = \begin{bmatrix} -\mathbf{h}\,\mathbf{h} & 0\,0 & 0\,0 \\ 0\,0 & -\mathbf{h}\,\mathbf{h} & 0\,0 \\ 0\,0 & 0\,0 & -\mathbf{h}\,\mathbf{h} \end{bmatrix}. \tag{7}$$

For an arbitrarily oriented interface element the matrix $\mathbf{B}_i$ subsequently has to be transformed to the local coordinate system of the integration point or node-set.

For analyses of fracture propagation that exploit interface elements, cohesive-zone models are used almost exclusively. In this class of fracture models, a discrete relation is adopted between the interface tractions $\mathbf{t}_i$ and the relative displacements $\mathbf{v}$:

$$\mathbf{t}_i = \mathbf{t}_i(\mathbf{v}, \kappa) \tag{8}$$

with κ a history parameter. After linearization, necessary to use a tangential stiffness matrix in an incremental–iterative solution procedure, one obtains:

$$\dot{\mathbf{t}}_i = \mathbf{T}\dot{\mathbf{v}} \tag{9}$$

with $\mathbf{T}$ the material tangent stiffness matrix of the discrete traction-separation law:

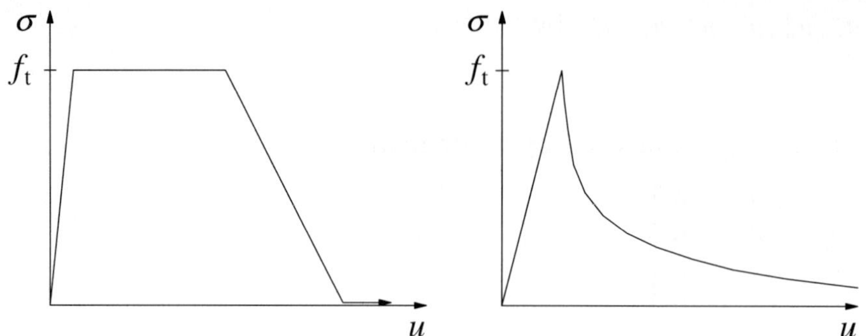

Figure 2. Stress-displacement curves for ductile separation (left) and quasi-brittle separation (right).

$$\mathbf{T} = \frac{\partial \mathbf{t}_i}{\partial \mathbf{v}} + \frac{\partial \mathbf{t}_i}{\partial \kappa}\ \frac{\partial \kappa}{\partial \mathbf{v}}. \tag{10}$$

A key element is the presence of a work of separation or fracture energy, $\mathcal{G}_c$, which governs crack growth and enters the interface constitutive relation (8) in addition to the tensile strength f_t. It is defined as the work needed to create a unit area of fully developed crack:

$$\mathcal{G}_c = \int_{v_n=0}^{\infty} \sigma\, \mathrm{d}v_n \tag{11}$$

with σ the stress across the *fracture process zone*. It thus equals the area under the decohesion curves as shown in Figure 2. Evidently, cohesive-zone models as defined above are equipped with an *internal length scale*, since the quotient $\mathcal{G}_c/E$, with E a stiffness modulus for the surrounding continuum, has the dimension of length.

Conventional interface elements have to be inserted in the finite element mesh at the beginning of the computation, and therefore, a finite stiffness must be assigned in the pre-cracking phase with at least the diagonal elements being nonzero. Prior to crack initiation, the stiffness matrix in the interface element therefore reads:

$$\mathbf{T} = \begin{bmatrix} d_n & 0 & 0 \\ 0 & d_s & 0 \\ 0 & 0 & d_t \end{bmatrix} \tag{12}$$

with d_n the stiffness normal to the interface and d_s and d_t the tangential stiffnesses.

With the material tangent stiffness matrix $\mathbf{T}$, the element tangent stiffness matrix can be derived in a straightforward fashion, starting from the weak form of the equilibrium equations, as:

$$\mathbf{K} = \int_{\Gamma_i} \mathbf{B}_i^{\mathrm{T}} \mathbf{T} \mathbf{B}_i\, \mathrm{d}\Gamma, \tag{13}$$

where the integration domain extends over the surface of the interface Γ_i. For comparison with methods that will be discussed in the remainder of this paper, we expand the stiffness matrix in the pre-cracking phase as, (cf. Schellekens and de Borst, 1992).

$$K = \begin{bmatrix} K_n & 0 & 0 \\ 0 & K_s & 0 \\ 0 & 0 & K_t \end{bmatrix} \tag{14}$$

with the submatrices K_π, $\pi = n, s, t$ defined as:

$$K_\pi = d_\pi \begin{bmatrix} h^T h & -h^T h \\ -h^T h & h^T h \end{bmatrix} \tag{15}$$

with d_π the (dummy) stiffnesses in the interface prior to crack initiation.

In cases where the direction of crack propagation is known *a priori*, interface elements equipped with cohesive-zone models have been used with considerable success. Figure 3 shows this for mixed-mode fracture in a single–edge notched concrete beam. In this example, the mesh has been designed such that the interface elements, which are equipped with a quasi–brittle cohesive-zone model, are exactly located at the position of the experimentally observed crack path (Rots, 1991).

Another example where the potential of cohesive-zone models can be exploited fully using conventional discrcte interface elements, is the analysis of delamination in layered composite materials (Allix and Ladevèze, 1992; Schellekens and de Borst, 1994) Since the propagation of delaminations is then restricted to the interfaces between the plies, inserting interface elements at these locations permits an exact simulation of the failure mode. Figure 4 shows an example of a uniaxially loaded laminate. Experimental and numerical results (which were obtained *before* the tests were carried out) show an excellent agreement, Figure 4, which gives the ultimate strain of the sample for different numbers of plies in the laminate (Schellekens and de Borst, 1994). A clear thickness (size) effect is obtained as a direct consequence of the inclusion of the fracture energy in the model.

To allow for a more arbitrary direction of crack propagation, Xu and Needleman (1994) have inserted interface elements equipped with a cohesive-zone model between *all* continuum elements. A related method, using remeshing, was proposed by Camacho and Ortiz (1996). Although such analyses provide much insight, they

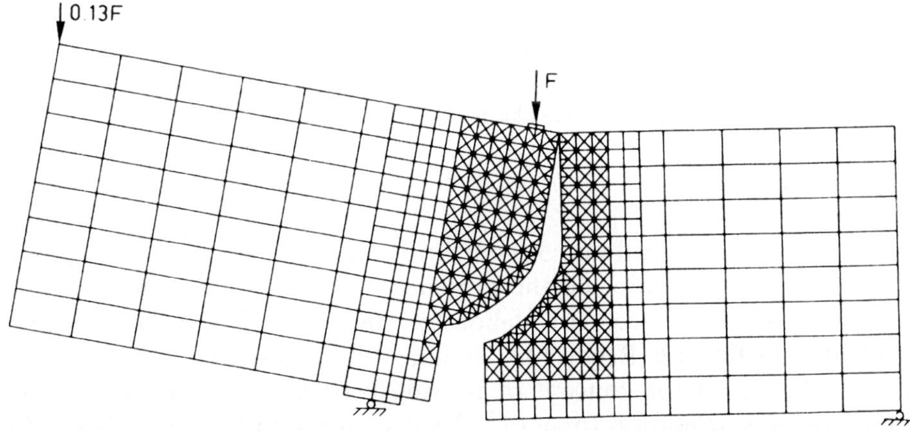

Figure 3. Deformed configuration of a single–edge notched beam that results from an analysis, where interface elements equipped with a quasi–brittle cohesive zone model have been placed *a priori* at the experimentally known crack path (Rots, 1991).

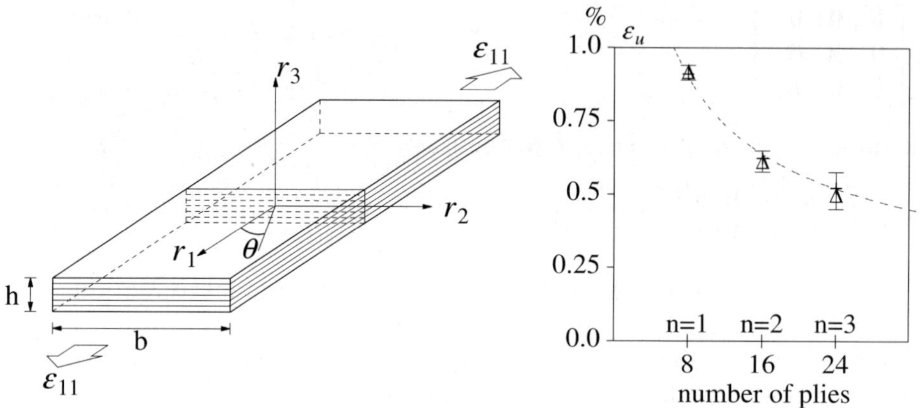

Figure 4. Left: Uniaxially loaded laminated strip. Right: Computed and experimentally determined values for the ultimate strain ϵ_u as a function of the number of plies (Schellekens and de Borst, 1994). Results are shown for laminates consisting of eight plies ($n=1$), 16 plies ($n=2$) and 24 plies ($n=3$). The triangles, which denote the numerical results, are well within the band of experimental results. The dashed line represents the inverse dependence of the ultimate strain on the laminate thickness.

suffer from a certain mesh bias, since the direction of crack propagation is not entirely free, but is restricted to interelement boundaries. This has been demonstrated in Tijssens et al., (2000) where the single–edge notched beam of Figure 3 has also been analysed, but now with a finite element model in which interface elements equipped with a quasi–brittle decohesion relation were inserted between all continuum elements, Figure 5.

As stipulated, conventional interface elements have to be inserted *a priori* in the finite element mesh. The undesired *elastic* deformations can be largely suppressed by choosing a high value for the stiffness d_n. However, the off-diagonal coupling terms of the submatrix $\mathbf{h}^{\mathrm{T}}\mathbf{h}$ that features in the stiffness matrix of the interface elements, cf. Equation (15), can lead to spurious traction oscillations in the pre-cracking phase for high-stiffness values (Schellekens and de Borst, 1992). This, in turn, may cause erroneous crack patterns. An example of an oscillatory traction pattern ahead of a notch is given in Figure 6. When analysing dynamic fracture, spurious wave reflections can occur as a result of the introduction of such artificially high-stiffness values prior to the onset of delamination. Moreover, the necessity to align the mesh with the potential planes of delamination, restricts the modelling capabilities.

3. Meshfree methods

In view of the above observations, discretization methods have been sought for that facilitate an improved resolution in the presence of stress singularities for *crack initiation* and that obviate the need for elaborate remeshing after *crack propagation*. In Nayroles et al., (1992) and Belytschko et al., (1994) interpolants have been introduced that are based on the concept of moving least squares. In such an interpolation scheme, the approximation function $u^h(\mathbf{x})$ is expressed as the inner product of a vector $\mathbf{p}(\mathbf{x})$ and a vector $\mathbf{a}(\mathbf{x})$,

$$u^h(\mathbf{x}) = \mathbf{p}^{\mathrm{T}}(\mathbf{x})\mathbf{a}(\mathbf{x}) \tag{16}$$

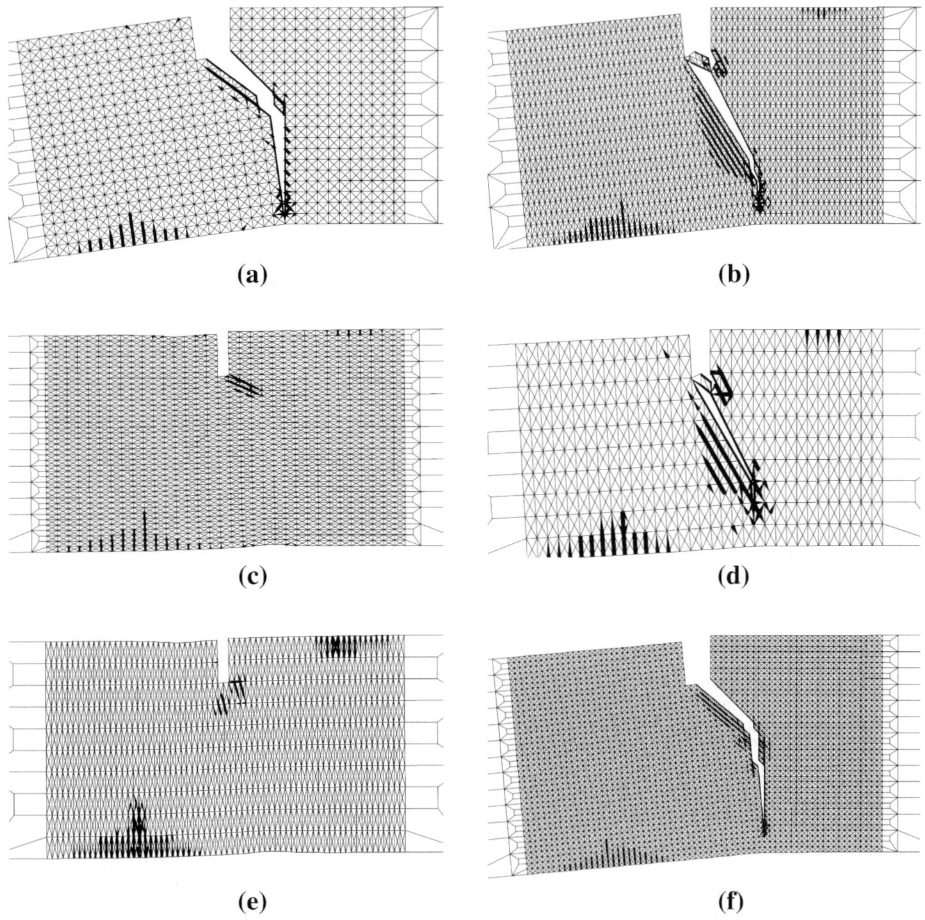

(a)   (b)

(c)   (d)

(e)   (f)

Figure 5. Crack patterns for different discretizations using interface elements between *all* solid elements. Only the part of the single–edge notched beam near the notch is shown (Tijssens et al., 2000).

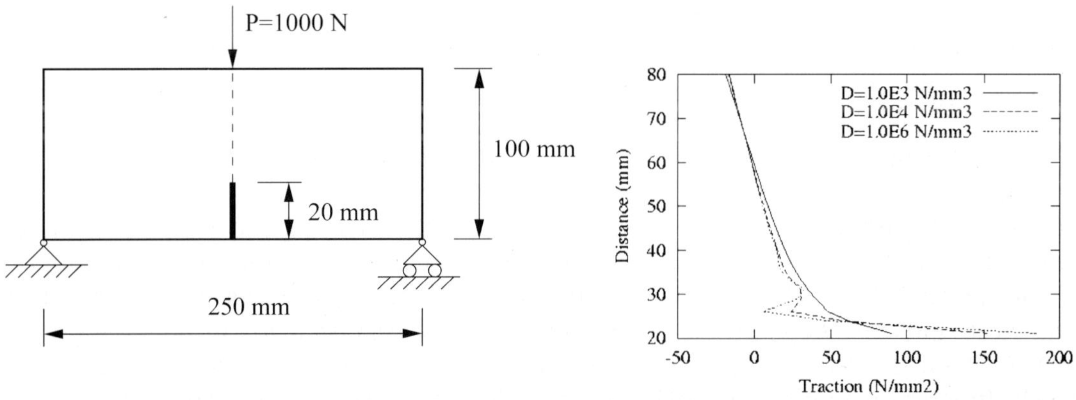

Figure 6. Left: Geometry of symmetric, notched three-point bending beam. Right: Traction profiles ahead of the notch using linear interface elements with Gauss integration. Results are shown for different values of the 'dummy' stiffness $D = d_n$ in the pre-cracking phase (Schellekens and de Borst, 1992).

in which $\mathbf{p}(\mathbf{x})$ contains basis terms that are functions of the coordinates $\mathbf{x}$. Normally, monomials such as $1, x, y, z, x^2, xy, \ldots$ are chosen, although also more sophisticated functions can be taken. The array $\mathbf{a}(\mathbf{x})$ contains the coefficients of the basis terms. In a moving least squares interpolation each node is assigned a weight function, which renders the coefficients $\mathbf{a}$ non-uniform. These weight functions w_i appear in the sum J^{mls} as:

$$J^{\mathrm{mls}} = \sum_{i=1}^{n} w_i(\mathbf{x}) \left(\mathbf{p}^{\mathrm{T}}(\mathbf{x}_i)\mathbf{a}(\mathbf{x}) - u_i \right)^2 \tag{17}$$

that has to be minimized with respect to $\mathbf{a}(\mathbf{x})$. Typical choices for the weight functions are Gaussian distributions or splines, whereby the domain of influence may take the shape of a disc (sphere) or rectangle (brick) in two (three)-dimensions. Elaboration of the stationarity requirement of J^{mls} with respect to $\mathbf{a}(\mathbf{x})$ gives:

$$\frac{\partial J^{\mathrm{mls}}}{\partial \mathbf{a}(\mathbf{x})} = \sum_{i=1}^{n} w_i(\mathbf{x}) \left[2\mathbf{p}(\mathbf{x}_i)\mathbf{p}^{\mathrm{T}}(\mathbf{x}_i)\mathbf{a}(\mathbf{x}) - 2\mathbf{p}(\mathbf{x}_i)u_i \right] = \mathbf{0}. \tag{18}$$

Thus, $\mathbf{a}(\mathbf{x})$ can be obtained as

$$\mathbf{a}(\mathbf{x}) = \mathbf{A}^{-1}(\mathbf{x})\mathbf{C}(\mathbf{x})\mathbf{u}, \tag{19}$$

where $\mathbf{u}$ contains all u_i, and

$$\mathbf{A}(\mathbf{x}) = \sum_{i=1}^{n} w_i(\mathbf{x})\mathbf{p}(\mathbf{x}_i)\mathbf{p}^{\mathrm{T}}(\mathbf{x}_i), \tag{20a}$$

$$\mathbf{C}(\mathbf{x}) = [w_1(\mathbf{x})\mathbf{p}(\mathbf{x}_1), w_2(\mathbf{x})\mathbf{p}(\mathbf{x}_2), \ldots, w_n(\mathbf{x})\mathbf{p}(\mathbf{x}_n)]. \tag{20b}$$

Equation (19) is substituted into Equation (16), which leads to:

$$u^h(\mathbf{x}) = \mathbf{p}^{\mathrm{T}}(\mathbf{x})\mathbf{A}^{-1}(\mathbf{x})\mathbf{C}(\mathbf{x})\mathbf{u}. \tag{21}$$

and the matrix that contains the shape functions $\mathbf{H}(\mathbf{x})$ can be identified as:

$$\mathbf{H}(\mathbf{x}) = \mathbf{p}^{\mathrm{T}}(\mathbf{x})\mathbf{A}^{-1}(\mathbf{x})\mathbf{C}(\mathbf{x}). \tag{22}$$

Shape functions which are generated in this manner, are usually not of a polynomial form, even though $\mathbf{p}(\mathbf{x})$ contains only polynomial terms. When moving least squares shape functions are used, the weight functions that are attached to each node determine the degree of continuity of the interpolants and the extent of the support of the node. A high degree of continuity can thus be achieved easily, so that steep stress gradients can be captured accurately, which is beneficial for the proper prediction of crack initiation. The fact that the extent of the support is determined by the weight function is in contrast with the finite element method. Consequently, there are no *elements* needed to define the support of a node. A mesh is not necessary and approximation methods based on moving least squares functions are often termed *meshfree* or *meshless methods*. However, the support of one node normally includes several other nodes and is therefore, less compact than with finite element methods and, therefore, leads to a larger bandwidth of the system of equations.

Discontinuous shape functions for use in fracture mechanics applications can be obtained in a straightforward manner by truncating the appropriate weight functions. Implicitly, the same procedure is applied as for nodes close to the boundary of the domain: the part of the domain of influence that falls outside the computational domain is simply not taken into account in the integration.

A different situation arises when the crack does not pass completely through a domain of influence, so that the crack tip lies inside the support. Figure 7 shows three different procedures how to truncate the domain of influence in the case of intersection by a crack, (see also Fleming et al., 1997). In the visibility criterion the connectivity between an integration point and a node is taken into account if and only if a line can be drawn that is not intersected by a nonconvex boundary. The resulting shape functions are not only discontinuous over the crack path, but also over the line that connects node and crack tip. Although convergent results can be obtained, the presence of discontinuities in the shape functions beyond the crack path is less desirable. As an alternative, it has been suggested to redefine the weight function. For instance, the line that connects node and integration point can be wrapped around the crack tip, in a similar way as light diffracts around sharp edges – hence the name diffraction criterion – or, the visibility criterion can be adapted such that some transparency is assigned to the part of the crack close to the crack tip. In either way, shape functions are obtained that are smooth and continuous for the part of the domain not intersected by the crack. In the see-through or continuous path criterion, truncation of the weight function only occurs when the domain of influence is completely intersected by the crack path. In this fashion, the effect of the crack propagation is delayed, and inaccuracies have been reported for this method (Fleming et al., 1997).

Another issue is the spatial resolution around the crack path and the crack tip. For linear-elastic fracture mechanics applications, the shape functions should properly capture the $r^{-1/2}$ – singularity near the crack tip in order to accurately compute the stress intensity factors. Apart from a nodal densification around the crack tip, this can be achieved by *locally* enriching the base vector **p** through the addition of the set

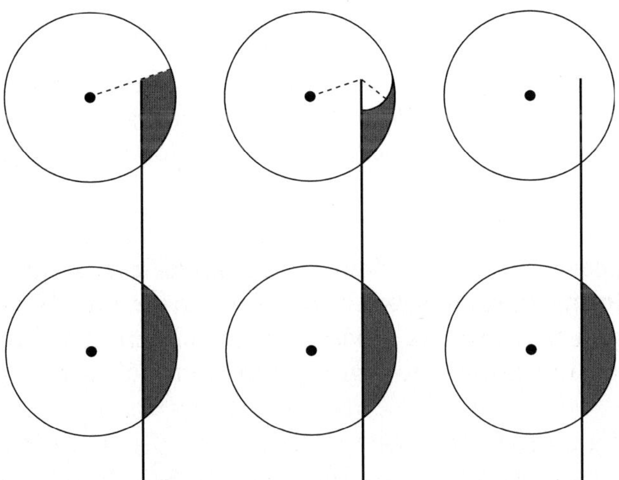

Figure 7. Domains of influence intersected by a crack or the crack tip: truncation of the weight function according to the visibility criterion (left), diffraction criterion (centre) and the see-through criterion (right)–shaded areas denote the neglected part of the domain of influence.

$$\boldsymbol{\psi} = \left(\sqrt{r}\cos(\theta/2), \quad \sqrt{r}\sin(\theta/2), \quad \sqrt{r}\sin(\theta/2)\sin(\theta), \quad \sqrt{r}\cos(\theta/2)\sin(\theta)\right)^{\mathrm{T}} \qquad (23)$$

where r is the distance from the crack tip and θ is measured from the current direction of crack propagation (Fleming et al., 1997). Alternatively, these functions can be added to the sum of Equation (17). This is possible by virtue of the fact that, similar to conventional finite element shape functions, shape functions obtained from a moving least squares approximation satisfy the partition-of-unity property, an issue to which we will return in Section 4.

As an example of a meshfree simulation of crack propagation using linear elastic fracture mechanics concepts, dynamic crack extension in a three-dimensional cube is considered (Krysl and Belytschko 1999). A penny-shaped crack is initially present, which extends internally in the cube. When the crack reaches the free surfaces of the cube, a full separation of the cube takes place. In Figure 8, the development of the crack is plotted for eight successive stages. It shows the ability of meshfree methods to describe not only cracks as line segments, but also as faces in three-dimensional analyses.

The high degree of continuity that is incorporated in meshfree methods makes them ideally suited for localization and failure analyses that adopt higher–order continuum models. Also, the flexibility is increased compared to conventional finite element methods, since there is no direct connectivity, which makes placing additional nodes in regions with high-strain gradients particularly simple. An example is offered in Figure 9 for a fourth-order gradient scalar damage model. This model can be summarized by the injective relation between the stress and strain tensors, $\boldsymbol{\sigma}$ and $\boldsymbol{\epsilon}$, respectively:

$$\boldsymbol{\sigma} = (1 - \omega)\mathbf{D}^{\mathrm{e}} : \boldsymbol{\epsilon} \qquad (24)$$

with ω a scalar damage variable, which grows from zero to one (at complete loss of integrity) and $\mathbf{D}^{\mathrm{e}}$ the fourth-order elastic stiffness tensor. This total stress–strain relation is complemented by a damage loading function f, which reads:

$$f = \tilde{\epsilon} - \kappa \qquad (25)$$

with the equivalent strain $\tilde{\epsilon}$ a scalar-valued function of the strain tensor, and κ a history variable. The damage loading function f and the rate of the history variable, $\dot{\kappa}$, have to satisfy the discrete Kuhn–Tucker loading–unloading conditions

$$f \leq 0, \quad \dot{\kappa} \geq 0, \quad \dot{\kappa} f = 0. \qquad (26)$$

The history parameter κ starts at a damage threshold level κ_i and is updated by the requirement that during damage growth $f = 0$. Damage growth occurs according to an evolution law such that $\omega = \omega(\kappa)$, which can be determined from a uniaxial test.

In a nonlocal generalization, the loading function (25) is replaced by

$$f = \bar{\epsilon} - \kappa, \qquad (27)$$

where the nonlocal equivalent strain $\bar{\epsilon}$ follows from the solution of the partial differential equation

$$\bar{\epsilon} - c_1 \nabla^2 \bar{\epsilon} - c_2 \nabla^4 \bar{\epsilon} = \tilde{\epsilon} \qquad (28)$$

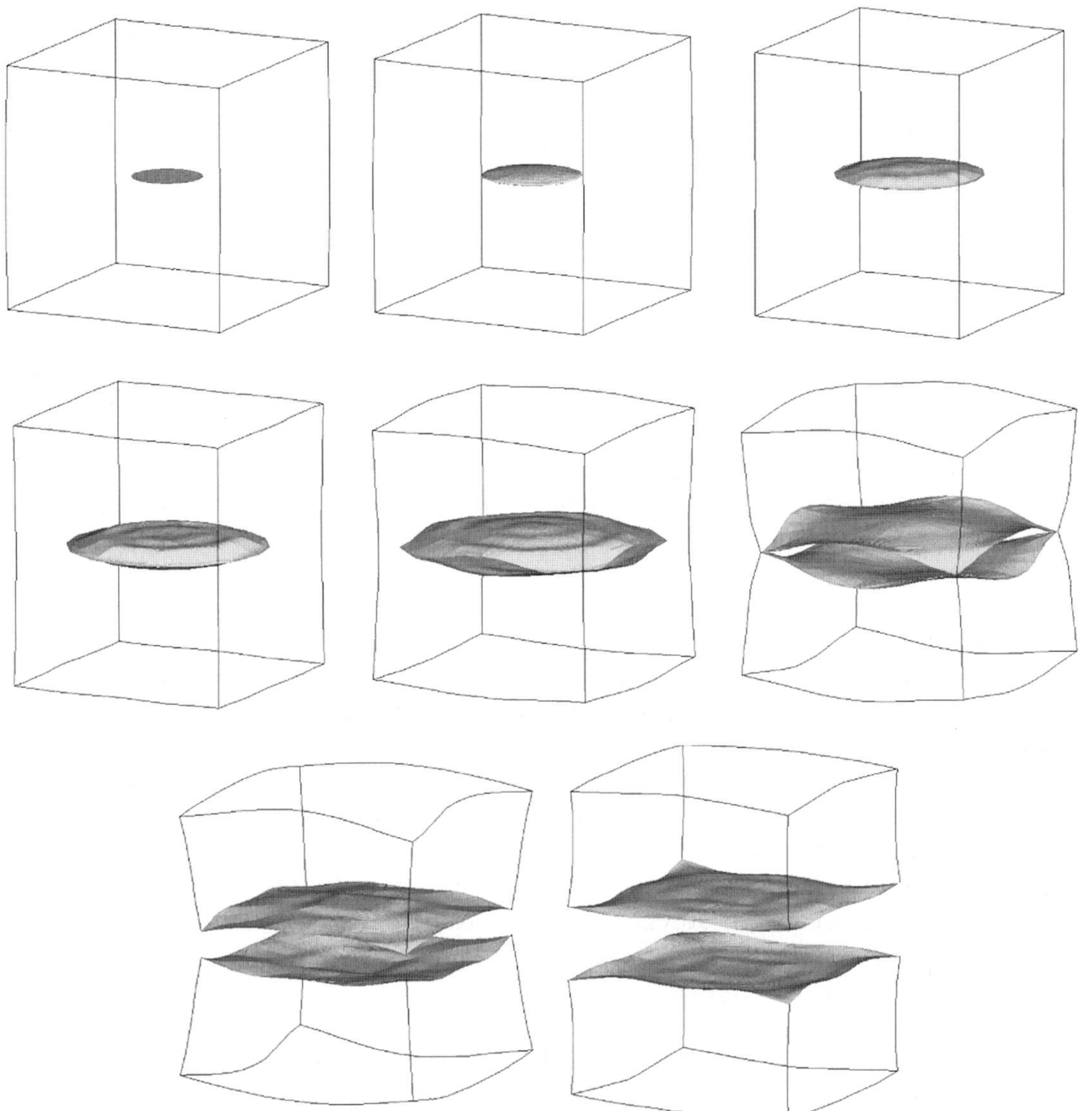

Figure 8. Cube with centered penny-shape crack – propagation of the crack towards the free surfaces of the specimen (Krysl and Belytschko, 1999).

with c_1 and c_2 two gradient constants, which is assumed to hold on the entire domain. Evidently, even after order reduction by partial integration, $\mathcal{C}^1$-continuous shape functions are necesary for the interpolation of the nonlocal strain $\bar{\epsilon}$, with all the computational inconveniences that come to it when finite elements are employed. Here, meshfree methods offer a distinct advantage, since they can be easily constructed such that they incorporate $\mathcal{C}^\infty$-continuous shape functions. In Figure 9, the element–free Galerkin method has been used to solve the damage evolution that is described by the fourth–order gradient scalar damage model of Equations (24), (26)–(28) to predict the damage evolution in a three–point bending beam (Askes et al., 2000). In both cases, a quadratic convergence behaviour was obtained when using a properly linearized tangent stiffness matrix. Interestingly, the differences between the fourth–order and the second–order ($c_2 = 0$) gradient damage models are almost negligible.

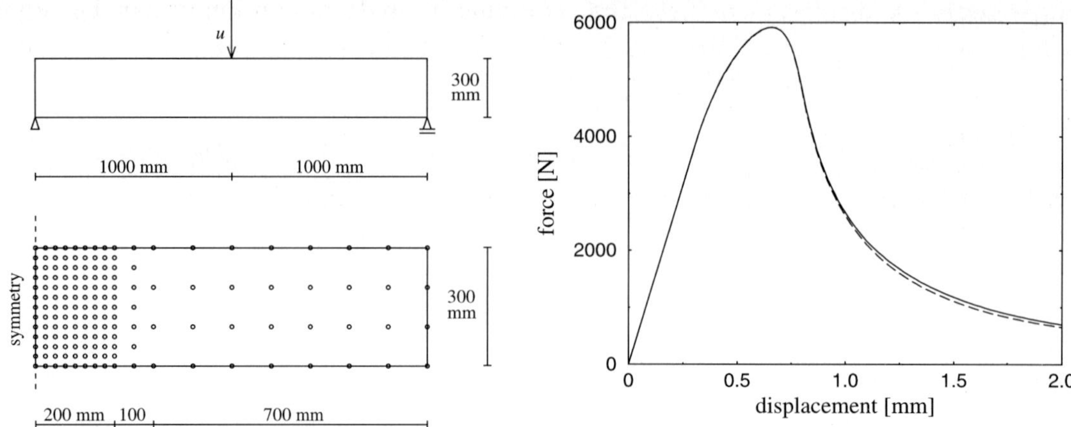

Figure 9. Left: Three–point bending beam and node distribution (for a symmetric half of the beam). Right: Load–displacement curves for three–point bending beam. Comparison between the second-order implicit gradient damage model (dashed line) and the fourth–order implicit gradient damage model (solid line) (Askes et al., 2000).

The size of the support of a node relative to the nodal spacing determines the properties of meshfree methods. When the support is made equal to the nodal spacing, shape functions as obtained in meshfree methods can become identical to finite element shape functions, thus showing that meshfree methods *encompass* finite element methods as a subclass. On the other hand, larger supports lead to shape functions in meshfree methods that are similar to higher–order polynomials, even if the base vector $\mathbf{p}(\mathbf{x})$ contains only constant and linear terms. Finally, it is noted that, although meshfree methods can straightforwardly accommodate cohesive–zone models, such computations seem absent in the literature.

4. The partition-of-unity approach

A unifying approach to domain–based discretization methods for crack initiation and propagation is enabled by the partition-of-unity concept (Babuska and Melenk, 1997). A collection of functions ϕ_ι, associated with nodes ι, form a partition of unity if $\sum_{\iota=1}^{n} \phi_\iota(\mathbf{x}) = 1$ with n the number of discrete nodal points. For a set of functions ϕ_ι that satisfy this property, a field u can be interpolated as follows:

$$u(\mathbf{x}) = \sum_{\iota=1}^{n} \phi_\iota(\mathbf{x}) \left(\bar{a}_\iota + \sum_{J=1}^{m} \psi_J(\mathbf{x}) \tilde{a}_{\iota J} \right) \tag{29}$$

with $\bar{a}_\iota$ the 'regular' nodal degrees-of-freedom, $\psi_J(\mathbf{x})$ enhanced basis terms, and $\tilde{a}_{\iota J}$ the additional degrees-of-freedom at node ι, which represent the amplitudes of the Jth enhanced basis term $\psi_J(\mathbf{x})$. In conventional finite element notation we can thus interpolate a displacement field as:

$$\mathbf{u} = \mathbf{H}(\bar{\mathbf{a}} + \mathbf{\Psi}\tilde{\mathbf{a}}), \tag{30}$$

where $\mathbf{H}$ contains the standard shape functions, $\mathbf{\Psi}$ the enhanced basis terms and $\bar{\mathbf{a}}$ and $\tilde{\mathbf{a}}$ collect the conventional and the additional nodal degrees-of-freedom, respectively. A displacement field that contains a single discontinuity can be represented by choosing, (e.g., Belytschko and Black, 1999; Moes et al., 1999; Wells and Sluys, 2001; Wells et al., 2002):

$$\boldsymbol{\Psi} = \mathcal{H}_{\Gamma_i}\mathbf{I} \tag{31}$$

with $\mathcal{H}_{\Gamma_i}$ the Heaviside function. Substitution into Equation (30) gives

$$\mathbf{u} = \underbrace{\mathbf{H}\bar{\mathbf{a}}}_{\bar{\mathbf{u}}} + \mathcal{H}_{\Gamma_i}\underbrace{\mathbf{H}\tilde{\mathbf{a}}}_{\tilde{\mathbf{u}}}. \tag{32}$$

Identifying the continuous fields $\bar{\mathbf{u}} = \mathbf{H}\bar{\mathbf{a}}$ and $\tilde{\mathbf{u}} = \mathbf{H}\tilde{\mathbf{a}}$ we observe that Equation (32) exactly describes a displacement field that is crossed by a single discontinuity, but is otherwise continuous. Accordingly, the partition-of-unity property of finite element shape functions can be used in a straightforward fashion to incorporate discontinuities in a manner that preserves their discontinuous character.

We take the balance of momentum

$$\nabla \cdot \boldsymbol{\sigma} + \rho\mathbf{g} = \mathbf{0} \tag{33}$$

as point of departure and multiply this identity by test functions $\boldsymbol{w}$, taking them from the same space as the trial functions for $\mathbf{u}$,

$$\boldsymbol{w} = \bar{\boldsymbol{w}} + \mathcal{H}_{\Gamma_i}\tilde{\boldsymbol{w}}. \tag{34}$$

Applying the divergence theorem and requiring that this identity holds for arbitrary $\bar{\boldsymbol{w}}$ and $\tilde{\boldsymbol{w}}$ yields the following set of coupled equations:

$$\int_{\Omega} \nabla^{\mathrm{sym}}\bar{\boldsymbol{w}} : \boldsymbol{\sigma}\,\mathrm{d}\Omega = \int_{\Omega} \bar{\boldsymbol{w}} \cdot \rho\mathbf{g}\,\mathrm{d}\Omega + \int_{\Gamma} \bar{\boldsymbol{w}} \cdot \mathbf{t}\,\mathrm{d}\Gamma, \tag{35a}$$

$$\int_{\Omega^+} \nabla^{\mathrm{sym}}\tilde{\boldsymbol{w}} : \boldsymbol{\sigma}\,\mathrm{d}\Omega + \int_{\Gamma_i} \tilde{\boldsymbol{w}} \cdot \mathbf{t}_i\,\mathrm{d}\Gamma = \int_{\Omega^+} \tilde{\boldsymbol{w}} \cdot \rho\mathbf{g}\,\mathrm{d}\Omega + \int_{\Gamma} \mathcal{H}_{\Gamma_i}\tilde{\boldsymbol{w}} \cdot \mathbf{t}\,\mathrm{d}\Gamma, \tag{35b}$$

where in the volume integrals the Heaviside function has been eliminated by a change of the integration domain from Ω to Ω^+. Interpolating the trial and the test functions in the same space,

$$\begin{aligned} \bar{\mathbf{u}} &= \mathbf{H}\bar{\mathbf{a}}, & \tilde{\mathbf{u}} &= \mathbf{H}\tilde{\mathbf{a}}, \\ \bar{\boldsymbol{w}} &= \mathbf{H}\bar{\mathbf{w}}, & \tilde{\boldsymbol{w}} &= \mathbf{H}\tilde{\mathbf{w}} \end{aligned} \tag{36}$$

and requiring that the resulting equations must hold for any admissible $\bar{\mathbf{w}}$ and $\tilde{\mathbf{w}}$, we obtain the discrete format:

$$\int_{\Omega} \mathbf{B}^{\mathrm{T}}\boldsymbol{\sigma}\,\mathrm{d}\Omega = \int_{\Omega} \rho\mathbf{B}^{\mathrm{T}}\mathbf{g}\,\mathrm{d}\Omega + \int_{\Gamma} \mathbf{H}^{\mathrm{T}}\mathbf{t}\,\mathrm{d}\Gamma, \tag{37}$$

$$\int_{\Omega^+} \mathbf{B}^{\mathrm{T}}\boldsymbol{\sigma}\,\mathrm{d}\Omega + \int_{\Gamma_i} \mathbf{H}^{\mathrm{T}}\mathbf{t}_i\,\mathrm{d}\Gamma = \int_{\Omega^+} \rho\mathbf{B}^{\mathrm{T}}\mathbf{g}\,\mathrm{d}\Omega + \int_{\Gamma} \mathcal{H}_{\Gamma_i}\mathbf{H}^{\mathrm{T}}\mathbf{t}\,\mathrm{d}\Gamma, \tag{38}$$

where $\mathbf{B} = \nabla\mathbf{H}$. After linearization, the following matrix–vector equation is obtained:

$$\begin{bmatrix} \mathbf{K}_{\bar{a}\bar{a}} & \mathbf{K}_{\bar{a}\tilde{a}} \\ \mathbf{K}_{\tilde{a}\bar{a}} & \mathbf{K}_{\tilde{a}\tilde{a}} \end{bmatrix} \begin{pmatrix} \mathrm{d}\bar{\mathbf{a}} \\ \mathrm{d}\tilde{\mathbf{a}} \end{pmatrix} = \begin{pmatrix} \mathbf{f}_{\bar{a}}^{\mathrm{ext}} - \mathbf{f}_{\bar{a}}^{\mathrm{int}} \\ \mathbf{f}_{\tilde{a}}^{\mathrm{ext}} - \mathbf{f}_{\tilde{a}}^{\mathrm{int}} \end{pmatrix} \tag{39}$$

with $\mathbf{f}_{\bar{a}}^{\mathrm{int}}, \mathbf{f}_{\tilde{a}}^{\mathrm{int}}$ given on the left–hand sides of Equations (35a) and (35b), $\mathbf{f}_{\bar{a}}^{\mathrm{ext}}, \mathbf{f}_{\tilde{a}}^{\mathrm{ext}}$ given on the right–hand sides of Equations (35a) and (35b) and

$$\mathbf{K}_{\bar{a}\bar{a}} = \int_{\Omega} \mathbf{B}^{\mathrm{T}} \mathbf{D} \mathbf{B} \, \mathrm{d}\Omega, \tag{40}$$

$$\mathbf{K}_{\bar{a}\tilde{a}} = \int_{\Omega^+} \mathbf{B}^{\mathrm{T}} \mathbf{D} \mathbf{B} \, \mathrm{d}\Omega, \tag{41}$$

$$\mathbf{K}_{\tilde{a}\tilde{a}} = \int_{\Omega^+} \mathbf{B}^{\mathrm{T}} \mathbf{D} \mathbf{B} \, \mathrm{d}\Omega + \int_{\Gamma_i} \mathbf{H}^{\mathrm{T}} \mathbf{T} \mathbf{H} \, \mathrm{d}\Gamma. \tag{42}$$

If the material tangential stiffness matrices of the bulk and the interface, $\mathbf{D}$ and $\mathbf{T}$, respectively, are symmetric, the total tangential stiffness matrix remains symmetric. It is emphasized that in this concept, the additional degrees-of-freedom cannot be condensed at element level, because it is node-oriented and not element-oriented. It is this property, which makes it possible to represent a discontinuity such that it is continuous at interelement boundaries.

When the discontinuity coincides with a side of the element, the traditional interface element formulation is retrieved. For this, we expand the term in $\mathbf{K}_{\tilde{a}\tilde{a}}$ which relates to the discontinuity as

$$\int_{\Gamma_i} \mathbf{H}^{\mathrm{T}} \mathbf{T} \mathbf{H} \, \mathrm{d}\Gamma = \begin{bmatrix} \mathbf{K}_n & \mathbf{0} & \mathbf{0} \\ \mathbf{0} & \mathbf{K}_s & \mathbf{0} \\ \mathbf{0} & \mathbf{0} & \mathbf{K}_t \end{bmatrix} \tag{43}$$

with $\mathbf{K}_{\pi} = d_{\pi} \mathbf{h}^{\mathrm{T}} \mathbf{h}$ (Simone, 2004) which closely resembles Equations (14) and (15). Defining the sum of the nodal displacements $\bar{\mathbf{a}}$ and $\tilde{\mathbf{a}}$ as primary variable $\mathbf{a}$ on the + side of the interface and setting $\mathbf{a} = \bar{\mathbf{a}}$ at the – side and rearranging then leads to the standard interface formulation. However, even though formally the matrices can coincide for the partition-of-unity-based method and for the conventional interface formulation, the former does *not* share the disadvantages of oscillating traction profiles and spurious wave reflections prior to the onset of decohesion, simply because the partition-of-unity concept permits the placement of cohesive surfaces in the mesh only at onset of decohesion, thereby by–passing the whole problem of having to assign a high-dummy stiffness to the interface prior to crack initiation.

To illustrate the potential of the method for linear elastic fracture mechanics problems, we consider the example of a block that contains a stationary, kinked crack. When utilizing linear elastic fracture mechanics, the nodes for which the support is crossed by a crack are enhanced with a Heaviside function and those for which the support contains a crack tip are enriched with near–tip terms as in Equation (23), similar to the procedure used in meshfree methods.

The problem has been analysed with a relatively coarse with a uniform element size, and with a finer, reference mesh, with the crack explicitly built in the mesh and refined around the crack tip. The bottom edge of the block is restrained and a uniformly distributed load p is applied to the top edge. For further details, the reader is referred to Askes et al. (2003). The deformed meshes that result from the computations are shown in Figure 10, which also gives the contours of the normal stress in the y-direction. The general form of the contour plot is the same for both computations. The resolution of the contours for the enhanced mesh is smaller, since stresses have been post–processed at nodal points only. This has a smoothing effect, with the stress singularity obvious only when the crack tip lies close to an element node. For completeness, Figure 11 shows the nodes which are enriched.

The partition-of-unity property of finite element shape functions is a powerful method to introduce discontinuities in continuum finite elements. Using the interpolation of Equation (32) the relative displacement at the discontinuity Γ_i is obtained as:

$$\mathbf{v} = \tilde{\mathbf{u}}\,|_{\mathbf{x} \in \Gamma_i} \,. \tag{44}$$

When using a cohesive-zone model, the tractions at the discontinuity can directly be derived from Equation (8). A key feature of the method is the possibility to extend a crack during the calculation in an arbitrary direction, independent of the structure of the underlying finite element mesh.

The objectivity of computations with this method for cohesive-zone formulations with respect to mesh refinement is now demonstrated for a three–point bending beam of unit thickness. The beam is loaded quasi-statically by means of an imposed displacement at the centre of the beam on the top edge. The geometric and material data can be found in Wells and Sluys (2001). Figure 12 show the crack after propagation throughout the beam. Two meshes are shown, one with 523 elements and the other with 850 elements. Clearly, in both cases the crack propagates from the centre at the bottom of the beam in a straight line towards the loading point, and is

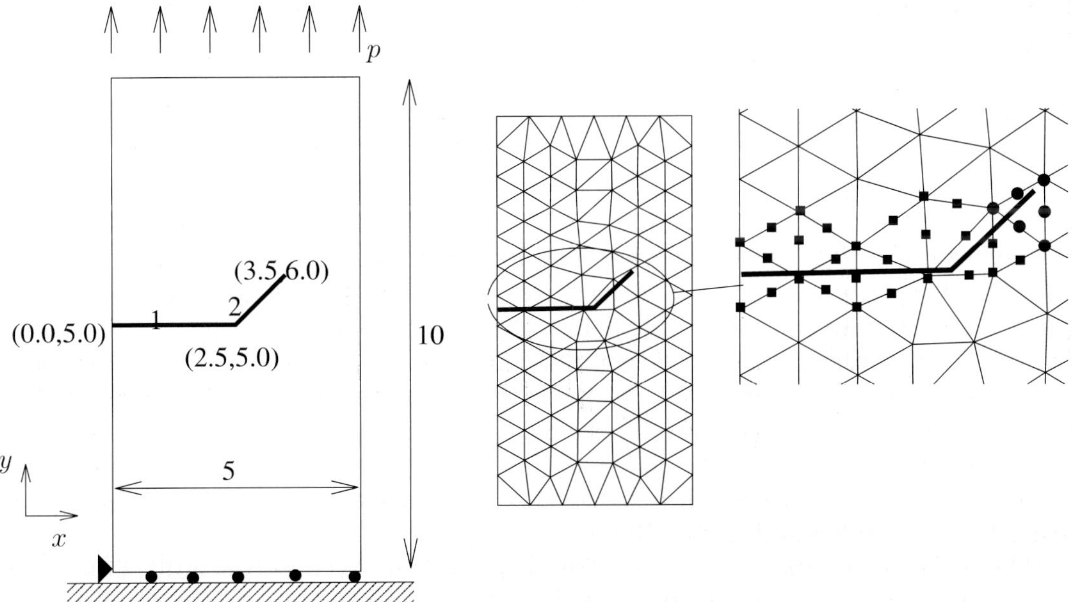

Figure 10. Geometry of a block that contains a stationary, kinked crack (left) and nodes enriched with jump functions (squares), or with crack–tip functions (circles) (right).

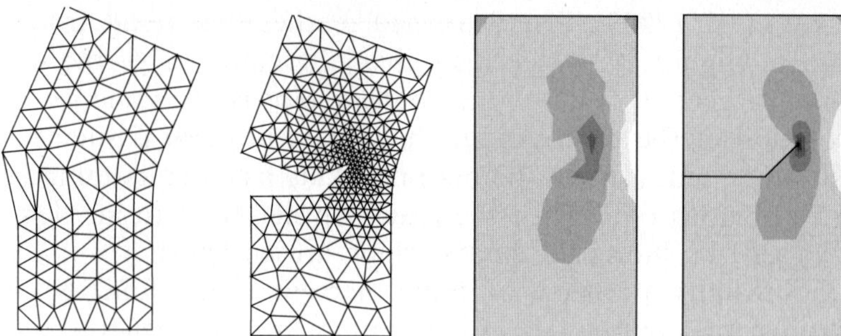

Figure 11. Deformed configurations for enhanced mesh (left) and reference mesh with explicit discontinuity (centre left), and contour plots of σ_{yy} for enhanced (centre right) and reference mesh with explicit discontinuity (right) (Askes et al., 2003).

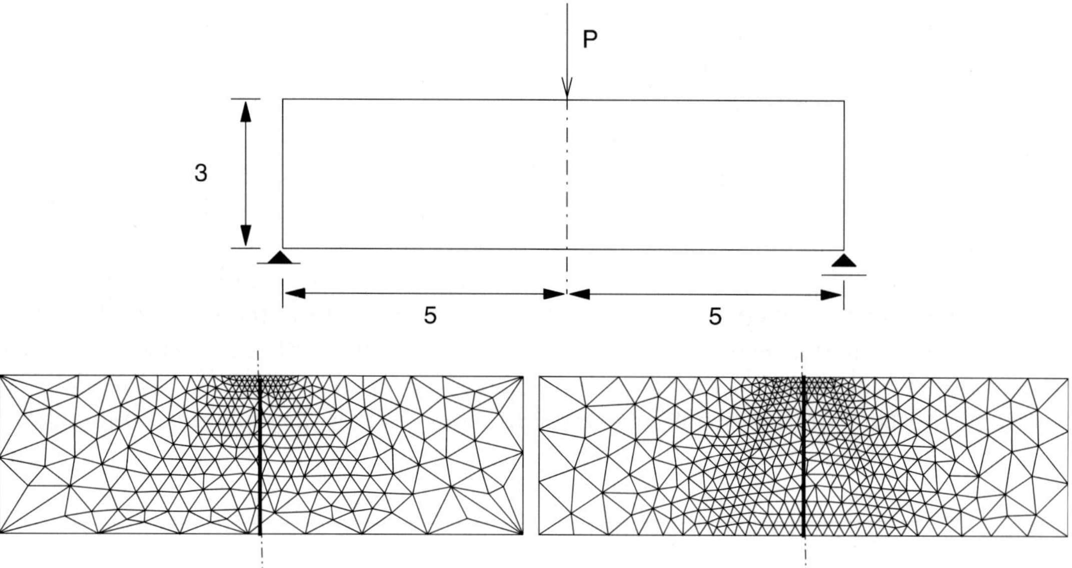

Figure 12. Crack path at the final stage of loading for the coarse mesh (523 elements) and the fine mesh (850 elements) (Wells and Sluys, 2001).

not influenced by the mesh structure. The load–displacement responses of Figure 13 confirm objectivity with respect to mesh refinement. From the curve for the coarser mesh the energy dissipation is calculated as 0.308 J, which only slightly exceeds the fracture energy multiplied by the depth and the thickness of the beam (0.3 J). Some small irregularities are observed in the load–displacement curve, especially for the coarser mesh. These are caused by the fact that in this implementation a cohesive zone is inserted entirely in one element when the tensile strength has been exceeded. A more sophisticated approach is to advance the crack only in a part of the element using level sets (Gravouil et al., 2002).

The requirement that the crack path is not biased by the direction of the mesh lines is normally even more demanding than the requirement of objectivity with respect to mesh refinement. Figure 14 shows that the approach also fully satisfies this requirement, since the numerically predicted crack path of the single–edge notched

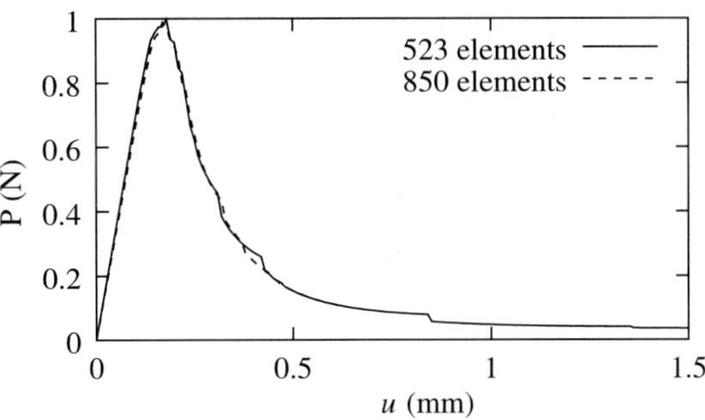

Figure 13. Load–displacement diagrams for the analysis of the symmetrically loaded beam using both meshes (Wells and Sluys, 2001).

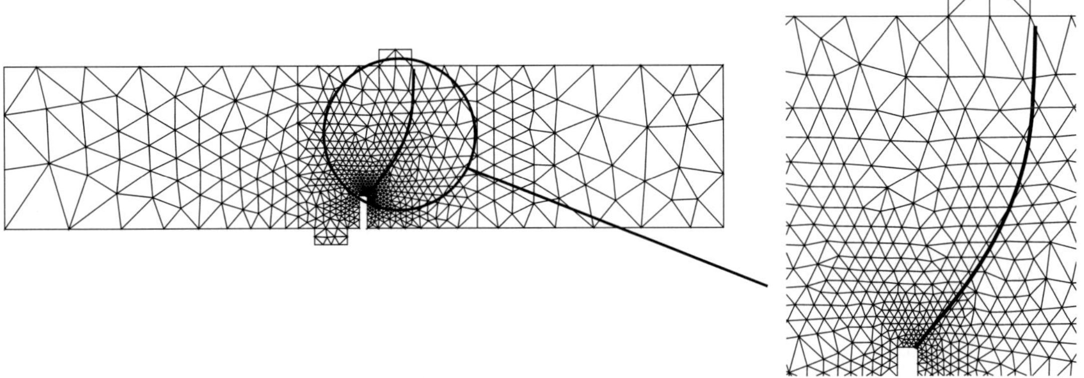

Figure 14. Crack path that results from the analysis of the single–edge notched beam using the partition-of-unity property of finite element shape functions (Wells and Sluys, 2001).

beam analysed before in Figures 3 and 5 is in excellent agreement with experimental observations.

In the above examples, a single, continuous cohesive crack growth was simulated. Crack propagation in heterogeneous materials and also fast crack growth in more homogeneous materials is often characterized by the nucleation of microcracks at several locations, which can grow, branch and eventually link up to form macroscopically observable cracks. To accommodate this observation, the concept of cohesive segments has been proposed in Remmers et al. (2003). Exploiting the partition-of-unity property of finite element shape functions, crack segments equipped with a cohesive law are placed over a patch of elements when a loading criterion is met at an integration point. Since the cohesive segments can grow and eventually coalesce, they can also simulate a single, dominant crack.

5. Discontinuous Galerkin methods

Discontinuous Galerkin methods have classically been employed for the computation of fluid flow (e.g. Cockburn, 2004). More recently, attention has been given to their potential use in solid mechanics, and especially for problems involving cracks (Mergheim et al., 2004), or for constitutive models that incorporate spatial gradients (Wells et al., 2004) such as gradient plasticity or gradient damage (e.g., de Borst and Gutiérrez, 1999; Askes et al., 2000). In the latter case, the fact that discontinuous Galerkin methods do not require interelement continuity *a priori*, by–passes the requirement of C^1–continuity on the damage or plastic multiplier field, which plague the implementation of many gradient models in a continuous Galerkin finite element method. In the former case, use of a discontinuous Galerkin formalism can be an alternative way to avoid traction oscillations in the pre-cracking phase (Mergheim et al., 2004).

For a discussion on the application of spatially discontinuous Galerkin to fracture it suffices by dividing the domain in two subdomains, Ω^- and Ω^+, separated by an interface Γ_i. In a standard manner, the balance of momentum (33) is multiplied by a test function w, and after application of the divergence theorem, we obtain:

$$\int_{\Omega/\Gamma_i} \nabla^{\mathrm{sym}} w : \sigma \, d\Omega - \int_{\Gamma_i} w^+ \cdot t_i^+ \, d\Gamma - \int_{\Gamma_i} w^- \cdot t_i^- \, d\Gamma = \int_{\Omega/\Gamma_i} w \cdot \rho g \, d\Omega$$

$$+ \int_{\Gamma} w \cdot t \, d\Gamma, \qquad (45)$$

where the surface (line) integral on the external boundary Γ has been explicitly separated from that on the interface Γ_i. Prior to crack initiation, continuity of displacements and tractions must be enforced along Γ_i, at least in an approximate sense:

$$\mathbf{u}^+ - \mathbf{u}^- = \mathbf{0}, \qquad (46a)$$

$$\mathbf{t}_i^+ + \mathbf{t}_i^- = \mathbf{0} \qquad (46b)$$

with $\mathbf{t}_i^+ = \mathbf{n}_{\Gamma_i}^+ \cdot \sigma^+$ and $\mathbf{t}_i^- = \mathbf{n}_{\Gamma_i}^- \cdot \sigma^-$. Assuming small displacement gradients, we can set $\mathbf{n}_{\Gamma_i} = \mathbf{n}_{\Gamma_i}^+ = -\mathbf{n}_{\Gamma_i}^-$, so that the expressions for the interface tractions reduce to $\mathbf{t}_i^+ = \mathbf{n}_{\Gamma_i} \cdot \sigma^+$ and $\mathbf{t}_i^- = -\mathbf{n}_{\Gamma_i} \cdot \sigma^-$.

A classical procedure to enforce conditions (46a) and (46b) is to use Lagrange multipliers. Then,

$$\lambda = \mathbf{t}_i^+ = -\mathbf{t}_i^- \qquad (47)$$

along Γ_i, and Equation (45) transforms into:

$$\int_{\Omega/\Gamma_i} \nabla^{\mathrm{sym}} w : \sigma \, d\Omega - \int_{\Gamma_i} (w^+ - w^-) \cdot \lambda \, d\Gamma = \int_{\Omega/\Gamma_i} w \cdot \rho g \, d\Omega + \int_{\Gamma} w \cdot t \, d\Gamma \qquad (48)$$

augmented with:

$$\int_{\Gamma_i} z \cdot (\mathbf{u}^+ - \mathbf{u}^-) d\Gamma = 0, \qquad (49)$$

where z is the test function for the Lagrange multiplier field λ. After discretization, Equations (48) and (49) result in a set of algebraic equations that are of a standard mixed format and therefore, give rise to difficulties when using solvers without pivoting. For this reason, alternative expressions are often sought, in which λ is directly expressed in terms of the interface tractions $\mathbf{t}_i^-$ and $\mathbf{t}_i^+$. One such possibility is to enforce Equation (46b) pointwise, so that Equation (47) is replaced by:

$$\lambda = -\mathbf{t}_i \qquad (50)$$

and one obtains:

$$\int_{\Omega/\Gamma_i} \nabla^{\text{sym}} \boldsymbol{w} : \boldsymbol{\sigma} \, d\Omega + \int_{\Gamma_i} (\boldsymbol{w}^+ - \boldsymbol{w}^-) \cdot \mathbf{t}_i \, d\Gamma = \int_{\Omega/\Gamma_i} \boldsymbol{w} \cdot \rho \mathbf{g} \, d\Omega + \int_{\Gamma} \boldsymbol{w} \cdot \mathbf{t} \, d\Gamma. \tag{51}$$

With aid of relation (6) between the relative displacements $\mathbf{v} = \mathbf{u}^+ - \mathbf{u}^-$ and the nodal displacements at both sides of the interface Γ_i, and the interface traction–relative displacement relation (9), the second term on the left–hand side can be elaborated in a discrete format as:

$$\int_{\Gamma_i} (\boldsymbol{w}^+ - \boldsymbol{w}^-) \cdot \mathbf{t}_i \, d\Gamma = \mathbf{w}^{\text{T}} \left(\int_{\Gamma_i} \mathbf{B}_i^{\text{T}} \mathbf{T} \mathbf{B}_i \, d\Gamma \right) \mathbf{a}, \tag{52}$$

which, not surprisingly, has exactly the same format as obtained for a conventional interface element.

Another possibility for the replacement of λ by an explicit function of the tractions is to take the average of the stresses at both sides of the interface:

$$\lambda = \frac{1}{2} \mathbf{n}_{\Gamma_i} \cdot (\boldsymbol{\sigma}^+ + \boldsymbol{\sigma}^-). \tag{53}$$

The surface integrals for the interface in Equation (48) can now be reworked as:

$$\int_{\Gamma_i} (\boldsymbol{w}^+ - \boldsymbol{w}^-) \cdot \lambda \, d\Gamma = \int_{\Gamma_i} \frac{1}{2} (\boldsymbol{w}^+ - \boldsymbol{w}^-) \cdot \mathbf{n}_{\Gamma_i} \cdot (\boldsymbol{\sigma}^+ + \boldsymbol{\sigma}^-) d\Gamma. \tag{54}$$

To ensure a proper conditioning of the discretized equations, one has to add Equation (49), so that the modified form of Equation (45) finally becomes:

$$\int_{\Omega/\Gamma_i} \nabla^{\text{sym}} \boldsymbol{w} : \boldsymbol{\sigma} \, d\Omega - \int_{\Gamma_i} \frac{1}{2} (\boldsymbol{w}^+ - \boldsymbol{w}^-) \cdot \mathbf{n}_{\Gamma_i} \cdot (\boldsymbol{\sigma}^+ + \boldsymbol{\sigma}^-) d\Gamma -$$
$$\alpha \int_{\Gamma_i} \frac{1}{2} (\nabla^{\text{sym}} \boldsymbol{w}^+ + \nabla^{\text{sym}} \boldsymbol{w}^-) : \mathbf{D} \cdot \mathbf{n}_{\Gamma_i} \cdot (\mathbf{u}^+ - \mathbf{u}^-) d\Gamma = \int_{\Omega/\Gamma_i} \boldsymbol{w} \cdot \rho \mathbf{g} \, d\Omega + \int_{\Gamma} \boldsymbol{w} \cdot \mathbf{t} \, d\Gamma. \tag{55}$$

To ensure symmetry, $\alpha = 1$, but then a diffusionlike term, $\int_{\Gamma_i} \tau (\boldsymbol{w}^+ - \boldsymbol{w}^-) \cdot (\mathbf{u}^+ - \mathbf{u}^-) d\Gamma$ has to be added to ensure numerical stability (Nitsche, 1970). The numerical parameter $\tau = \mathcal{O}(|k|/h)$, with $|k|$ a suitable norm of the diffusion like matrix that results from elaborating this term and h a measure of the grid density. For the unsymmetric choice $\alpha = -1$, addition of a diffusion-like term may not be necessary (Baumann and Oden, 1999).

With a standard interpolation on both Ω^- and Ω^+ and requiring that the resulting equations hold for any admissible $\mathbf{w}$, we obtain the discrete format:

$$\int_{\Omega^-} \mathbf{B}^{\text{T}} \boldsymbol{\sigma} \, d\Omega + \int_{\Gamma_i} \frac{1}{2} \mathbf{N}^{\text{T}} \mathbf{n}_{\Gamma_i}^{\text{T}} (\boldsymbol{\sigma}^+ + \boldsymbol{\sigma}^-) d\Gamma - \alpha \int_{\Gamma_i} \frac{1}{2} \mathbf{B}^{\text{T}} \mathbf{D} \mathbf{n}_{\Gamma_i}^{\text{T}} (\mathbf{u}^+ - \mathbf{u}^-) d\Gamma$$
$$= \int_{\Omega^-} \mathbf{B}^{\text{T}} \rho \mathbf{g} \, d\Omega + \int_{\Gamma} \mathbf{N}^{\text{T}} \mathbf{t} \, d\Gamma,$$
$$\int_{\Omega^+} \mathbf{B}^{\text{T}} \boldsymbol{\sigma} \, d\Omega - \int_{\Gamma_i} \frac{1}{2} \mathbf{N}^{\text{T}} \mathbf{n}_{\Gamma_i}^{\text{T}} (\boldsymbol{\sigma}^+ + \boldsymbol{\sigma}^-) d\Gamma - \alpha \int_{\Gamma_i} \frac{1}{2} \mathbf{B}^{\text{T}} \mathbf{D} \mathbf{n}_{\Gamma_i}^{\text{T}} (\mathbf{u}^+ - \mathbf{u}^-) d\Gamma$$
$$= \int_{\Omega^+} \mathbf{B}^{\text{T}} \rho \mathbf{g} \, d\Omega + \int_{\Gamma} \mathbf{N}^{\text{T}} \mathbf{t} \, d\Gamma \tag{56}$$

with $\mathbf{n}_{\Gamma_i}$ now written in a matrix form:

$$\mathbf{n}_{\Gamma_i}^{\mathrm{T}} = \begin{bmatrix} n_x & 0 & 0 & n_y & 0 & n_z \\ 0 & n_y & 0 & n_x & n_y & 0 \\ 0 & 0 & n_z & 0 & n_z & n_x \end{bmatrix}, \tag{57}$$

where n_x, n_y, n_z are the components of the vector $\mathbf{n}_{\Gamma_i}$. After linearization, one obtains:

$$\begin{bmatrix} \mathbf{K}^{--} & \mathbf{K}^{-+} \\ \mathbf{K}^{+-} & \mathbf{K}^{++} \end{bmatrix} \begin{pmatrix} \mathrm{d}\mathbf{a}^- \\ \mathrm{d}\mathbf{a}^+ \end{pmatrix} = \begin{pmatrix} \mathbf{f}_{\mathbf{a}^-}^{\mathrm{ext}} - \mathbf{f}_{\mathbf{a}^-}^{\mathrm{int}} \\ \mathbf{f}_{\mathbf{a}^+}^{\mathrm{ext}} - \mathbf{f}_{\mathbf{a}^+}^{\mathrm{int}} \end{pmatrix} \tag{58}$$

with $\mathbf{a}^+, \mathbf{a}^-$ arrays that contain the nodal values of the displacements at the minus and the plus side of the interface, respectively, with $\mathbf{f}_{\mathbf{a}^-}^{\mathrm{ext}}, \mathbf{f}_{\mathbf{a}^+}^{\mathrm{ext}}$ and $\mathbf{f}_{\mathbf{a}^-}^{\mathrm{int}}, \mathbf{f}_{\mathbf{a}^+}^{\mathrm{int}}$ on the right and left–hand sides of Equation (56), respectively, and the submatrices defined by:

$$\mathbf{K}^{--} = \int_{\Omega^-} \mathbf{B}^{\mathrm{T}} \mathbf{D} \mathbf{B} \, \mathrm{d}\Omega + \frac{1}{2} \int_{\Gamma_i} \mathbf{N}^{\mathrm{T}} \mathbf{n}_{\Gamma_i}^{\mathrm{T}} \mathbf{D} \mathbf{B} \, \mathrm{d}\Gamma + \frac{1}{2}\alpha \int_{\Gamma_i} \mathbf{B}^{\mathrm{T}} \mathbf{D} \mathbf{n}_{\Gamma_i} \mathbf{N} \mathrm{d}\Gamma, \tag{59a}$$

$$\mathbf{K}^{-+} = \frac{1}{2} \int_{\Gamma_i} \mathbf{N}^{\mathrm{T}} \mathbf{n}_{\Gamma_i}^{\mathrm{T}} \mathbf{D} \mathbf{B}, \, \mathrm{d}\Gamma - \frac{1}{2}\alpha \int_{\Gamma_i} \mathbf{B}^{\mathrm{T}} \mathbf{D} \mathbf{n}_{\Gamma_i} \mathbf{N} \mathrm{d}\,\Gamma, \tag{59b}$$

$$\mathbf{K}^{+-} = \frac{1}{2}\alpha \int_{\Gamma_i} \mathbf{B}^{\mathrm{T}} \mathbf{D} \mathbf{n}_{\Gamma_i} \mathbf{N} \, \mathrm{d}\Gamma - \frac{1}{2} \int_{\Gamma_i} \mathbf{N}^{\mathrm{T}} \mathbf{n}_{\Gamma_i}^{\mathrm{T}} \mathbf{D} \mathbf{B} \, \mathrm{d}\Gamma, \tag{59c}$$

$$\mathbf{K}^{++} = \int_{\Omega^+} \mathbf{B}^{\mathrm{T}} \mathbf{D} \mathbf{B} \, \mathrm{d}\Omega - \frac{1}{2} \int_{\Gamma_i} \mathbf{N}^{\mathrm{T}} \mathbf{n}_{\Gamma_i}^{\mathrm{T}} \mathbf{D} \mathbf{B} \, \mathrm{d}\Gamma - \frac{1}{2}\alpha \int_{\Gamma_i} \mathbf{B}^{\mathrm{T}} \mathbf{D} \mathbf{n}_{\Gamma_i} \mathbf{N} \, \mathrm{d}\Gamma. \tag{59d}$$

6. Concluding remarks

An overview has been given of various domain-based discretization methods for damage and fracture. Where possible, differences and similarities have been pointed out between the methods discussed, viz. conventional interface elements, meshfree methods, the partition-of-unity approach. At present, the application of discontinuous Galerkin methods to damage and fracture is yet in its infancy, which is why the discussion of the merits and disadvantages of this class of methods is rather scanty.

References

Allix, O. and Ladevèze, P. (1992). Interlaminar interface modelling for the prediction of delamination. *Composite Structures* **22**, 235–242.

Askes, H., Pamin, J. and de Borst, R. (2000). Dispersion analysis and element–free Galerkin solutions of second and fourth–order gradient–enhanced damage models. *International Journal for Numerical Methods in Engineering* **49**, 811–832.

Askes, H., Wells, G.N. and de Borst, R. (2003). Novel discretization concepts. In *Comprehensive Structural Integrity, Vol III. 'Numerical and Computational Methods'* (edited by de Borst, R. and Mang H.), Elsevier, Oxford, 377–425.

Babuska, I. and Melenk, J.M. (1997). The partition of unity method. *International Journal for Numerical Methods in Engineering* **40**, 727–758.

Barenblatt, G.I. (1962). The mathematical theory of equilibrium cracks in brittle fracture. *Advances in Applied Mechanics* **7**, 55–129.

Barsoum, R.S. (1976). On the use of isoparametric finite elements in linear fracture mechanics. *International Journal for Numerical Methods in Engineering* **10**, 225–237.

Baumann, C.E. and Oden, J.T. (1999). A discontinuous *hp* finite element method for the Euler and Navier–Stokes problems. *International Journal for Numerical Methods in Fluids* **31**, 79–95.

Belytschko, T. and Black, T. (1999). Elastic crack growth in finite elements with minimal remeshing. *International Journal for Numerical Methods in Engineering* **45**, 601–620.

Belytschko, T., Lu Y.Y. and Gu, L. (1994). Element–free Galerkin methods. *International Journal for Numerical Methods in Engineering* **37**, 229–256.

Camacho, G.T. and Ortiz, M. (1996). Computational modeling of impact damage in brittle materials. *International Journal of Solids and Structures* **33**, 2899–2938.

Cockburn, B. (2004). Discontinuous Galerkin methods for computational fluid dynamics. In *The Encyclopedia of Computational Mechanics*, (edited by Stein, E., de Borst, R. and Hughes Wiley, T.J.R.) Vol III, Chap. 4 Chichester.

de Borst, R. and Gutiérrez, M.A. (1999). A unified framework for concrete damage and fracture models with a view to size effects. *International Journal of Fracture* **95**, 261–277.

de Borst, R. and Nauta, P. (1985). Non–orthogonal cracks in a smeared finite element model. *Engineering Computations* **2**, 35–46.

Duarte, C.A. and Oden, J.T. (1996). $H - p$ clouds – an $h - p$ meshless method. *Numerical Methods in Partial Differential Equations* **12**, 673–705.

Dugdale, D.S. (1960). Yielding of steel sheets containing slits. *Journal of the Mechanics and Physics of Solids* **8**, 100–108.

Fleming, M., Chu, Y.A., Moran, B. and Belytschko, T. (1997). Enriched element-free Galerkin methods for crack tip fields. *International Journal for Numerical Methods in Engineering* **40**, 1483–1504.

Gravouil, A., Moës, N. and Belytschko, T. (2002). Non-planar 3D crack growth by the extended finite element and level sets – Part I: Mechanical model and Part II: Level set update. *International Journal for Numerical Methods in Engineering* **53**, 2549–2586.

Hansbo, A. and Hansbo, P. (2004). A finite element method for the simulation of strong and weak discontinuities in solid mechanics. *Computer Methods in Applied Mechanics and Engineering*, **193**, 3523–3540.

Henshell, R.D. and Shaw, K.G. (1975). Crack tip finite elements are unnecessary. *International Journal for Numerical Methods in Engineering* **9**, 495–507.

Hillerborg, A., Modéer, M. and Petersson, P.E. (1976). Analysis of crack formation and crack growth in concrete by means of fracture mechanics and finite elements. *Cement and Concrete Research* **6** 773–782.

Ingraffea, A.R. and Saouma, V. (1987). Numerical modelling of discrete crack propagation in reinforced and plain concrete. *In Fracture Mechanics of Concrete*. Martinus Nijhoff Publishers. Dordrecht, 171–225.

Krysl, P. and Belytschko, T. (1999). The element–free Galerkin method for dynamic propagation of arbitrary 3-D cracks. *International Journal for Numerical Methods in Engineering* **44**, 767–800.

Lemaitre, J. and Chaboche, J.L. (1990). *Mechanics of Solids Materials*. Cambridge University Press, Cambridge.

Liu, W.K., Jun, S. and Zhang, Y.F. (1995). Reproducing kernel particle methods. *International Journal for Numerical Methods in Fluids* **20**, 1081–1106.

Mergheim, J., Kuhl, E. and Steinmann, P. (2004). A hybrid discontinuous Galerkin/interface method for the computational modelling of failure. *Communications in Numerical Methods in Engineering* **20**, 511–519.

Mergheim, J., Kuhl, E. and Steinmann, P. (2005). A finite element method for the computational modelling of cohesive cracks. *International Journal for Numerical Methods in Engineering* **63**, 276–289.

Moës, N., Dolbow, J. and Belytschko, T. (1999). A finite element method for crack growth without remeshing. *International Journal for Numerical Methods in Engineering* **46**, 131–150.

Nayroles, B., Touzot, G. and Villon, P. (1992). Generalizing the finite element method: diffuse approximations and diffuse elements. *Computational Mechanics* **10**, 307–318.

Nitsche, J.A. (1970). Über ein Variationsprinzip zur Lösung Dirichlet–Problemen bei Verwendung von Teilraümen, die keinen Randbedingungen unterworfen sind. *Abhandlungen des Mathematischen Seminars Universität Hamburg* **36**, 9–15.

Rashid, Y.R. (1960). Analysis of reinforced concrete pressure vessels. *Nuclear Engineering and Design* **7**, 334–344.

Remmers, J.J.C., de Borst, R. and Needleman, A. (2003). A cohesive segments method for the simulation of crack growth. *Computational Mechanics* **31**, 69–77.

Rots, J.G. (1991). Smeared and discrete representations of localized fracture. *International Journal of Fracture* **51**, 45–59.

Schellekens, J.C.J. and de Borst, R. (1992). On the numerical integration of interface elements. *International Journal for Numerical Methods in Engineering* **36**, 43–66.

Schellekens, J.C.J. and de Borst, R. (1994). Free edge delamination in carbon-epoxy laminates: a novel numerical/experimental approach. *Composite Structures* **28**, 357–373.

Simone, A. (2004). Partition of unity–based discontinuous elements for interface phenomena: computational issues. *Communications in Numerical Methods in Engineering* **20**, 465–478.

Tijssens, M.G.A., Sluys, L.J. and van der Giessen, E. (2000). Numerical simulation of quasi-brittle fracture using damaging cohesive surfaces. *European Journal of Mechanics: A/Solids* **19**, 761–779.

Wells, G.N., de Borst, R. and Sluys, L.J. (2002). A consistent geometrically non-linear approach for delamination. *International Journal for Numerical Methods in Engineering* **54**, 1333–1355.

Wells, G.N., Garikipati, K. and Molari, L. (2004) A discontinuous Galerkin formulation for a strain gradient–dependent damage model. *Computer Methods in Applied Mechanics and Engineering* **193**, 3633–3645.

Wells, G.N. and Sluys, L.J. (2001). A new method for modeling cohesive cracks using finite elements. *International Journal for Numerical Methods in Engineering* **50**, 2667–2682.

Xu, X.P. and Needleman, A. (1994). Numerical simulations of fast crack growth in brittle solids. *Journal of the Mechanics and Physics of Solids* **42**, 1397–1434.

Zienkiewicz, O.C., Taylor, T.L., Sherwin, S.J. and Peíro, J. (2003). On discontinuous Galerkin methods. *International Journal for Numerical Methods in Engineering* **58**, 1119–1148.